AA003486

14th Topical on Refinery Processing 2011

Topical Conference at the 2011 AIChE Spring Meeting and 7th Global Congress on Process Safety

Chicago, Illinois, USA
13-17 March 2011

ISBN: 978-1-61782-895-9

Printed from e-media with permission by:

Curran Associates, Inc.
57 Morehouse Lane
Red Hook, NY 12571

Some format issues inherent in the e-media version may also appear in this print version.

Printed by Curran Associates, Inc. (2011)

For permission requests, please contact AIChE
at the address below.

AIChE
3 Park Avenue
New York, NY 10016-5991

Phone: (203) 702-7660
Fax: (203) 775-5177

www.aiche.org

Additional copies of this publication are available from:

Curran Associates, Inc.
57 Morehouse Lane
Red Hook, NY 12571 USA
Phone: 845-758-0400
Fax: 845-758-2634
Email: curran@proceedings.com
Web: www.proceedings.com

TABLE OF CONTENTS

(10a)

This welcoming session sponsored by Fuels & Petrochemicals Division (F&PD) is a new initiative organized by the AIChE's national and F&PD technical programming team as a preamble to the division's sponsored technical sessions. This will be opened by a short introduction by the chair of the division. Besides recognizing leadership in his introductory short presentation, the chair will highlight the key elements of value additions the division brings to its members. The presentation will be concluded with a special request to join the division where professional opportunities are limited only by members' imagination, to positively impact academia and industry and thereby society at large. For details of the division and its many facets of activities and functions please follow the link www.aiche-fpd.org.

(10b) What Are We Going to Do with All This Natural Gas? - Ramifications of the Dramatic Expansion of Natural Gas Available to the Market Due to North American Shale Gas Production and LNG Liquefaction

Sam Thomas, Chart Industries, Inc., Garfield Heights, OH

The success in Shale and other unconventional formations has dramatically expanded the N. American Natural Gas reserves and promises to also do so in many other parts of the world. New LNG liquefaction capacity in the Middle East and Asia along with additional large scale capacity from Australia and Papua New Guinea coming on-line in the next 5 years opens the opportunity for a global trade in Natural Gas stabilizing historically volatile pricing. How will this important energy and petrochemical feedstock be utilized? The success in improving techniques to allow unconventional gas reserves to be proven and significantly increase the available reserves has not as yet been fully appreciated by markets and policy makers. This presentation reviews the ramifications of natural gas becoming a larger part of the global energy supply for which there are numerous competing interests. Natural Gas holds the ability to significantly reduce carbon emissions by substitution for coal in power generation, it can be used as a transportation fuel either as LNG, CNG or GTL clean diesel increasing energy independence of some of the largest economies in the world simultaneously reducing carbon emissions. It could enable the USA to maintain its position as a leading petrochemical producer if used for ethylene feedstock. How will these competing interests play out and how will governments respond?

See more of this Session: Fuels & Petrochemicals Division Welcome Session and Energy Leadership Presentation
See more of this Group/Topical: Topical 7: 14th Topical on Refinery Processing

(14a) The Energy Challenge: Providing Sustainable Supplies to Meet Growing Demand

F. Emil Jacobs, ExxonMobil Research and Engineering Company, Annandale, NJ

As economies grow and living standards improve, the world faces a dual energy challenge: meeting the growing demand for energy while mitigating emissions and their impact on the environment. ExxonMobil believes that addressing this dual challenge requires an integrated set of solutions which includes increasing the supply of all commercially viable energy sources, reducing emissions, and improving energy efficiency in all sectors of the economy. Technology will be a key enabler of continued progress in all of these areas.

Many experts agree that the immense scale of global energy demand necessitates a continuing role for oil and gas as a substantial portion of the energy supply mix for at least the next several decades. These experts also forecast that the overall energy mix will shift toward a relatively higher level of sustainability through a combination of increased efficiency across the economy, growth in lower-carbon sources of energy, and development and deployment of technologies that help mitigate emissions.

ExxonMobil has developed a number of technologies that contribute to increased supplies of energy, reduced emissions, and higher efficiency in its operations and in consumer use of its products. The presentation will provide an overview of ongoing research and development activity in a broad range of technology areas including algae-based biofuels. It will also highlight some of company's current technologies including lithium-ion battery separator films, lightweight plastics that improve automotive efficiency, and processes that could advance carbon capture and storage.

See more of this Session: Spring Meeting Keynote Address - Sponsored by the AIChE Fuels & Petrochemicals Division
See more of this Group/Topical: Liaison Functions

(17a) Biobutanol Production In Recombinant Escherichia Coli BL21

Seunghyun Ryu, Department of Chemical Engineering, Texas Tech University, Lubbock, TX and M. Nazmul Karim, Chemical Engineering, Texas Tech University, Lubbock, TX

In this study, *Escherichia coli* BL21 was engineered for butanol production. The mutant *E. coli* BM322 was constructed by disrupting phosphotransacetylase (*pta*), lactate dehydrogenase A (*ldhA*), bifunctional acetaldehyde/ethanol dehydrogenase (*adhE*$_B$), fumarate reductase (*frdBC*), and pyruvate oxidase B (*poxB*) genes. Additionally, the transcriptional regulator FNR protein gene was also deleted. The recombinant plasmid pBUOH was constructed by cloning the key enzyme coding genes that are involved in the butanol production pathway in *Clostridium acetobutylicum*. These genes include: thiolase (*thl*), 3-hydroxybutyryl-CoA dehydrogenase (*hbd*), crotonase (*crt*), butyryl-CoA dehydrogenase (*bcd*), electron transfer flavoprotein (*etfAB*), and the bifunctional butyraldehyde/butanol dehydrogenase (*adhE*$_C$). The recombinant *E. coli* BM322/pBUOH produced 1.2 g/L butanol with 10.2 g/L lactate as a by-product for 60 h.

See more of this Session: Biodiesel and Renewable Fuels
See more of this Group/Topical: Topical 7: 14th Topical on Refinery Processing

(17b) Biorenewable Diesel and Jet Fuel Production

Geoffrey Fichtl, Renewable Energy and Chemicals, UOP - Honeywell, Des Plaines, IL

Conversion of natural plant and animal-derived oils to hydrocarbon fuels has been established using hydroprocessing technologies to produce renewable jet and diesel products by hydrotreating and then isomerizing/cracking the natural plant and animal-derived oils via a two-stage process. Investigation of a range of conversion and cracking/isomerization levels has shown that an alternative design could provide better economics than just running at maximum jet yield. This mode of operation also leads to an initial capital investment similar to the maximum diesel yield mode while allowing the refiner to produce a significant amount of jet fuel by operating at higher conversion levels with a modified fractionation scheme. The positive relationship between conversion and isomerization requires optimization of fractionation to meet all critical jet fuel specifications, especially freeze point, at different levels of conversion. However, operation at increased conversion risks cracking some of the high-value distillate range material into low-value naphtha range. Additionally, the cold flow properties of the diesel product are impacted by fractionating out a jet product. Therefore, maximizing product yields according to market demand while meeting all required product specifications requires careful management of cracking and fractionation.

See more of this Session: Biodiesel and Renewable Fuels
See more of this Group/Topical: Topical 7: 14th Topical on Refinery Processing

(17c) Integrated Hydropyrolysis and Hydroconversion Process (IH^2) for Production of Gasoline and Diesel Fuel From Wood, Algae, Cornstover, Lemna, and Bagasse Feedstocks

Terry Marker, Larry Felix, Martin Linck and Michael Roberts, Gas Technology Institute, Des Plaines, IL

Experimental results and a process design will be presented for an integrated process which uses hydropyrolysis plus hydroconversion to convert biomass into gasoline and diesel. The hydrogen needed for the hydropyrolysis and hydroconversion process is produced by reforming the light gases so that no external H_2 is required. Finished hydrocarbon products with less than 2% oxygen are directly produced using this processing approach.

Yield and product quality data from IH^2 testing of a variety of diverse biomass feedstocks will be presented, including data from wood, algae, corn stover, lemna and bagasse feeds. All the biomass feeds show good yields. In all cases the IH^2 hydrocarbon product has less than 1% oxygen, with TAN number less than 1. The IH^2 product is primarily gasoline and diesel. Product quality is affected by the type of feed and also the catalyst and conditions used. Algae produced the best diesel, while wood produced the highest octane gasoline.

Some data will also be shown, on the effects of catalyst and process variables on the IH^2 process. The relative amount of gasoline and diesel is affected by both the biomass feedstock and the processing conditions, particularly the catalyst.

Future IH^2 experimental and commercialization plans will also be discussed.

See more of this Session: Biodiesel and Renewable Fuels
See more of this Group/Topical: Topical 7: 14th Topical on Refinery Processing

(17d) Investigations of Mass Transfer and Kinetic Effects During Transesterification of Canola Oil Using a Novel Technique

Krupal D. Pal, Chemical/Biochemical Engineering, University of Western Ontario, London, ON, Canada

Abstract

Transesterfication of a triglyceride (i.e. vegetable oil) with alcohol is commonly used to produce biodiesel. However, the triglyceride and alcohol phases are nearly immiscible and high degree of mixing is required to have a better contact between the reactants. The kinetics of the system can be explained by three regimes: initial mass transfer controlled regime followed by chemical reaction controlled regime and final regime close to equilibrium. The initial lag is observed due to the immiscibility of oil and alcohol. A study has been conducted to eliminate the mass transfer control regime and reduce the reaction time by means of a new method developed for the transesterfication of canola oil and ethanol using KOH and NaOH as base catalyst. The kinetics of the reaction parameters of the new method were also studied and compared with available literature data. The new method gave good results as compared to the conventional method for transesterfication. The concentration of acylglycerols (triglycerides, di-glycerides and mono-glycerides) was quite low in the final product by applying this method. The quality and yield of biodiesel obtained by this were much better than conventional method in less reaction time. The new method developed is cost effective, compact and easily upgraded from conventional method. A pilot plant of 50L capacity has been developed to further demonstrate the benefits of new approach.

See more of this Session: Biodiesel and Renewable Fuels

See more of this Group/Topical: Topical 7: 14th Topical on Refinery Processing

Instrumentation, Automation and Control Applications in a Supercritical Biodiesel Process

Marc Privitera

Biodiesel Process Chemistry

Biodiesel is typically produced by reacting a triglyceride with alcohol traditionally in the presence of a base catalyst, and most recently using supercritical fluid process conditions. The reaction is known as transesterification. When using methanol as the reacting alcohol, the resulting desired product is a fatty acid methyl ester (FAME). When using ethanol, the resulting product is fatty acid ethyl ester (FAEE). The FAME or FAEE is the biodiesel. The base catalyzed biodiesel systems need low free fatty acid (FFA) feedstock to avoid the formation of soap. The supercritical system can use higher levels of free fatty acid feedstock as the competing soap reaction is not a factor. Other methods of converting high free fatty acid feedstocks have been successful by first esterifying all the FFA using an acid catalyst then transesterifying the remaining triglycerides in the feedstock using the base catalyst.

Traditional Batch Process

A batch biodiesel process system is usually a series of tanks where various blends of feedstock, alcohol, and catalyst are reacted. The primary variables of concern are liquid tank level and temperatures. The process uses the liquid phase solubility differences of the various products to allow gravity separation in batch tanks. After the water and oil phase separations, each phase must then be separated into its constituent components for purification of the product or recycle of unused reagents. The most challenging of these separations is the recovery and drying of the methanol. The efficient use, separation, and reuse of the methanol is a key economic driver of biodiesel production. Dry methanol is critical as water interferes with transesterification in many different ways.

Continuous Intensified Process

A continuous system using a high shear mixing device is commonly referred to as an intensified mixing system. The reaction kinetics are limited by the solubility limits of the feedstock in the alcohol. The creation of the optimum liquid droplet dispersed phase is the critical control parameter driving an increased reaction rate. Flow ratio control becomes critical with a continuous intensified system and usually is accomplished using PID loops with the feedstock flow loop being a wild stream and the methanol slaving its set point from either the total flow or the feedstock flow. Attention to developing a stable flow measurement drives the system economics as the molar ratios are key to both reaction efficiency and separation efficiency.

Supercritical Process

The supercritical process uses temperature and pressure to drive both the alcohol and the feedstock beyond their critical points so that the reaction occurs in one phase. The critical point for methanol is 1415 psia and 480 °F. These reaction conditions allow the methanol to lose its polar nature and solvates both the triglyceride and free fatty acid. This method is able to use very high free fatty acid (FFA) feedstocks as there is no sodium hydroxide saponification reaction that competes with the transesterification with this method. High FFA feedstocks are commonly found available as waste streams that have no other economic purpose. The use of ethanol instead of methanol at supercritical conditions presents what some may term a more complete "green" solution since ethanol can be fermented from renewable feedstocks whereas most methanol is currently produced from the hydrolysis of natural gas. The greatest advantage with supercritical process is that the lowest value feedstocks may be used and the conversion has very few detrimental byproducts. As with all transesterification, water is a concern and must be controlled carefully.

Overall Instrument Installation and Operational Concerns:

Most of the instruments are installed in classified electrical locations due to the presence of methanol or ethanol in the systems. Many details regarding intrinsic safety and hazardous environment instrument installation methods have been part of each system's installation. The instruments installed must have wetted parts that are compatible with the various process fluids in the systems. A tricky detail with biodiesel system process fluids is conflicting materials of compatibility (MOC) information. Compounding the many factors involved in the MOC selection includes the fact that many process unit operations see changing concentrations and characterization of various active materials that cause compatibility problems. One of the more interesting examples is the level of free fatty acid (FFA) in the feedstock. The higher the FFA, then the more it tends to require more chemical resistant materials than its base composition components. FAME or FAEE and methanol or ethanol have been found to have mutually exclusive materials of compatibility concerns. In general, stainless steel for hard wetted surfaces and PTFE for elastomeric surfaces have been successful. Different grades and brands of FKM in differing service have exhibited different results. The most significant being the conflicting compatibility data and experience regarding the use of FKM for biodiesel applications. In some applications it is quite successful, and in others with just slightly different temperature and minor constituent in process components has caused failure of the material.

Temperature

A wide variety of temperature measuring devices have been installed in all three types of systems. Local indicators are as simple as a bimetallic dial gauge installed in a common thermowell. Unique to the application is the need for considering the thermal insulating properties of the feedstocks due to the varying melt points, viscosity versus temperature profiles and the low thermal conductivity of the solidified feedstocks. Each feedstock is made up of various triglyceride components that can have a varying affect on the bulk properties of the feedstock itself. Much data exists on the properties of neat oils and for the low free fatty acid feedstocks the information published in literature predicted well the expected properties of the feedstocks. The area that became a challenge and one to pay close attention to is when the feedstocks had high levels of water or the mixtures of triglycerides included significant portions of saturated components. Typical oils for biodiesel feedstock are a blend of C16:0, C18:0, C18:1 and C18:2 oils. This gives insight as to how the different components can drop out of solution and form coatings on the surfaces of instrument measurement devices.

Another interesting aspect of this application is that as the commonly known phenomena of the separation of the oil/water mixture as temperature increases. The water coalesces and drops to the bottom of the various feedstock tanks. Since the tanks are filled with various feedstocks with varying moisture levels, the interface region and the measurement point composition will vary as they are used and refilled. Depending upon the placement of the temperature probe or gauge, the actual temperature of the material in the tank will show variations versus catch samples due to thermal conductivity of the material now closest to the thermowell that contains the temperature element or gauge probe. If a film is formed on the thermowell due to the material changing its flow properties as the concentration of both moisture and triglyceride mass fraction vary, this also leads to erroneous reading in the bulk temperature of the material in the tank.

In the supercritical process, the temperature measuring devices have to be installed with the requirements of the high pressure piping in mind. Thermowells have to be rated consistent with the high pressures in the piping system. The pressures used in the system require high pressure flanges, so the connection details are critical to safety and success. Keep in mind the need to consider the classification of the electrical installation. The most important temperature measurement applications of the system are the dual temperature measurements for the control of the discharge of the process heater to insure that no over temperature failure can occur in the fired process heater coil. The common configuration of a local PID controller backed up by a separate high temperature relay contact is installed. The system follows the usual NFPA best practices for burner control, with the not so common twist of the supercritical state of the process fluid.

A series of flash columns are installed downstream for the separation of the various reaction products. The temperature measuring devices in the columns are installed in the common arrangement of: overhead, feed, and bottoms . The installation uses a thermowell extending deep into the most turbulent flow path for the feed and overhead vapors. The bottoms measurement is placed in the headspace below the bottom column packing but above the liquid collection still. The most interesting detail of this application is the affect of the glycerin. Glycerin is a reaction byproduct that remains in the bottoms with the FAME and it is immiscible in the FAME unless in the presence of the methanol. This little solubility detail has derailed many biodiesel systems' ability to meet the rather tight ASTM 6751 specification both on glycerin and methanol contamination in the fuel. The glycerin may begin to coalesce on the surface of the bottoms thermowell and give an insulating affect on the thermowell due to its different thermal and flow properties than the methanol and FAME also in the system. Since the glycerin is the most viscous material at the lowest temperature, it tends to insulate the thermowell and give a reading in the bottoms that has a bias from sample temperatures taken from the sample port nearby. This problem is solved by careful control of the bottoms make up during the start up and instrument maintenance data runs to keep a good understanding of the bias presented by the glycering coating in the bottoms thermowell.

Pressure

A series of pressure transmitters and gauges are installed throughout the process. Some key elements for each device is the MOC details of the tubes, sockets, and in some cases the isolation diaphragms.

The range applied to the desired measurement area seems simple, but in a new system start up the establishment of the process operator measurement unit "lingo" is important to consider. Vacuum and pressure can become confusing and the pressure ranges in the various unit operations of the system have a wide range. A plant vernacular has to be either established, if new, or if existing, the terms of use of the operators must conform to the adopted manner of the plant to prevent confusion in communications. Since these were grassroots plant start ups with no historical local nomenclature, the following was established as practice: For pump discharges and any pressure above 0 gauge, all terms are in psig. For the local gauge readouts on the vacuum systems, the use of dual range gauges lead to the local plant term of "inches" which is actually inches of Hg vacuum. For the pressure transmitters in vacuum service, the use of torr absolute was established as the terms for the units of measure of the pressure in the system. It seems a simple detail, but in training and troubleshooting, and in discussions with many people new to the system, many errors and misunderstandings can be avoided if the basis of unit "lingo" is established beforehand.

Flow

The use of a high pressure coriolis meter as the flow control element on the fired heater control system allowed the rangeability required for the needed stability in the control loop for the different start up, steady state, shutdown, and emergency shutdown process conditions. One of the more interesting applications in the systems is the concentration control of the recycled methanol. The coriolis mass flow meters have been some of the most reliable and repeatable devices in the system.

Level

The system has both radar and guided wave radar level devices in the system. The radar level sensor is placed on the top of the various tanks. Once calibrated for the tank geometry, the devices have given excellent service through the varying headspace conditions of temperature and moisture to which the tank has been exposed. The more interesting application has been the guided wave radar devices installed through the system in the smaller tanks and the side chambers in the system. A culprit that had originally caused problems was the glycerin and its tendency to form films or enter the small tanks and side chambers in small globs versus a steady stream. These mainly occur due to the environmental conditions changing between night and day and cold spots forming in the piping where the heat tracing or insulation has been exposed due to repairs or other operations. As the glycerin hits the probe, it has been observed that false signals are reflected back to the transmitter which result in erroneous level readings. The problem can be contained by using the software functions of the transmitter to block out and ignore areas that are repeated problems like the entrance ports to some of the various side chambers. A changing value comparison trap in the PLC logic was helpful to keep these random errors from affecting the stability of the various level control loops in the system.

Controlling the Reactors at High Temperature and Pressure

The supercritical reactors were run at high temperature and high pressure. As mentioned earlier, the critical point of methanol is 1174 psia and 462 °F. A globe valve was used to control the pressure in the reactors. The valve had to be sized such that the needed steady state conditions were controllable as at supercritical the fluid properties change dramatically. The valve had to be sized to allow for the sub-critical conditions during start up and the warm hold recycle and the emergency full flow reactor pressure let down conditions. The reactor system was protected by the required ASME safety relief valve, but the design basis included capability for a rapid controlled release of the pressure (that was ultimately protected mechanically by the safety relief system). As the pressure drops, the supercritical fluid will choke flow more rapidly than when at supercritical conditions as the velocity of the fluid can quickly reach its sonic limit depending upon the detail of the orifice design. This is a challenging design basis for a control valve. A digital positioner was used as an inner control loop to insure that the forces from the changing high temperature and high pressure fluid conditions were compensated for by the stem so that a stable pressure set point could be maintained. The inner loop had a very short time constant and the outer pressure loop was detuned so to not over run the inner loop response.

Methanol Water Separation with Varying Reactor Produced Water

One of the most critical parameters in the supercritical reaction is removing the water produced due to the high free fatty acid conversion. One mol of FFA reacted with one mol of MeOH produces one mol of water along with every mol of FAME produced. As with all methods of transesterification, the water can inhibit the reaction. As the feedstock varies in FFA, the resultant glycerol and water produced also vary. This adds a bit of a control challenge to the distillation since the height of the separation column cannot be changed and strict temperature control at different levels of the column will drive the separation. Too hot and fuel goes overhead, too cold and the water stays in the fuel.

A two column strategy is employed, the first separating the water and methanol overhead leaving the FAME and the glycerol in the bottoms, then the second column to dehydrate the methanol. A coriolis flow meter is used to measure the density of the first and second distillation column reflux and the second column distillation bottoms. The density combined with the temperature gives a very stable concentration of the water in the methanol to apply to control the separation.

Vacuum System Control via Temperature

A wiped film evaporator is used to separate the fame from the low level impurities. This vacuum distillation step of the fuel is the first step in the control of various constituents that come into the system with the feedstock. A 1 torr vacuum is required for the vacuum distillation of the fuel. The vapor pressure of methanol varies with temperature. In a liquid ring pump sealed with methanol, the temperature control of the fluid in the ring loop determines the vacuum that the pump can create based upon the temperature absolute pressure relationship. Chilled water is circulated through a heat exchanger to remove the heat picked up by the vacuum fluid due to the condensation of the methanol being separated from the fuel, or the heat of compression from the two subsequent vacuum stages. The liquid ring pulls the first stage to 100 torr. Two blowers in series then complete the vacuum drawing the wiped film evaporator down to 1 torr to enable the biodiesel distillation to occur around 400 °F.

Processing Heavy Crudes in Existing Refineries

Prepared for 2011 AIChE Spring Meeting, Chicago, IL

Kevin Turini
James Turner
Arthur Chu
Shankar Vaidyanathan
Fluor Enterprises, Inc., Sugar Land, Texas, USA

Introduction

As world demand for petroleum products increases, and production in existing light crude fields declines, heavier, high sulfur crude oils are increasing in market share. Since heavy sour crudes typically trade at $5-15 per barrel discount to light sweet crudes, it may be profitable for an existing refinery to increase their throughput of heavy sour crudes. Furthermore, Canadian heavy crude bitumen production has increased from 500,000 BPD production in 1996 to over 1 million BPD in 2005, and production is predicted to increase to over 3 million BPD in 2020 (1). As these crudes currently have transportation logistics challenges to get on the open market, existing refineries that are logistically advantaged may be able to purchase these crudes at even higher discounts to light sweet crudes, and can improve their operating margins if they can process more of the lower value heavy sour crude.

The heaviest crudes, often called bitumen in their raw form, have viscosities at ambient temperatures that prevent economical transportation. Therefore, these crudes are typically blended either with a diluent, which is typically naphtha to light kerosene boiling range material, or with a full boiling range syncrude. The combination of diluent and bitumen crudes are called "dilbit", and the combination of synthetic crude and bitumen are called "synbit".

Processing these crudes in existing refineries may require significant refinery modifications. This paper presents some of the major issues that may be encountered by refiners wanting to process more of these crudes.

Feasibility Study

A refiner considering modifying their refinery to process incremental heavy sour crude will typically hire an external team to study the options. While there are many different approaches that may be successful, a proven approach is to engage a small team of experienced professionals from an Engineering Contractor who can provide the necessary technical expertise and project cost certainty.

Processing Heavy Crudes in Existing Refineries

This team can perform the following:

- Organize and manage the study
- Coordinate information from licensors and catalyst suppliers
- Check existing equipment and determine optimum revamp strategies
- Determine optimum refinery stream mappings for various cases
- Provide cost estimates for unit revamps and new units
- Estimate utility consumption and OSBL impacts
- Assist with economic calculations
- Prepare study reports and assist with owner stage gate documentation

Fluor has been involved in many such studies in the last 20 years, and can utilize this experience to execute these studies quickly and efficiently.

At the beginning of a study, several different crude types and refinery capacity options may need to be considered. The typical approach is to determine the primary options under consideration, and do the work required to determine the economics for each case. However, the team should be flexible enough to recognize where major investment breakpoints may occur – this may result in the desire to adjust the design basis for one or more cases to reflect what may be a more optimum solution than the original cases that were defined.

Incremental economics should be calculated to determine the incremental return for the differential investment between cases. OSBL costs may be factored, but a sense check should be done to determine major investment points between cases – for example, OSBL costs for a new unit may be much higher than for a small revamp to an existing unit.

Crude Design Basis

The type of heavy crude specified may have a large impact on the investment required to process it. Below is a brief discussion of the major design basis considerations for crude selection. Typical Alberta bitumen is a 8.5 deg API, 4.8% sulfur, and 3.3 TAN material.

<u>Dilbit Crude</u>

As discussed above, a dilbit crude is one in which a light/mid range hydrocarbon is blended with the raw crude to allow it to be transported economically via pipeline. The crude and diluent are normally blended to a viscosity specification. A density specification may also apply, but usually this specification is not controlling. The pipeline grade is typically 19 deg API and 350 Cst @ 40 deg C. Most of the shipped crudes are in the 20-21 deg API range. Often the diluent is purchased by the crude supplier, so its quality and composition may not be as well understood or well defined as the heavy crude itself.

Depending on pipeline transport logistics, the heavy crude may be "contaminated" by lighter crudes transported in the same pipeline. While this is likely an economic benefit fro the refinery, it may overload a bottlenecked naphtha and light ends processing system, possibly reducing the total crude throughput.

Depending on the refinery location and logistics, it may be economic for the refinery to ship diluent back to the production supplier. Purchased diluent typically has an RVP and olefins specification to allow storage in atmospheric tanks and minimize gum formation.

Dilbit crudes are sometimes called "dumbbell" crudes, since if one plots the volume of each cut, the naphtha/light kero range and resid range are high, whereas the kerosene and diesel range are low. In the Crude/Vac unit, dumbbell crudes stress the resid and naphtha processing areas of the unit. The vac resid handling includes the vacuum column system, including the heater, column, vac resid pump, and quench and heat recovery system. The naphtha handling system includes the top of the atmospheric column and the atmospheric column overhead cooling and reflux system.

Figure 1 - Oil Distributions of Baseline Crudes and Canadian Synbit and Dilbit

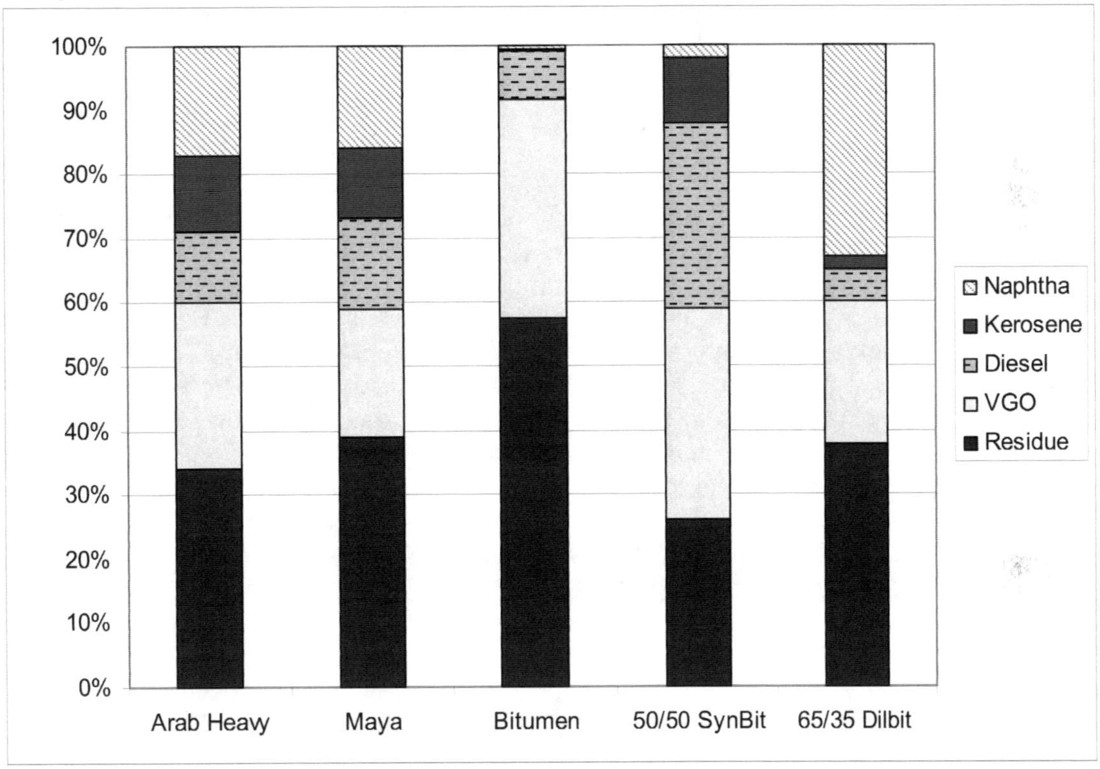

The affected downstream units, such as the coker, naphtha hydrotreater, naphtha reformer, and other naphtha processing equipment, are similarly affected. Gasoline product specifications, such as RVP and octane, may also be adversely affected by the diluent, particularly if it is a light naphtha that is lighter than the feed range of the reformer unit.

<u>Synbit Crude</u>

As discussed above, a Synbit crude is one where synthetic crude, which is typically produced by processing bitumen in an upgrader, is used as the "diluent" to get the blended crude to meet the viscosity specifications. Typical SynCrude from the upgrader is a 36 deg API, 0.05% sulfur material. Typical SynBit blend is 50/50 bitumen/diluent to a

20 deg API, 2.5% sulfur, 1.8 TAN and 350 Cst viscosity grade for pipeline shipment from Alberta. The netback on SynBit is lower than DilBit since the bitumen content is lower.

A Synbit crude is not typically a "dumbbell" crude therefore, may be a better fit for existing Crude/Vac units. However, Synbit crudes can have their processing difficulties as well. These crudes were essentially refined and hydrotreated in the upgrader and blended back together. Bitumen based crudes tend to be high in aromatic content. When these crudes are processed in a delayed coker, the aromatic concentration increases. When the coker products are hydrotreated, the "easiest" sulfur is removed, but much of the sulfur bound in highly aromatic compounds will not be removed. The resulting Syncrude appears to be low in sulfur in the diesel and gas oil range, but the sulfur that remains is the hardest sulfur to remove. Therefore, the Synbit fractions are much harder to desulfurize than straight run crude fractions. The fraction from the bitumen itself is also very difficult to hydroprocess, so this compunds the difficulty of processing the crude. Gas oil from Synbit or Dilbit typically requires significant hydrotreating before being sent to an FCC unit, or else the FCC yield will be very poor, and the product qualities will also be poor.

Naphthenic Acid Content of Crude

Both Dilbit and Synbit crudes have shown high Total Acid Numbers (TAN), which indicates high naphthenic acid contents. These acids are known to cause corrosion at high temperatures, and the corrosion rates are affected by concentration, temperature, flow characteristics, H_2S content, etc. Depending on whether the unit is alloyed for these acids, a crude change may require significant material upgrades in the Crude/Vac unit, and may also require material upgrades in downstream coker and hydrotreater units, as well as in OSBL piping (this is discussed further in later sections of the paper). A lot of research has been performed in recent years on predicting naphthenic acid corrosion rates, so it may be worthwhile to investigate using recent analysis, instead of historical guidelines. For example, a recent Joint Industry Research Program has performed laboratory testing of naphthenic acid corrosion under many conditions, and compiled the data into a proprietary computer program (6).

Sulfur

Higher sulfur in the crude will impact many systems in the refinery. The sulfur plant, amine and sour water systems are obviously affected by higher sulfur content in the crude, since almost all of the sulfur has to be removed from the refinery liquid products. Hydrotreaters will also be impacted from higher sulfur levels in the crude.

Higher sulfur in the crude and crude cuts may also increase corrosion in the Crude/Vac, coker and hydrotreater units if the refinery was not previously alloyed for handling sour feed stocks. This should be considered in the analysis of existing systems.

The gas oil and vacuum resid transfer temperatures between units (i.e. OSBL piping) may have an impact on whether existing piping and equipment can withstand the naphthenic acid or sulfidation corrosion from the proposed crudes – it may be economical to change the heat integration to reduce the transfer temperature instead of replacing a large amount of equipment or piping.

Processing Heavy Crudes in Existing Refineries

Vacuum Column Cutpoint

A key design basis consideration for a heavy crude project is the vacuum column cutpoint. The vacuum column cutpoint is the nominal final boiling point temperature for the heavy vacuum gas oil cut, and the nominal initial boiling point for the vacuum resid cut. The vacuum column cutpoint is actually defined by taking the volumetric vacuum resid product rate that corresponds to the cutpoint temperature on a TBP distillation curve of the whole crude. The actual vacuum gas oil and vacuum resid cuts will overlap, since there is not a perfect cut in the vacuum column.

A higher vacuum column cutpoint requires some combination of a lower flash zone pressure in the vacuum column (i.e. higher vacuum), more stripping steam, or a higher vacuum feed inlet temperature. For standard crudes, vacuum column cutpoints above 1000°F are typical, and units have been designed and successfully operated up to 1050°F. Vacuum heater coil outlet temperatures (COT) above 800°F are common for standard crudes, and typically correspond to cutpoints above 1000°F. However, for vacuum columns processing heavy crudes, cutpoints this high may not be practical. This may be especially true for Canadian heavy crudes, which are known to be more problematic in operating units.

In theory, a higher vacuum column cutpoint appears to be advantageous, since the lower value vacuum resid product rate is minimized, and more "higher value" gas oil is recovered. Some of the incremental heavy gas oil fed to the Coker will convert to coke, lowering the liquid yield from the refinery. However, a closer look reveals that a higher cutpoint, even if technically feasible, may not be the most economic option in the long run.

Heavy crudes are known to have a higher cracked gas make in the vacuum heater, and likewise produce more coking in the vacuum heater. Like all crudes, the vacuum gas rate and coking rate increase as the heater outlet temperature is increased. However, for heavy crudes, and in particular Canadian heavy crudes, the rate of increase in gas make and coking with increased temperature may be particularly steep, and may occur at much lower temperatures. A survey of vacuum columns operating on mostly Canadian heavy crudes have operating COTs of 690-740°F, which correspond to much lower vacuum cutpoints.

Thinking through all of the impacts for higher cutpoints can explain why it may not be economic, even if technically feasible. The three ways to increase cutpoint are to lower the flash zone operating pressure, increase the velocity/stripping steam rate, or increase the flash zone temperature. However, all of these can have negative impacts on the vacuum column system. A lower pressure requires a higher vacuum, which puts a higher load on the overhead vacuum system. This can significantly increase the capital and operating cost of the vacuum system. The vacuum column diameter also needs to be larger to process the same feed rate at lower pressure. Likewise, higher steam rates require a larger column diameter and a larger overhead vacuum ejector/condenser system. Finally, a higher heater outlet temperature can lead to a higher gas make rate, which also stresses the vacuum ejector system, and increases heater coking, which may require more frequent heater shutdowns for decoking.

As the vacuum cutpoint is increased, the incremental material is recovered into the heavy vacuum gas oil cut. In refineries processing heavy crudes, this cut is typically fed to a gas oil cat feed hydrotreater or a hydrocracker. Since this incremental cut is of sloppy distillation quality and is highly aromatic, it has a higher entrainment of nitrogen, metals and con carbon. Desulfurization of this material may be very difficult, leading to higher design severity or shorter catalyst run length in the hydroprocessing unit.

If the cutpoint is not increased, this material instead stays in the vacuum resid, which is typically fed to a Coker unit. Studies have indicated that the increased coke make in the Coker for this incremental heavy gas oil may be relatively small. Since the Coker unit includes a bottoms recycle from the Coker Main Fractionator, the cutpoint in the Coker can be specified to keep the worst heavy gas oil material out of the gas oil draw.

Unit Impacts

Figure 2 below depicts a typical refinery block flow diagram. Units are color coded based on the likelihood that a refinery designed for a medium sweet crude will require significant investment to convert to a heavy sour crude. Red units are highly likely to require major revamp and/or new units, yellow units are fairly likely to require investment, and white units are less likely to require investment. However, depending on the design basis chosen, it is possible that every unit in the refinery will be impacted in some way.

Figure 2 - Simplified Refinery Block Flow Diagram with Impacts

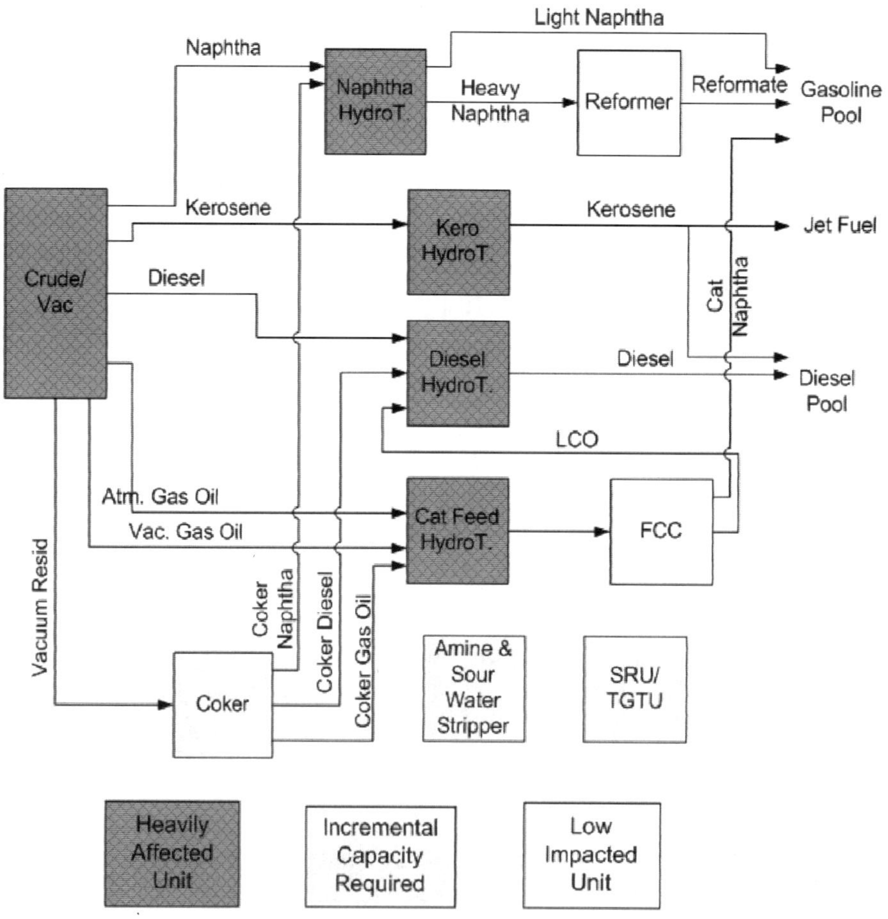

Processing Heavy Crudes in Existing Refineries

Crude/Vac

There are several key areas that must be addressed in the Crude/Vac unit design when considering running Canadian crude in a refinery.

Metallurgy

The metallurgy impacts on the Crude/Vacuum unit can be some of the most significant in the refinery. As mentioned above one of the first concerns of heavy sour crudes can be the metallurgy. There are two major causes for concern, sulfidic attack due to increased sulfur content and naphthenic acid corrosion. The sulfidation corrosion rates are typically addressed with the Modified McConomy Curves (7). However, most Canadian crudes have an additional problem of high naphthenic acid content commonly measured by TAN. The most common solution to the naphthenic acid corrosion problem is increased metallurgy in the affected equipment to 317L SS or better alloys with ≥3% molybdenum.

However, some studies have shown that not all Canadian crude naphthenic acids corrode at easily predictable rate. There are theories that the corrosion rates are lower than predicted in some cases based on some forms of naphthenic acids being less aggressive than others. (8, 9) There were reports that corrosion in units processing synbit or dilbit is exacerbated by an increased solids content (i.e. erosion-corrosion), but this has been mitigated by solids removal. Corrosion rate studies should be investigated before considering replacement of existing equipment and piping. These studies are not as critical for new installations since the incremental cost of upgraded metallurgy will not impact the overall costs as much.

One key parameter that can be very expensive is the transfer temperature for Crude/Vac products to the downstream units. If the OSBL transfer piping is carbon steel, it is important to maintain the unit transfer temperature below the temperature at which naphthenic acid corrosion is a concern, typically about 400-450°F.

Cracked Gas Make

The Vacuum Heater Cracked Gas Make is a potential major impact from running heavy sour crudes. Canadian crudes, especially bitumen derived crudes, have a very high cracking propensity. These crudes have been known to produce cracked gas at more than double, and sometimes triple, the rate of common conventional crudes. Revamps to increase the amount of Canadian crude in the feed slate to a Crude/Vacuum unit usually include improved Vacuum Systems with either new ejector condenser systems or new liquid ring vacuum pumps.

Coking and Vacuum Unit Design

Along with the high propensity for cracking, processing heavy sour crudes can lead to a high rate of coke formation, typically concentrated in the Vacuum Heater furnace. To minimize coke formation in the Vacuum Heater, several key mitigation strategies should be incorporated into the design. The Vacuum Heater should definitely have velocity steam (2, 3), which will maintain a high velocity in the heater tubes. A Double Fired

Vacuum Heater design is also recommended since it will reduce the peak heat flux in the tubes to minimize the coking potential.

Additionally, stripping steam in the vacuum column is recommended (2, 3). Studies show that this will minimize the required vacuum heater outlet temperature for a fixed vacuum resid cutpoint and vacuum column diameter.

One option to help minimize the impact of all the additional steam on the vacuum system would be the installation of a precondenser upstream of the ejectors or liquid ring vacuum pump. However, the installation of a precondenser adds pressure drop to the system, typically increasing the flash zone pressure.

Crude Preheat Exchanger Train

Several key design issues can affect the design and operation of the crude preheat exchanger train. First and foremost, the Canadian crudes typically have a higher crude viscosity that significantly impairs the heat transfer in the cold preheat exchangers. Unfortunately, there are not many options to improve the heat transfer in existing heat exchangers. There are options to improve heat transfer with unique baffle configurations, or twisted tube bundles with new exchanger designs.

The second major issue is the vaporization of the preheat train. Due to the high naphtha content of many of the dilbit crudes, this can lead to premature vaporization in the crude preheat train ahead of the crude heater pass valves if the unit does not have a preflash drum/column. The simplest solution to this problem is installation of the aforementioned preflash drum/column.

Desalting

Desalting of heavy Canadian crudes is much more challenging, and most likely the existing desalter system will need to be enlarged or replaced with one utilizing a newer desalting technology. Desalter performance can be hampered by factors, such as the increased salt content in the Canandian crudes, as well as the high crude density and viscosity, that make oil-water separation more difficult. Moreover, the combination of asphaltene precipitation and high naphthenic acid concentration will increase the tendency to form stable water-in-oil emulsion (rag layer), and potentially cause high oil entrainment in brine water and difficulty in maintaining the desired BS&W and salt removal at the desalter outlet.

One of the critical aspects of preventing desalter upsets is the ability to detect and monitor the emulsion/rag layer. The use of conventional level measurement devices, such as guided wave radar and displacer float column, has not been proven as accurate and reliable in stabilizing the desalter interface levels. The poor rag layer detection would also often result in relatively high chemical injection rates to control the rag layer. Recent advances in level instrumentation for 3-phase (oil, rag, water) level detection have been tested and implemented to accurately monitor the rag layer to optimize the desalter performance.

Processing Heavy Crudes in Existing Refineries

Kerosene Product Sulfur Specification

With a heavy sour crude slate, the Crude/Vac unit Kerosene may not meet the Jet specifications for smoke point or sulfur. Therefore the Kerosene may require further hydroprocessing to meet the Jet fuel specifications or blended into the Diesel pool ahead of the Diesel hydrotreater.

<u>Coker/Resid Processing</u>

Although the Delayed Coker is one of several options for vacuum resid processing, it is the most common unit in the United States for heavy ends processing due to economic considerations. Therefore, the discussion will be focused on that unit. Since the Coker's function is to upgrade the heavy vacuum resid into more valuable liquid products, and heavy sour crudes have significantly higher amounts of Vacuum resid in the feed, the Coker is typically one of the most overwhelmed in terms of capacity.

Although debottlenecking is usually possible by improvements to drum cycle time, most refiners will find that the capacity of their existing Coker is soon overwhelmed by the amount of vacuum resid that needs to be processed. Therefore, additional coking capacity in the form of a new unit is usually required to significantly increase the amount of heavy sour crude processed by the refinery.

Most Cokers are alloyed against sulfidation corrosion, but if the feed increases significantly in TAN, some metallurgy reviews are needed. Luckily, the heater temperatures cause the naphthenic acids to decompose, and there have been no reports of corrosion in the heaters or downstream. Hence, it is only the feed area after some preheating that has this corrosion risk.

<u>Hydrotreaters</u>

Hydrotreaters are one of the most impacted areas of the refinery as the amount of heavy sour crude increases. Typically, there is significantly more sulfur and aromatics in the Crude/Vac products that need hydroprocessing. Additionally, the Coker also makes more Coker distillate and gas oil that also must be hydroprocessed. The most significant impact is the catalyst cycle length if no additional hydroprocessing capacity is added.

First, there is incremental sulfur in the feed to the unit that must be treated. This incremental sulfur requires additional catalyst volume or increased reactor temperatures to achieve the same product specifications. If the crude includes a synthetic crude, the material has been previously hydroprocessed. Therefore, the remaining sulfur molecules are more difficult to treat.

Second, the metals and silicon are typically higher in heavy sour crudes than conventional crudes.

Third, with the increase of Coker capacity mentioned above, there will be more Coker stocks that must be hydroprocessed. In addition to containing a high level of sulfur, nitrogen, and aromatics compared to straight run Crude/Vac products, the coker stocks also may contain high levels of silicon. The silicon is usually related to the operation of the Coker and the amount of anti-foam that is injected. The silicon usually requires additional guard catalyst.

With the incremental changes in the feed quality to the unit, the demand for hydrogen to achieve the product specifications increases. This is because the bitumen derived stocks are naturally deficient in hydrogen content. It must be considered if the units have sufficient makeup hydrogen capacity and recycle hydrogen capacity.

One of the first concerns with the hydroprocessing units can be the metallurgy. As with the crude unit, the naphthenic acid corrosion can be a concern while the material is hot. However, many refineries transport the oils between units at moderate temperatures (<400°F) that alleviate the need for upgraded unit interconnecting pipe. However, in the hydroprocessing unit, the oil is usually heated as part of the process. It has been shown that once the oil is mixed with hydrogen, this prevents 300 series SS alloys (even those without significant molybdenum contents) from experiencing the acidic attack of the naphthenic acid. (10) Therefore, a careful review of the unit configuration is required to ensure that either upgraded metallurgy is provided, or that hydrogen is mixed with the oil before heating.

Another key unit design issue can be the reactor effluent wash water rate. The increased sulfur and nitrogen will lead to increased demand for wash water to minimize the corrosion potential by keeping the ammonium bisulfide (NH_4HS) concentration in the sour water below acceptable corrosion levels. The desired concentration will vary depending on the alloys and design, as discussed below under Sour Water.

Finally, one other parameter that must be considered is the amine loading on the recycle gas scrubber, if the hydrotreater in question has one. With increased sulfur liberated to H_2S, higher rich amine loadings are inevitable if the amine rate remains unchanged. The loadings must be maintained below industry acceptable norms to minimize corrosion potential.

Amine System

As a result of the increased H_2S produced in the hydroprocessing, coking and crude/vac units, the amine treating system can quickly become overloaded. One solution to improve capacity is to evaluate the amine selection and its concentration. If the refinery currently runs a DGA or MEA, additional capability might be achieved by considering MDEA. However, the selectivity of MDEA towards H_2S in the presence of CO_2 should be considered regarding the impact on fuel gas system metallurgy and fuel gas pipeline specifications. Also, a complete study of the amine system can ensure that no spare capacity is going unused.

Sour Water System

As with the amine system, the increased H_2S and ammonia production in the hydroprocessing units may quickly overload the regeneration system capacity. One possible solution to help debottleneck the system is to optimize the wash water in the various source units. Many times refiners are not recycling wash water up to the maximum limits of the metallurgy in individual units. Depending on the metallurgy of the system, the limits on ammonium bisulfide (NH_4HS) can range from 2 wt% for carbon steel systems to up to 6-8 wt% with upgraded metallurgy. However, more often than not the optimization of the wash water will not get the sour water system back in balance and additional sour water strippers will be needed.

One further optimization of the sour water system that may be considered would be the reduction of stripped sour water that is sent to the Waste Water plant. This can be done if stripped sour water is recycled back to the hydroprocessing units as part of the makeup wash water supply.

SRU/TGTU

One of the classic ways to improve the capabilities of the SRU is adding oxygen enrichment to the system. This is a very low cost solution in a refinery that has easy access to an air separation unit or a pipeline oxygen supply. Depending on the amount of oxygen enrichment, this may require from very simple modifications up to installing a complete new furnace.

Another possibility for an existing refinery is to change the sulfur unit sparing philosophy. Many refineries historically operated with a “spare unit” philosophy, where the sulfur throughput was kept below the total available sulfur refinery capacity with any one sulfur unit down for maintenance. In this case, all the units are normally operated below their maximum capacity, but with the capability to turn up to their maximum capacity quickly if any one unit goes down. However, if it is possible to store sour water or quickly shed sour water production, this may be a way of managing sulfur plant capacity during unit upsets instead of maintaining spare unit capacity.

Desalter Underflow Wastewater Treating

Effluent water from Crude unit desalters processing heavy crude may have more hydrocarbon carry-under during normal and mudwash operation, specifically brine “rag layer” emulsions with higher BOD/COD loads than from typical crudes. While every effort should be taken to optimize the performance of the desalters in the Crude unit, consideration may also be given to either upgrading the refinery wastewater treating system, or installing a unit to segregate and pretreat the desalter brine, prior to routing to refinery water treating system.

Fluor has developed a desalter water treating system to accomplish this task, which has performed very well in laboratory tests. Other suppliers also have proprietary technology that reportedly will be able to pre-treat this stream, although there do not appear to be any operating units at this time.

Naphtha Processing

By its very nature, a dilbit based crude has a significant amount of naphtha that needs to be processed. There are two basic options. First, return it back to the oil production site, or second, upgrade the naphtha quality to salable gasoline. The issue with returning the naphtha to the production site is a very complex question related to the logistics of the shipment of the naphtha. In areas with a short distance between the refinery and the production site, it makes a lot of sense to return the naphtha to the site via pipeline. However, for many Canadian dilbit processing projects in the United States, it is not reasonable to ship the diluent naphtha back to Canada, since pipeline capacity from the refinery to the northbound pipelines is limited at many refinery sites. Therefore, many refineries will require incremental naphtha hydrotreating and reforming capacity to upgrade the naphtha octane to meet gasoline requirements.

Processing Heavy Crudes in Existing Refineries

Conclusion

As world demand for petroleum products increases, and production in existing light crude fields declines, heavier, high sulfur crude oils are increasing in market share. Canadian crudes are becoming more commonplace in the market and present a number of challenges to process due to their nature. Processing these crudes in existing refineries may require significant refinery modifications. Many process units are affected with key units to consider being the Crude/Vac, Hydrotreaters, Coker/Resid processing, Amine, Sour Water, and SRU/TGTU.

References

1. McGinnis, J. and Confrotte, E., "Oil sands: Vision resource for the energy industry", *Hydrocarbon Processing*, September 2007
2. Golden, S. et al, "Vacuum unit design for high metals crudes", *Process Technology Quarterly*, Q1 2007
3. Grande, M., "Designing vacuum distillation for deep-cut bitumen service", *Hydrocarbon Processing*, February 2010
4. Golden, S. "Canadian crude processing challenges", *Process Technology Quarterly*, Q1 2008
5. Henderson, R. et al, "Consider modifying your refinery to handle heavy opportunity crude oils", *Hydrocarbon Processing*, September 2005
6. Chambers, B., "Assessing Refinery Crude Corrosivity," Joint Industry Research Proposal, Honeywell web site, 2006, http://hpsweb.honeywell.com/NR/rdonlyres/ED973E17-7EEB-4D82-B44C-B10CC55F5324/89839/NAP_JIP_Prospectus_021809.pdf
7. "Overview of Sulfidic Corrosion in Petroleum Refining," NACE International Publication 34103, NACE International, Houston, TX, Feb. 2004.
8. Messer, B. et al, "New Theory for Naphthenic Acid Corrosivity of Athabasca Oilsands Crudes," NACE CORROSION/04, Paper 04634, NACE International, Houston, TX, Mar. 2004.
9. Dettman, H. et al, "Refinery Corrosion, Organic Acid Structure, and Athabasca Bitumen," NACE CORROSION/09, Paper 09336, NACE International, Houston, TX, Mar. 2006.
10. Shargay, C. et al, "Survey of Materials in Hydrotreater Units Processing High TAN Feeds," NACE CORROSION/07, Paper 07573, NACE International, Houston, TX, Mar. 2007.
11. Janjua, R. "Wastewater Treatment for Refineries Using Canadian Tar Sands SAGD Crude as Feedstock", 2011 AIChE Spring Meeting, Chicago, IL, Mar. 2011

بسم الله الرحمن الرحيم

Two years after the financial crisis: Where are we? Was it all bad?

Ahmed Khogeer, PhD, PE, PMP
Mohamed Soliman
Saudi Aramco

Mar 17, 2011

ارامكو السعودية
Saudi Aramco

Outline

- Review of the effects of the financial crisis
  - Current situation for NOC
  - Current situation for IOC
- Other side effects
- Outlook for the markets
- Outlook for the construction industry
- Why are most gulf countries and china different?
- Benefits for these counties:
- Conclusion

Incremental Oil Demand 2003-2015
(before crisis)

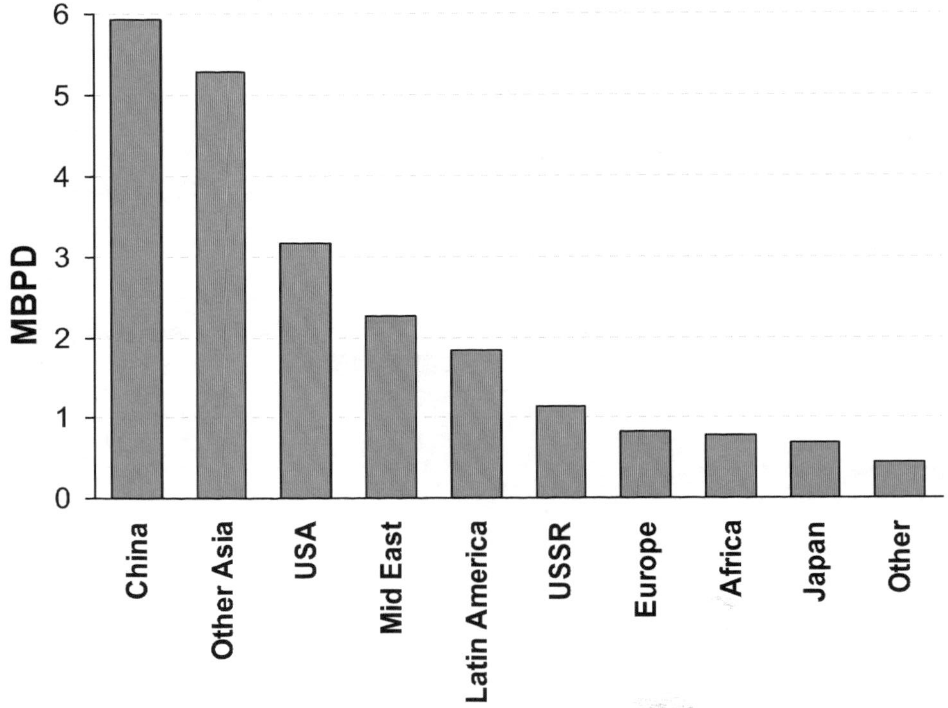

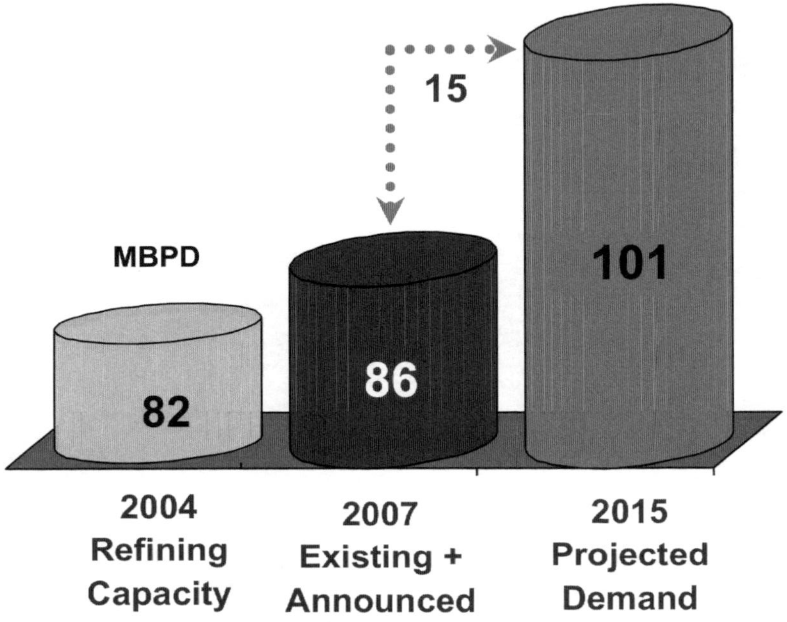

Sources: O&GJ, PGI, Woodmac

Significant growing refined products deficit (before crisis)

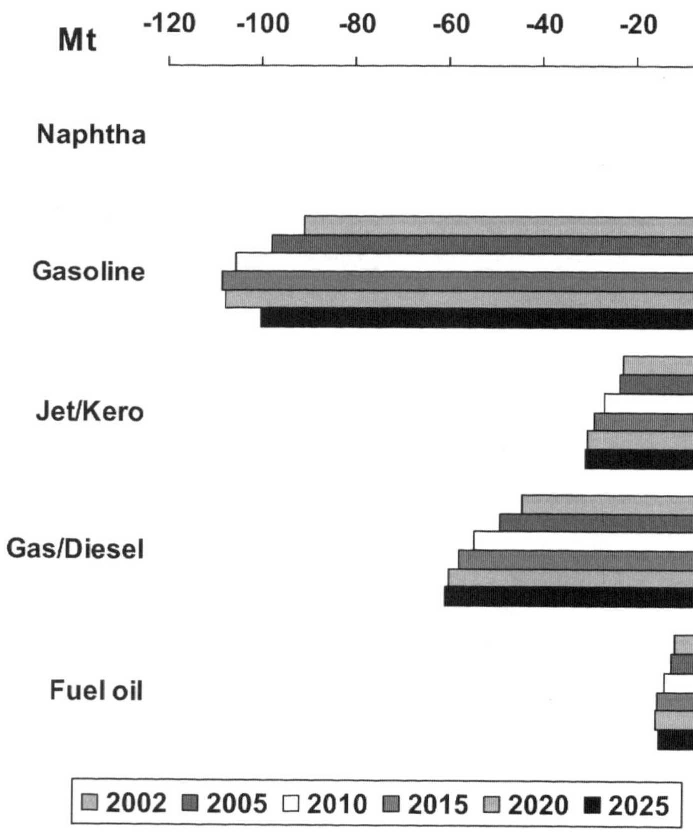

PADD-1

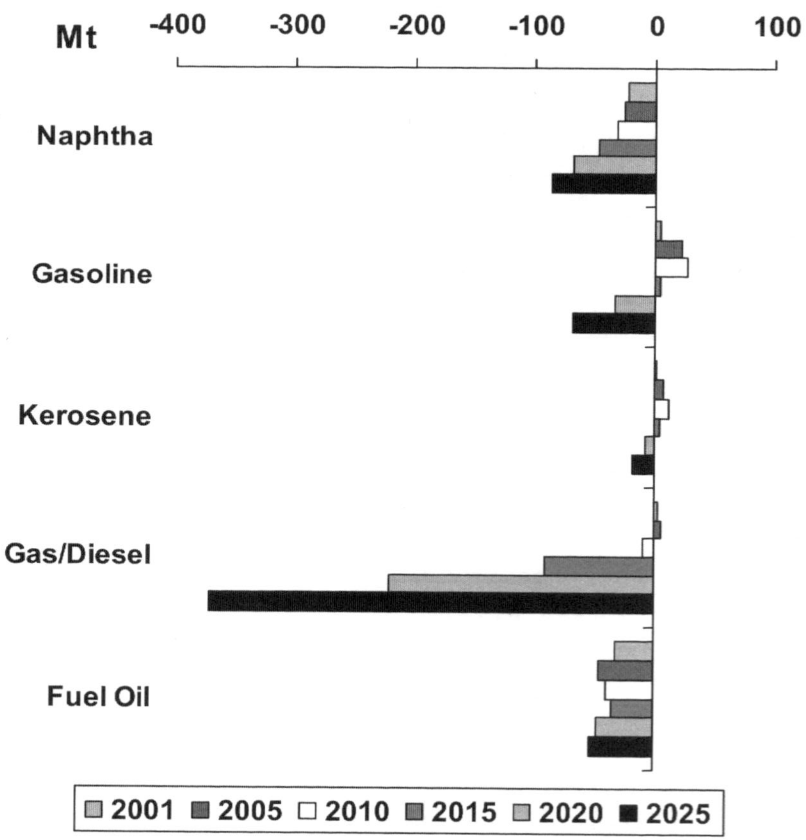

Asia

Source: Woodmac

Saudi Aramco

Effect of crisis on crude and products

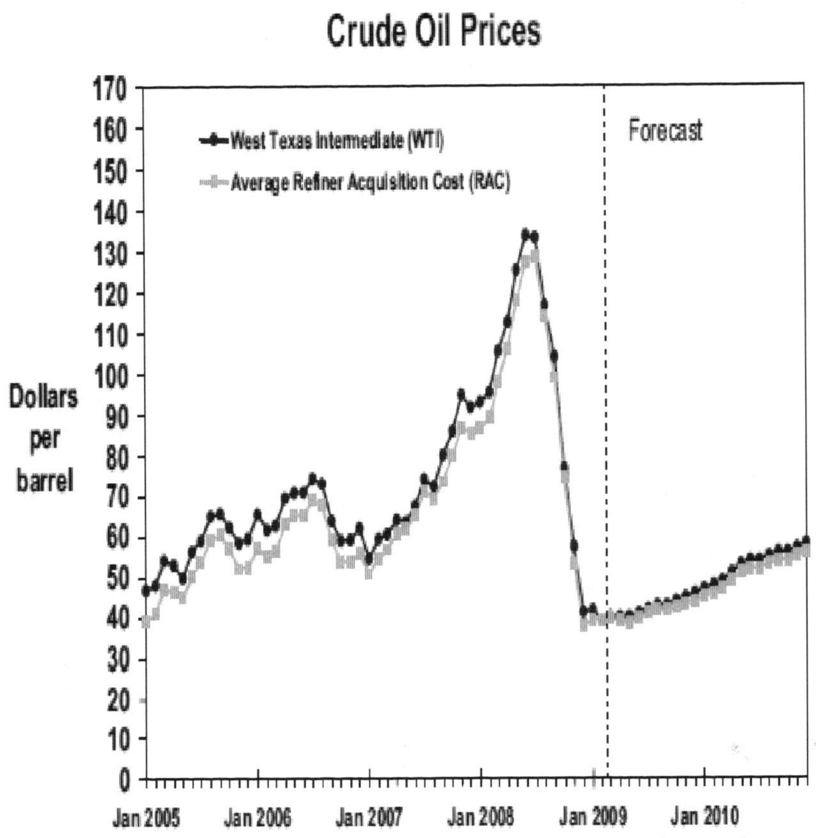

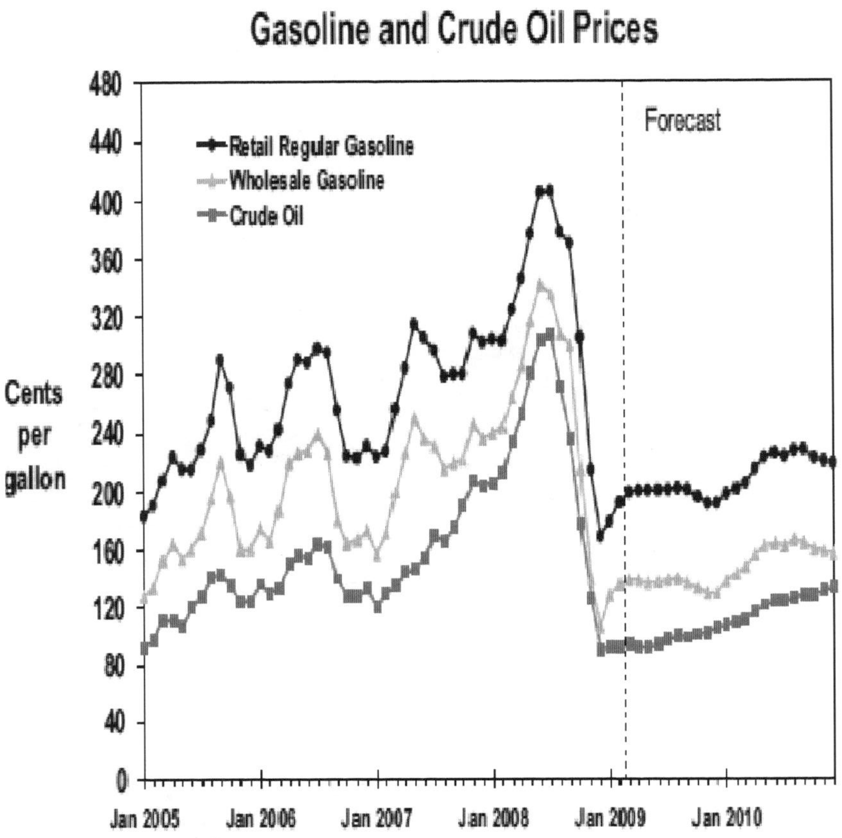

Source: EIA

...and jobs

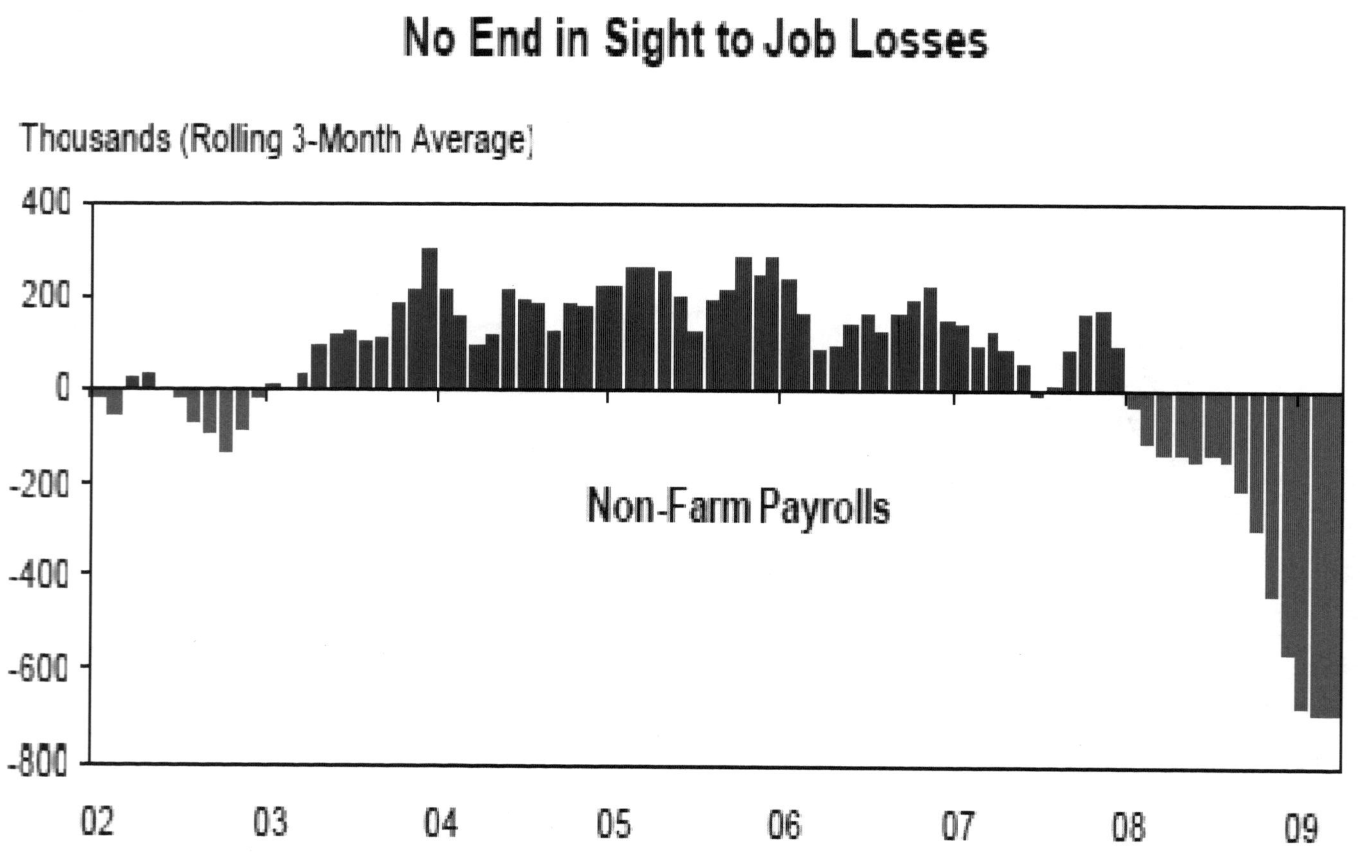

Sources: Bloomberg and BBH Analysis

Current situation for international oil companies IOC

- IOC market value depreciated as a result of :

1. Low oil prices which means low assets
2. Low demand on all fuels such as transportation, power, industrial
3. The believe that time economy recovers, more efficient vehicles will relax the demand

- IOC now have much lower credit rating and line of credits to finance multi billion projects
- Most Refinery projects are cancelled
- Refining margins are at near historical low

أرامكو السعودية
Saudi Aramco

Current situation on national oil companies NOC

- Most NOC (and thus governments) are in better shape due to cash reserve they made in past years
- For smaller countries such as Qatar, the main emphasis is on export. National demand is negligible.
- For Saudi Arabia with highly developed petrochemical industries, domestic demand can be 25 % of production
- Problem with low oil demand, less associated gas for petrochemicals
- Need for domestic refineries ~ 800 K is going forward, may benefit form lower costs. Simple not full conversion due to need of fuel oil
- With population increase rate of 3.7 % and 70 % under age 30, more need for gasoline
- What if all women drive??

Effects on oil exporting countries, china, India

- In most countries where Islamic financial laws (sharieah) are practiced, financial markets and housing markets particularly where not effected much by the credit crisis.
- No selling on margin and no variable interest rates, no fine prints.
- Makah showed great appreciation in the real estate prices.
- Only 25 % of population own houses
- Most banks showed profit in last two quarters
- Gulf countries: $ trillion in power, oil, refining, and infrastructure.
- Export oriented projects effected much due to international demand
- Infrastructure projects can benefit from lower construction costs

أرامكو السعودية
Saudi Aramco

Some mega projects in Saudi Arabia.. Which one you think will be cancelled??

Project	Cost, $ B	After crisis
Aramco- ConocoPhillips 400 k export refinery	12-15	10
Aramco- Total 400 k export refinery	12	9
Aramco-Dow RTIP petrochemical complex	30	~20 with scope reduction
Oil production expansion	??	
King Abdullah economic city	25	stalled
Jazan economic city	25	stalled
Makah real estate projects	50	
New Jeddah airport	10	7
Jeddah sewer system expansion	2	?
Desalination expansion projects	20	?
Education system upgrade	5	
King Abdullah university for science and tech.	10	
Saudi Airlines Fleet modernization	10	

Effect on EPC

- Many economies have now entered periods of recession.
- Steel prices increased by more than 60% at their peak but fell to 11% higher than they were on January 1, 2008.
- Copper prices increased 23%, drop to 47% lower than first of 2008.
- Manufactured equipment prices continued to remain high, reflecting existing commitments and high manufacturer

أرامكو السعودية
Saudi Aramco

2008 Financial Crisis:

- The 2008 turmoil has lead to Lehman **Brothers** become the first major victim of the financial crisis and began the **largest bankruptcy** proceedings in world history, when it went bankrupt on September 15, 2008 with assets worth $691 billion and debt of $613 billion. It was soon followed by the country's largest savings and loan holding company WaMu, which went bankrupt on September 26, 2008, and its $327 billion asset was immediately sold to J P Morgan Chase for mere $1.9 billion. Wamu was the second largest bankruptcy. The auto sector started falling apart a few months later and Chrysler filed for bankruptcy on April 30, 2009, with assets worth $ 39.3 billion. The very next day, Thornburg Mortgage followed with assets of $ 36.5 billion. GM has become the latest to join the list as fourth biggest bankruptcy with assets worth $91 billion. General Motors Corp filed for bankruptcy on Monday June 1st, 2009 forcing the 100-year-old iconic car maker which has been one of the most visible symbols of American capitalism into uncharted waters of nationalization or government ownership. This was the largest-ever US bankruptcy in the manufacturing arena. GM, Chrysler and Thornburg Mortgage have found place among the 10 biggest ever bankruptcies in the American history.

أرامكو السعودية
Saudi Aramco

Causes of the financial crisis:

- Many factors directly and indirectly caused the 2008 financial crisis (which started with the US subprime mortgage crisis), with experts placing different weights upon particular causes.
- The U.S. **Financial Crisis Inquiry Commission** reported its findings in January 2011. It concluded that "the crisis was avoidable and was caused by: Widespread failures in financial regulation, including
  - The Federal Reserve's failure to stem the tide of toxic mortgages;
  - Dramatic breakdowns in corporate governance including too many financial firms acting recklessly and taking on too much risk;
  - An explosive mix of excessive borrowing and risk by households and Wall Street that put the financial system on a collision course with crisis;
  - Key policy makers ill prepared for the crisis, lacking a full understanding of the financial system they oversaw;
  - Systemic breaches in accountability and ethics at all levels.
- The U.S. Financial Crisis Inquiry Commission findings clearly indicate that the financial institutions were behind the 2008 financial earthquake.

The impact of 2008 Financial Meltdown on Chemical and oil& gas industries

- Because of the current globalization, the crisis was not limited to the financial sector alone but spread to affect every business not only in America alone but also around the globe. Many industries that were performing very well before the crisis were dramatically affected by the financial crisis. Some of these industries were petrochemical, oil& gas, mining, .. industries.
-
- Chemical, petrochemical and oil & gas industries were not isolated from the financial meltdown and were dramatically affected primarily due to the credit squeeze as follows.

- Many projects has been cancelled or postponed as a result.
  - Yanbu Export refinery (YERP), a joint venture between saudi aramco and Conoco-philips was put on hold. Later, conocophilips withdrew from the project
  - Oil sand projects were cancelled or put on hold
- Royal Dutch Shell PLC delayed their Canadian oil sands projects like its 100,000 barrel per day Carmon Creek project. Along with Shell, Suncor Energy Inc , Nexen Inc , Petro-Canada , Canadian Natural Resources Ltd. and others delayed or deferred projects in the region. http://www.financialpost.com/story.html?id=1000660
-
  - Dow chemical joint venture with saudi aramco was put on hold
  -
- Companies disappeared from the global map
  - Nortel Networks
  - Wamu
  -
- Laying off

أرامكو السعودية
Saudi Aramco

Was it really bad?

- Many rich cash companies and countries has benefited from the financial turmoil around the globe. Out of the top 15 mergers and acquisition in 2009, 50 % were chemical and oil & gas related companies as shown in the table below.
- **China**
- *China's Powerful State-Owned Enterprises Scour the Globe during and after the financial crisis to buy assets to Quench the Country's Insatiable Thirst for Oil & Gas. Dome of the mergers and acquisition are listed in the graph below. China managed to take control of major oil & gas assets that they will never to achieve during normal stable environment.*
- http://seekingalpha.com/article/166968-2009-a-chinese-energy-acquisition-odyssey

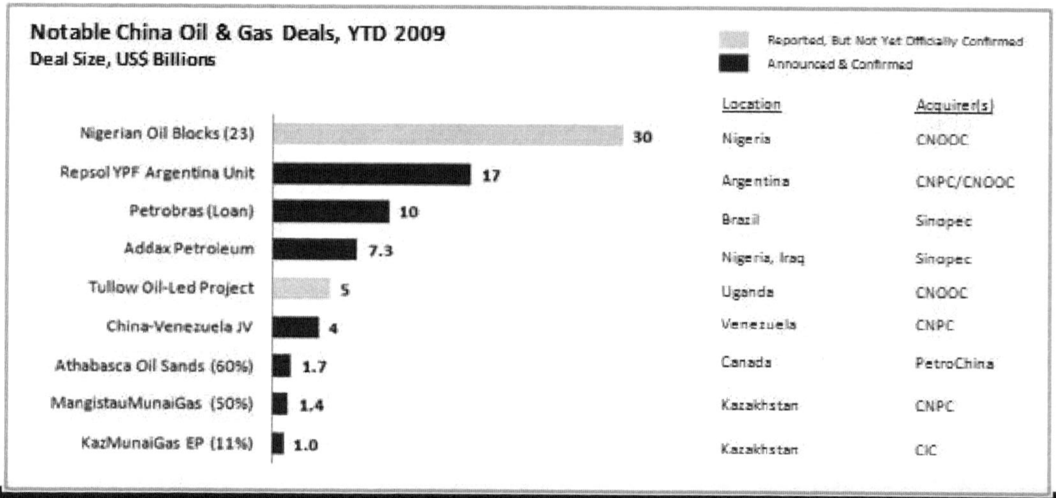

- **Gulf states**
- **Saudi Arabia**
- Saudi Aramco
  - YERP Refinery
  - Manifa Project
  - Dow Chemical
  - Hiring top qualified employees at cheaper rate
- Sabic
- Others
- **Qatar**

- **Global Top 15 Mergers and acquisition in 2009:**
- The table below shows that 30% of top M&A in 2009 was in the oil& gas industries. The aquiror got the target companies for up to 50% discount to their average value at the beginning of 2008. For example, Suncor energy acquired petrocanada for 18.2 billions while its value was more than 26.0 billions just before the financial meltdown. The table below shows that more than 50 % of the top 11 acquisitions were directly related to companies where chemical engineers play an important role. These sectors are Pharmaceuticals, Energy and mining.

Rank	Date	Target (%Sought/Acquired)	Acquiror	Rank Value (US$m)	Target Macro / Mid industry
1	1/26/2009	Wyeth (100%)	Pfizer Inc	64,479.70	Pharmaceuticals
2	6/5/2009	Rio Tinto PLC-WA Iron Ore Asts (100%)	BHP Billiton-WA Iron Ore Asts	58,000.00	Materials / Metals & Mining
3	3/9/2009	Schering-Plough Corp (100%)	Merck & Co Inc	45,704.40	Pharmaceuticals
4	6/1/2009	General Motors-Cert Assets (100%)	Vehicle Acq Holdings LLC	42,979.50	Industrials / Automobiles & Components
5	2/26/2009	RBS	HM Treasury	41,878.70	Financials / Banks
6	12/14/2009	XTO Energy Inc (100%)	Exxon Mobil Corp	40,659.10	Energy and Power / Oil & Gas
7	11/3/2009	Burlington Northern Santa Fe (78.1%)	Berkshire Hathaway Inc	35,948.00	Industrials / Transportation & Infrastructure
8	7/30/2009	Citigroup Inc (38.81%)	Preferred Shareholders	28,078.30	Financials / Banks
9	11/2/2009	Cenovus Energy Inc (100%)	Shareholders	23,312.90	Energy and Power / Oil & Gas
10	9/7/2009	Cadbury PLC (100%)	Kraft Foods Inc	19,255.40	Consumer Staples / Food and Beverage
11	3/23/2009	Petro-Canada (100%)	Suncor Energy Inc	18,245.80	Energy and Power / Oil & Gas
12	5/4/2009	Liberty Entertainment Inc (100%)	DirecTV Group Inc	14,499.10	Media and Entertainment / Cable
13	12/3/2009	NBC Universal Inc (100%)	Comcast Corp	14,400.00	Media and Entertainment / Broadcasting
14	2/20/2009	Endesa SA (25.01%)	Enel SpA	13,470.00	Energy and Power / Power
15	6/11/2009	Barclays Global Investors Ltd (100%)	BlackRock Inc	13,345.20	Financials / Asset Management

How to overcome and benefit from the crisis

Flexibility in Execution Strategies

- Utilize cost reimbursable contracting strategies.
- Utilize “limited” converted LSTK contracting strategies.
- Include provisions in contracts for commodity price indexing.
- Ensure use of appropriate negotiating parameters
- Optimize contract schedules.
- Increase use of early funding.
- Optimize engineering standards.

Relax Financial Constraints

- Include provisions in contracts (as an option) for advance payments.
- Optimize progress measurement systems towards neutral cash flow.

أرامكو السعودية
Saudi Aramco

Utilize cost reimbursable contracting for EPC

Company pay actual costs for materials and services.

Benefits:

1. Pay no more than the actual market costs for material and services.
2. More contractor participation and competition due to less risks.
3. Can achieve lowest cost and shortest schedule.

Risks:

1. More administrative efforts than normal LSTK.
2. Project cost can increase if prices rise in future.
3. Project budget is uncertain due to future price unknowns.
4. Contractor's risk for performance is shifted to Company.
5. Contractor limited incentive to control costs and meet schedule.

- Applies to new projects where contracts have not been awarded.

Utilize Limited Convertible LSTK

Modify LSTK to allow portions of lump sum price be determined after award based on actual costs.

Eg: leaving out the cost for a single key equipment PO and take offs for bulk materials, and/or the cost of executed subcontracts.

Benefits:

1. Minimizes contractor's risk and thus his contingency due to high volatility of commodity and materials.
2. Company can realize cost savings when future prices drop; significant price reductions are expected under existing conditions.

Risks:

1. Project cost can increase if future prices rise.
2. Company budget is uncertain due to future price unknowns.
3. Some additional administrative effort will be required

Include commodity price indexing in LSTK

- **Benefits:**

1. Minimize contractor's risk due to high volatility of commodity prices, thus reducing contractor's contingency in the bid price.
2. Company can realize cost saving when future prices drop. Significant price reductions are expected under the existing market conditions.

- **Risks:**

1. Project cost can increase if future prices rise.
2. Project budget is uncertain due to future price unknown.

Optimize schedule completion based on project criticality and revenue-generating status

Select the scheduled completion date that results in best balance between contractor price and Company costs and revenue.

Benefits:

1. Possible early completion resulting in increased revenue and reduced overheads.
2. Possible cost savings resulting from lower contractor price for an extended schedule.

Risks:

1. Possible additional cost for advance delivery.

Utilize early funding for purchasing materials before ERA

When prices are expected to rise, early funding for early procurement of project materials at a favorable price.

Risks:

1. Cost savings may not be achieved if prices do not rise as expected.
2. If a project is cancelled, there could be costs for abandoned engineering, cancelation of purchase orders, purchase of unneeded material, and/or abandoned construction work.
3. In a situation of falling oil prices and reduced production, likelihood of cancelled or deferred projects increase.
4. Purchased material must be suitable for use on the project.
5. Company cash flow could be advanced

ارامكو السعودية
Saudi Aramco

increase the amounts of advance payment for LSTK and LSPB Contractors

During tight credit market, number of interested bidders may be limited due to lack or high financing costs. Offering to provide or increase advance payments to contractors may attract more bidders.

Benefits:

More available bidders and, therefore, more competition thus potential cost savings from lower bids

summary

- The financial crisis has most negative effect on IOC and export oriented projects.
- NOC, had made good reserve of cash during the past w a few years but market is still unclear.
- Domestic demand projects, and infrastructure projects has to continue and can benefit much from the lower EPC cost .
- Cancellation of international projects can help some domestic oriented projects be more competitive to export surplus products if specs allow.

Thank you
أرامكو السعودية
Saudi Aramco

(20c) Recovery of High Purity Propylene From Refinery LPG

R.E. (Ed) Palmer, Mustang Engineering, L.P., Houston, TX

Many refiners are investigating options for manufacturing higher value specialty products from traditional refinery product and intermediate streams as a means of improving weak cash margins that currently prevail. One potential opportunity is the recovery of high purity propylene from refinery-mixed propane/propylene (PPs) product. Normally, the bulk of this material originates from an FCC Unit and is processed along with butane/butylenes to make gasoline via Alkylation or Polymerization (or Dimerization). However, with the projected declining demand for gasoline in the US many refiners are placing increasing scrutiny on the disposition of this valuable material.

In this paper, a financial analysis is presented for an onsite facility to recover high purity propylene from refinery PPs The critical variables involved in the optimization of equipment sizes and operating costs for the recovery of high purity propylene by conventional distillation and distillation utilizing a heat pump are examined. This is followed by a capital cost estimate for each scheme and ranking based on an investment plus operating cost formula. The scope and capital costs for the associated offsites facilities including prefractionation, treating, storage and shipping are also included. Finally, a financial analysis is presented showing the discounted cash flow rate of return (DCF) as a function of propylene value, utilities cost and facility capacity. The value of the PP feed stream is also calculated as a function of the rate of return for the project and the other variables noted above.

See more of this Session: New Projects, Expansion, Revamps, Debottlenecking, and Reliability
See more of this Group/Topical: Topical 7: 14th Topical on Refinery Processing

(20d) Naphtha Normal Paraffin Separation Using a Dividing Wall Column and Simulated Moving Adsorption Bed

S. Thomas King[1], Kurt J. Cenek[1], Lawrence W. Miller[1], Joe R. Haas[2] and Cynthia K. Zimmerman[1], (1) Olefins, Detergents, and Alkylation Technology Center, Engineering Department, UOP / Honeywell, Des Plaines, IL, (2)Optimization Services, Engineering Department, UOP / Honeywell, Des Plaines, IL

Improving the separation of normal paraffin from naphtha can increase ethylene yield from naphtha crackers. This is achieved by increasing the concentration of normal paraffin (n-paraffin) in the naphtha cracker feedstock. The n-paraffin depleted naphtha byproduct can then, in turn, be fed to a reforming unit to increase the octane value of the refiner's gasoline pool. This presentation will, first, discuss the application of Dividing Wall Column design using high capacity trays to decrease naphtha prefractionation utility consumption and reduce a two column system to a single column without sacrificing product purity and recovery. Second, it will discuss how a simulated moving adsorption bed can be used to separate prefractionated naphtha normal and non-normal paraffins to increase the respective naphtha cracker and reforming unit performance.

See more of this Session: New Projects, Expansion, Revamps, Debottlenecking, and Reliability
See more of this Group/Topical: Topical 7: 14th Topical on Refinery Processing

ADAPTING THE NAPHTHA COMPLEX TO A CHANGING ENVIRONMENT

Jeff M. Bray and Steven L. Kleinman, UOP LLC, A Honeywell Company - Des Plaines, IL

It seems that all of the activity in the refining business is in the areas of the refinery that process distillate and heavier material, and much of this emphasis is justified. Worldwide, the market for motor fuels is moving from gasoline towards diesel, while at the same time the demand for the heavier fuel oils is decreasing. But this does not mean that the naphtha complex can be ignored. Indeed, the above changes and others all impact how the naphtha complex needs to be configured and operated.

While not every refinery has all of the same process units, the naphtha complex generally includes hydrotreating, reforming, alkylation, isomerization, benzene saturation, and any associated separation units. The design and operation of this equipment is determined by both the process unit feeds and product needs. The feed is affected by the types of crude being purchased, which are trending heavier, as well as the material being produced by other units outside the naphtha block, including any hydrocrackers, cokers, and other residue processing units. The product specifications are determined by customer preferences, as well the ever tightening environmental requirements – lower sulfur, lower volatility, and less aromatics, especially benzene. In addition, the rest of refinery is becoming more dependent on the naphtha block's reformer hydrogen for its hydrotreating and hydrocracking needs.

This environment makes it increasingly difficult and increasingly important to consider the naphtha processing units as a complex, rather than individually. The output of one unit influences the operation of the next. Cut-point choices and separation efficiency affect the next two processes. The ability of the refiner to integrate accurate process technology models with representative separation models, product blending models, and current demand and pricing data is crucial to optimizing the overall complex operations.
The tools and skills needed to conduct this type of study may not be available in every refinery, or within every refining corporation.

To insure that the maximum value is being captured from the processing of the refinery naphtha, a study of this portion of the refinery needs to evaluate changes including:

1. Feed changes – This could include cut point changes in naphtha streams, processing purchased streams, use of hydrocracked or FCC naphtha, or changing routing of naphtha between units.
2. Equipment – How the processing can be changed by the debottlenecking of existing equipment or addition of other processing equipment?
3. Catalyst changes – As objectives and catalyst technology improves it is important to understand the effects of changes.
4. Conditions – Changes in fractionation efficiency or catalytic unit conditions should be evaluated.
5. Different Product Mix – The ability of the naphtha block to adjust to changing gasoline specifications needs to be understood well.

To achieve this range of evaluation, an effective naphtha study needs to have the following characteristics:

1. Component based modeling of C12 and lighter materials
2. Fairly rigorous modeling of separation and equipment
3. Strong reaction modeling that use compositional data and can show feed composition effects as well as operating conditions and catalyst change effects
4. Effective modeling of gasoline blending – This is needed to show the effects in the final product production and properties so that when combined with any petrochemical feedstocks (aromatics or naphtha sales), the overall economics of any change can be measured with a minimum of pricing of intermediate streams

The authors have conducted numerous refinery studies for a range of refinery clients, from small independent, single plant refiners to some of the world's largest multi-facility, integrated plant operators. Two case studies will be presented within this range of services. The first case will discuss a study commissioned to evaluate the benzene reduction alternatives needed to bring a refinery into compliance with the most recent Mobile Source Air Toxics regulations (MSAT2). The second case study will investigate configuring a naphtha block that is designed to produce material for both gasoline and aromatics production, and the conflicting and complex factors that must be accounted for in developing an optimal and flexible configuration.

(20f) Separation of Butane Isomer Mixtures Using Dense 6FDA-DAM Membranes

Omoyemen Esekhile and W.J. Koros, School of Chemical & Biomolecular Engineering, Georgia Institute of Technology, Atlanta, GA

SEPARATION OF BUTANE ISOMER MIXTURES USING DENSE 6FDA-DAM MEMBRANES

Omoye Esekhile, William J. Koros

oesekhile3@gatech.edu, bill.koros@chbe.gatech.edu

School of Chemical and Biomolecular Engineering

Georgia Institute of Technology, Atlanta, GA

The glassy 6FDA-DAM polymer has a relatively high fractional free volume, thus allowing for high permeabilities of butane isomers compared to most other polymers. The separation performance of this membrane has been studied previously under ideal conditions of single gas feeds with upstream pressures up to 2atm at 100°C, and downstream under vacuum. These ideal conditions may give overly optimistic separation performance estimates.

Mixture gas permeation studies introduce factors such as sorption and transport competition, and so-called bulk flow or "frame of reference" nonidealities. These factors may significantly influence the separation performance of the membrane even in the absence of plasticization. The dual mode transport model has been extended to account for competition in the Langmuir environment of the glassy polymer, and modified to account for the bulk flow effect. Current work with butane isomers, however, further suggests a possible breakdown in local equilibrium in the transport of this mixture system. This adds a previously unobserved effect that may be a new feature for glassy membrane-based separations of large penetrants

The breakdown in local equilibrium is hypothesized to result from the slow jump rate of isobutane from the Langmuir to dissolved environments, relative to the diffusion time scale of nC4 into the Langmuir environment. An *apparent* affinity constant for the faster moving n-butane isomer can be used to describe this breakdown.

In this presentation, I will discuss the behavior of this mixture system over various feed compositions and discuss further the possible breakdown of local equilibrium. I will also suggest ways of enhancing the separation performance of the membrane.

(22a) Fundamentals of Alarm Management

Peter G. Herena, Kenexis Consulting Corporation, Elgin, IL

When the information era began thirty years ago, the focus for the refinery control specialist was placing as much data as possible into the hands of engineers and operators. The philosophy remained prevalent though the volume of transferred data leaped by orders of magnitude every decade. Eventually refinery operators began to realize that the sheer volume of data was, in part, causing the problem; and thus the field of Alarm Management was born.

But how does one go through the process of identifying the most critical alarms, and once that is known, what are the methods used to get the most important information in the hands of the correct person? In this presentation the fundamentals of Alarm Management will be discussed, including assessing hazards, identifying data transmission bottlenecks and understanding how our brains are wired to recognize and respond to information. Specific examples will be cited and current good practices will be described.

See more of this Session: Process Control Optimization and Tutorial
See more of this Group/Topical: Topical 7: 14th Topical on Refinery Processing

(22b) Who's Controlling Your Industrial Control System? and What Makes You So Sure? An Information Security Primer for Systems Professionals

Todd Haverkos, Iconium Security, Elgin, IL

In the wake of Stuxnet – a game changing piece of malware specifically targeting air gapped industrial control systems infrastructure – electronic security has rocketed up the worry list for control systems professionals. Todd will discuss the tools in the current arsenal of modern attackers, how they evade detection, and the ecosystem surrounding the underground economy that fuels for-profit malicious software and hackers as well as discussing specifics of Stuxnet that reverse engineers are discovering as they work to decompile and decrypt the intent behind the highly advanced class of threat that Stuxnet represents. Todd will discuss what specifically a control systems professional should be thinking about with respect to information security in their day-to-day work, and how to keep systems from becoming the next victim (or at least detect whether a system is a current victim).

See more of this Session: Process Control Optimization and Tutorial
See more of this Group/Topical: Topical 7: 14th Topical on Refinery Processing

(22c) A Look Into Advanced Controls for Automotive Engines

Aaron Matthews, University of Wisconsin-Madison, Chicago, IL

This presentation features an in-depth look at engine controls for gasoline, diesel, and hybrid drivetrains. The most important control tasks are then identified and their implications on engine performance are analyzed. Examples are given showing driver input and controller output sensitive to environmental conditions. There are also examples of advanced hardware configurations with a brief list of pros/cons for each. The most important control tasks are then identified and their implications on engine performance are analyzed. The presentation also discusses emissions testing procedures for transient and stationary engines.

See more of this Session: Process Control Optimization and Tutorial
See more of this Group/Topical: Topical 7: 14th Topical on Refinery Processing

Importance of Data Reconciliation on Improving Performances of Crude Refinery Preheat Trains

Edward M. Ishiyama[1,2], Simon J. Pugh[1], D. Ian Wilson[2], William R. Paterson[2] and Graham T. Polley[3], (1) Process Engineering Technology, IHS-ESDU, London, United Kingdom, (2) Chemical Engineering and Biotechnology, University of Cambridge, Cambridge, United Kingdom, (3) Chemical Engineering, Universidad de Guanajuato, Guanajuato, Mexico

Abstract

Performance of a preheat train is evaluated through plant monitoring data (temperature and flow measurements); here the reliability of data reconciliation plays a significant role. Most preheat trains do not monitor all stream temperatures (or flow rates) and frequently guess the missing data through 'short-cut' methods. Wrong choice of 'short-cut' methods such as the use of linear interpolation is shown to result in misleading conclusions.

This paper introduces a systematic data reconciliation approach through three steps. 1) Generation of missing data through a modified preheat train simulator based on *Energy Fuels*, 2009, *23*(3), pp 1323–1337). 2) Filtering unreliable data through a 'trusted' heat balance. 3) Grouping heat exchanger monitoring data into different time sections to identify trends under different operating periods.

The superiority of the new data reconciliation methodology is illustrated through a series of case studies. The reconciled data are then utilized to make key operational decisions to improve preheat train performance.

1. Introduction

Preheat trains are networks of heat exchangers, usually complex, with flow splitters and bypass streams, which transfer heat between the distillation tower products and its crude oil feed. These units are prone to fouling. Fouling is the formation of unwanted deposits on the heat transfer surfaces with time. The fouling behaviour is dynamic and is a combination of several factors (*i.e.* a combination of particulate, chemical reaction, crystallization and corrosion fouling in the preheat train). The complex nature of the fouling behaviour was illustrated by Ishiyama *et al.* (2010), where refinery fouling analysis and laboratory experimental reports were summarized (reproduced as Figure 1). Figure 1 is a plot between actual plant heat exchanger fouling rates against wall shear stress. Note that all exchangers in this plot are shell-and-tube units with crude on the tube-side. The solid line is the correlation proposed by Joshi *et al.* (2009) based on plant fouling data collected by Shell Global Solutions. This correlations show that the average rate of fouling decreased with increasing tube-side wall shear and the influence of temperature was not significant to the observed fouling rates. On the same plot are hollow symbols (two data sets) and filled symbols (5 data sets), representing fouling rate observed in pilot plant experiment and 5 other refineries. Each data point present exchangers operated at different temperatures (detailed in Ishiyama *et al.*, 2010), and would raise the concern that the use of shear stress alone is insufficient in evaluating fouling behaviours in certain exchangers in various refineries.

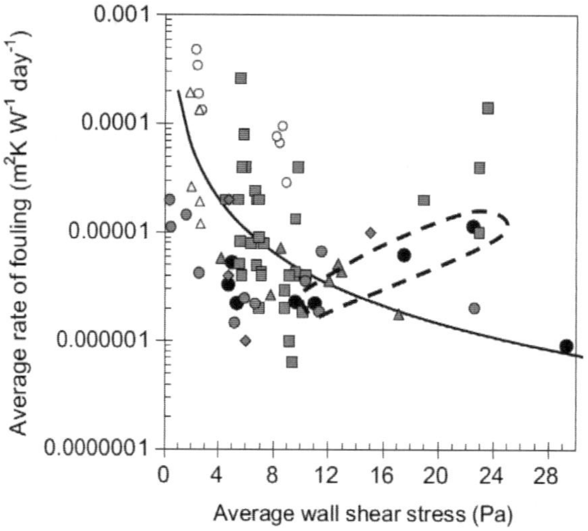

Figure 1: Plot of average fouling rate against estimated wall shear stress. Solid locus: correlation reported by Joshi *et al.* (2009). Solid symbols are the fouling rates obtained from 5 refineries; open triangle and open square are pilot plant fouling data obtained from Wood River tube flow and Exxon tube flow, respectively. (both pilot plant data sets reported by Knudsen *et al.*, 1999). (Reproduced from Ishiyama *et al.*, 2010).

Evaluations of fouling rates are based on the temperature and flow measurements available in the plant, and the accuracy and the reliability of these plants monitoring data reflects the accuracy of the calculated fouling rates. Most of the time, not all stream temperatures are measured and/or the measured stream temperatures are erroneous. In this paper we denote data reconciliation as fitting plant monitoring data to specific models in order to get data that follow mass and energy conservation laws.

The fouling resistance of an exchanger, R_f, is given by,

$$R_f = 1/U - 1/U_{cl} \tag{1}$$

Here U is the overall heat transfer coefficient and U_{cl} is the overall heat transfer coefficient at the clean state, respectively. Both U and U_{cl} are based on the tube external surface area. Note that U_{cl} is calculated for each operating condition, not the design (clean) value based on the design operating condition.

Exchanger heat duty, Q, is evaluated by

$$Q \quad = UA_oF\,\text{LMTD} \tag{2a}$$

$$= m_c\, C_{pc}\,(T_{c,out} - T_{c,in}) \tag{2b}$$

Here, Q, is the heat duty, A_o is the heat transfer area (based on the tube external surface), F is the temperature correction factor and LMTD is the logarithmic mean temperature difference. m is the mass flow rate, C_p is the specific heat capacity and $(T_{c,out} - T_{c,in})$ is the temperature driving force. Subscript 'c' denote crude stream.

With missing temperature or flow measurements it is impossible to evaluate the value of U from equation (1, 2a,b) alone. The main aim of this document is to demonstrate the methodology practiced in the data reconciliation and network simulation software tool (*smart*PM™) which is being developed at IHS ESDU, UK. The method is based on guessing the fouling resistances of each exchanger; the guessed fouling resistances are used in the network simulator to obtain the reliable plant measurements. The initial guess for the fouling resistance is very important to obtain an accurate data reconciliation; this is based on obtaining reliable temperature guesses for missing/unreliable data sets.

2. Missing temperatures and/or flow measurements – assuming available measurements are reliable

When the plant consists of missing measurements, the missing values are generated through a constructed simulator, which could either be used for a single heat exchanger or a heat exchanger network of any structure. The steps are divided into four parts.

1. Initial mass balance is performed to generate any missing stream mass flow rates.
2. Initial guess for the missing temperature is made.
3. Then these values are used to calculate the fouling resistance guesses.
4. These fouling resistance values are then used as initial values and the temperatures of the streams generated through the simulation. The steps are summarized in Figure 2.

The initial temperature guess is done through linear interpolation of the known temperatures associated with the exchanger group.

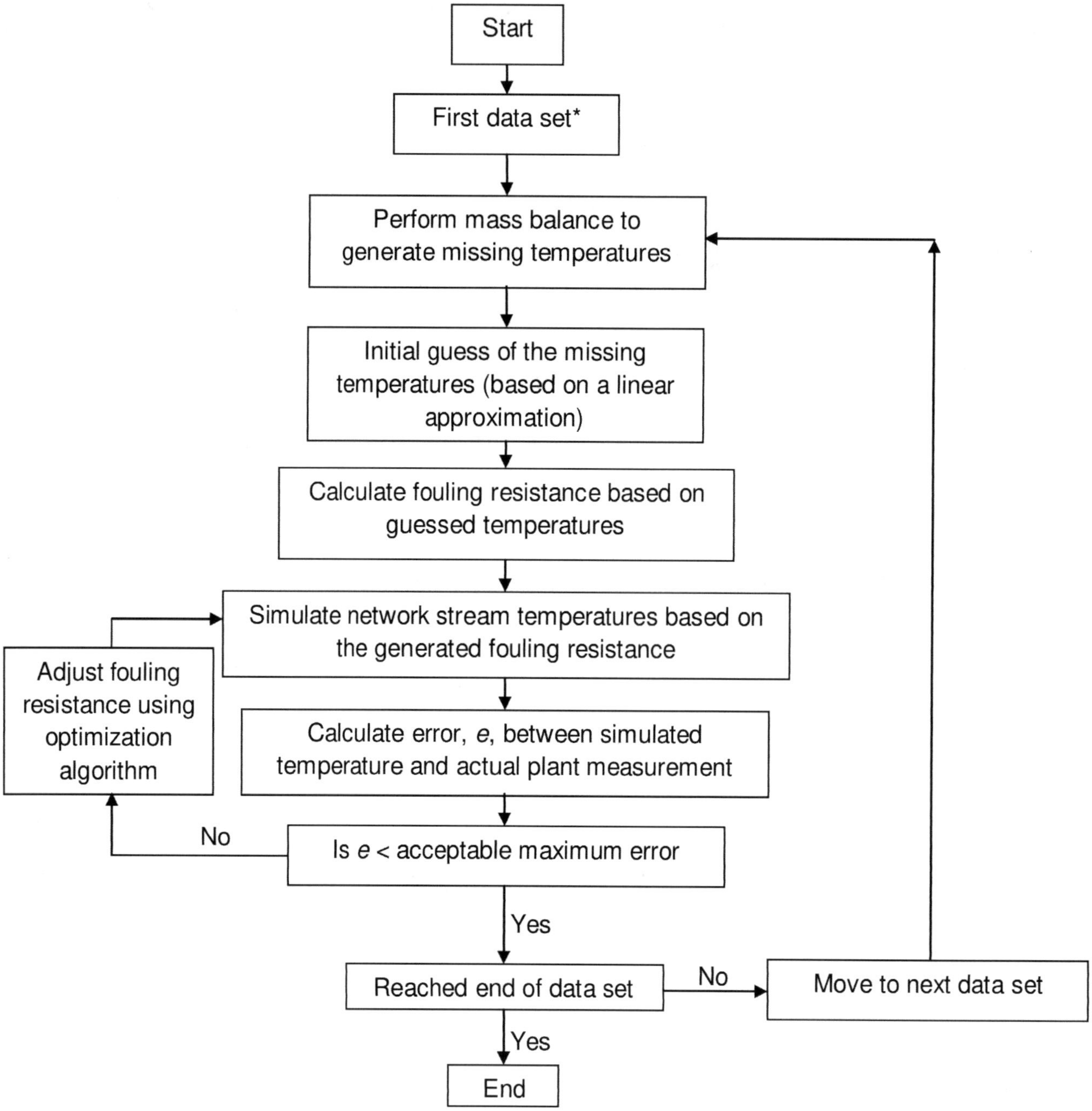

Figure 2: Data reconciliation steps with missing intermediate flow/temperature measurements. *'Data set' refers to plant monitoring data

3. Missing temperatures and/or flow measurements – some available measurements are unreliable

In cases where the available measurements could be unreliable (this is likely to be the case in most refineries), the data reconciliation process needs further analysis. It is necessary to have information on what measurements could be trusted. For instances, measurements at the coil inlet, desalter inlet/outlet, storage tank and product/pump-around streams coming from the distillation column are considered reliable.

The reliable data are then used to provide the basis for mass and energy balances. One approach is to assume that there is linear variation between temperatures of the reliable streams and get an approximate initial guess. If there is not more than one reliable temperature on a stream then the temperature is inferred comparing the clean operating condition (based on the simulated value) to the current operating condition. For example, assume a preheat train consisting of reliable measurements at the preheat train inlet and the coil inlet. The clean operating condition could be obtained through performing a network simulation taking all exchangers are clean.

Unmeasured temperatures are guessed based on an idealized location of the stream between the crude storage tank and the furnace inlet (both temperatures are always measured). Assume that the temperature increment linearly with distance from the tank to the furnace, represented by the linear form, $y = m x + c$, where x gives the location and y the temperature at location 'x'. The value of x is not important in guessing the stream temperature at the fouled state as this could be eliminated as

$$y_2 = \left(\frac{m_2}{m_1}\right) y_1 + \left(c_2 - \frac{m_2}{m_1} c_1\right)$$

Here subscripts 1 and 2 represent the clean and the fouled states, respectively.

By using the initial temperature guess and obtaining the initial fouling resistance, the results are then fine tuned to minimized measured and simulated temperature drop differences across heat exchangers using *smart*PM simulation. The approximate search method assumes the temperature profile to change linearly as it moves along the length of the preheat train. The rigorous search then uses an optimization algorithm to find the missing/unreliable data based on reliable measurements.

Case study

Figure 4 is a section of the preheat train of a well instrumented refinery processing heavy crude oil. All temperatures are measured. In the first case study the temperatures labelled T_1 to T_8 in the streams are assumed to be missing and in the second case study the temperatures labelled T_1 to T_{10} in the streams are assumed to be missing. Data reconciliation process carried out in two ways. 1) data reconciliation is based on linear approximation of temperature and local heat balances. 2) data reconciliation is carried out using the methodology described under sections 2 and 3.

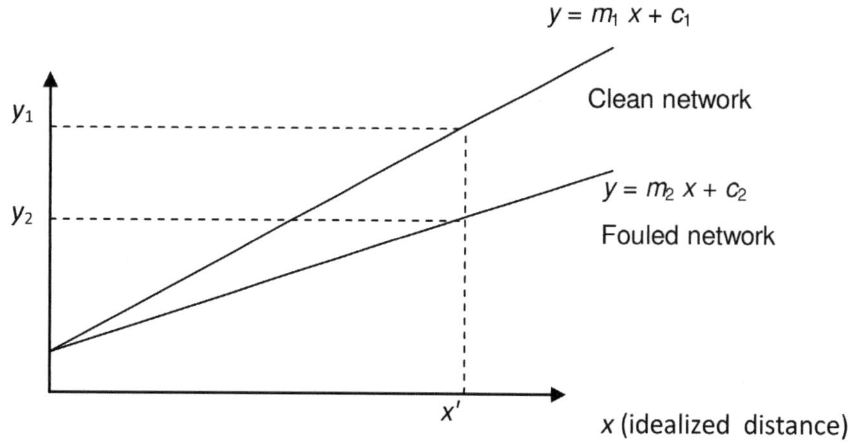

Figure 3: Illustration of evaluating the initial guess for the unmeasured stream temperature

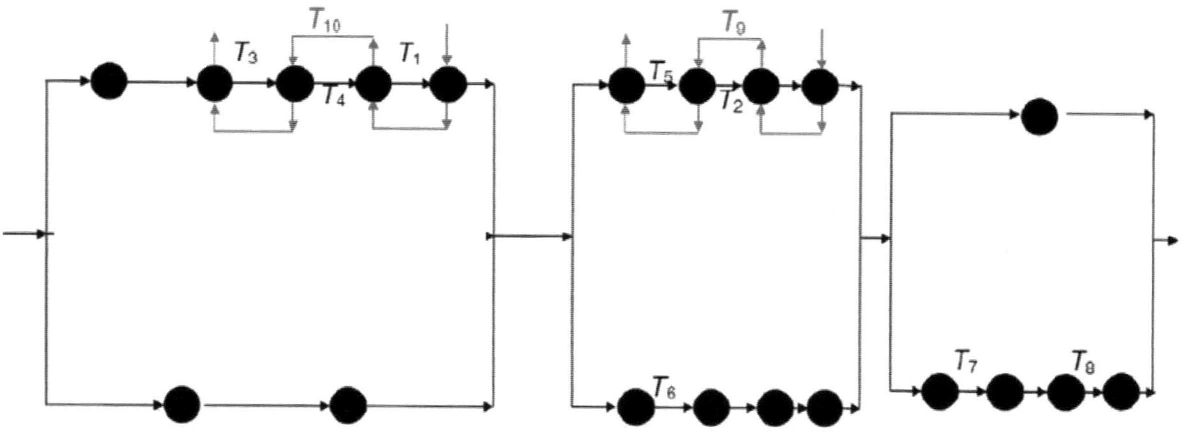

Figure 4: Case study preheat train. The refinery is well instrumented. The labels T_1 to T_{10} are the stream temperature measurements that are the subject of this study.

Discussion

The fouling resistance profile at a given time instance is compared in Figure 5 with the actual plant fouling resistance. The actual plant fouling resistance could be obtained as all plant measurements are available. The assumption of having temperatures T_1 to T_{10} missing is for illustration purpose only.

In Figures 5 (a,b), the data set appearing as 'I' shows an over-prediction of the fouling resistance. Thus telling the user to clean an exchanger earlier than it should have. For the data set denoted as 'II' fouling is under predicted, hence masking the actual effect of fouling in the unit. This would provide misleading information not to clean a unit, when it should have been cleaned.

(a)

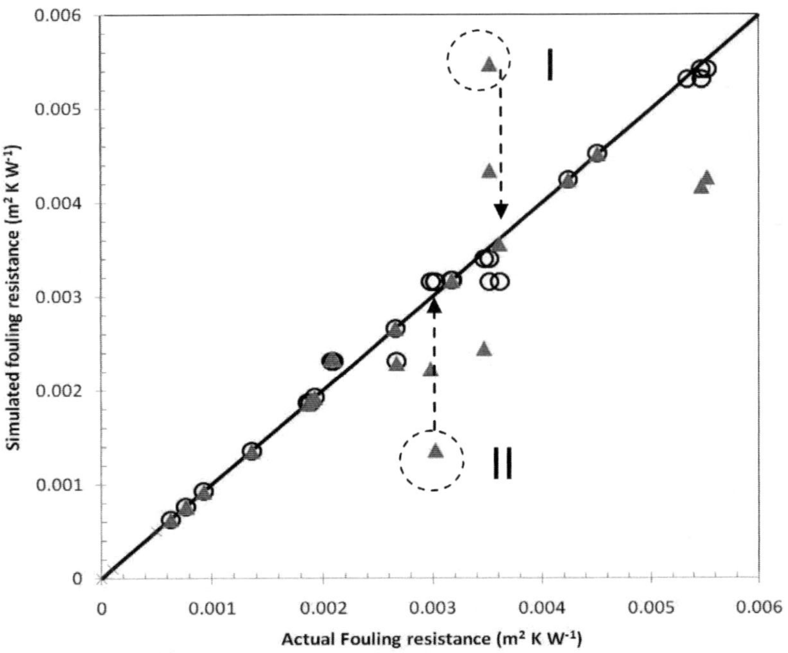

(b)

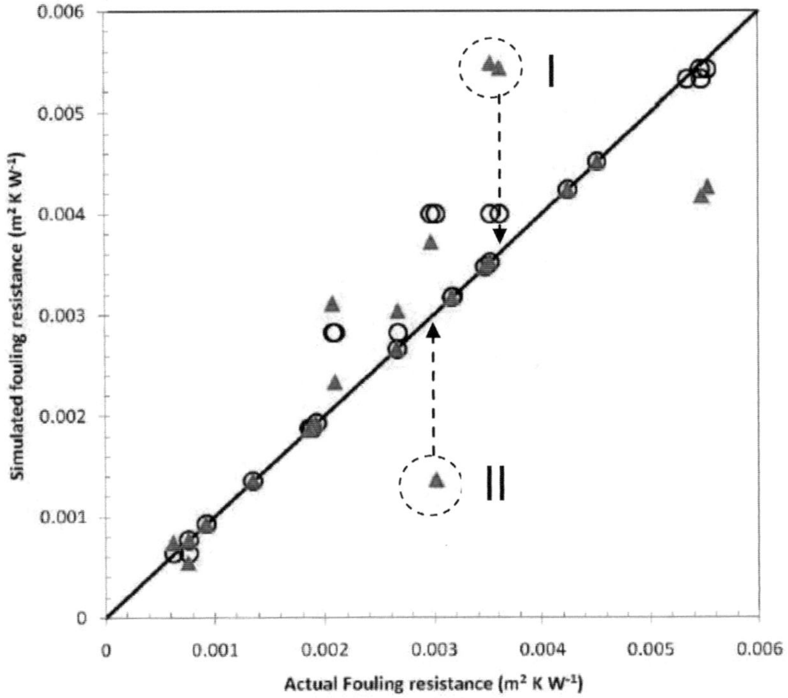

Figure 5: Fouling resistance at a given time instance for the case study exchangers. Filled triangles present results based on simple linear temperature approximation alone and the hollow circles present result obtained through this study. (a) assumes temperatures T_1 to T_8 are missing and (b) assumes temperatures T_1 to T_{10} are missing.

Conclusions

Data reconciliation is a vital process before analysing plant monitoring data. A method of reconciling data with missing/unreliable stream temperatures and flow rates is illustrated through a newly developed software tool (*smart*PM^{TM}) at IHS based on the work done at Cambridge University. The importance of data reconciliation is demonstrated through a case study network where incorrect data reconciliation is illustrated to result in wrong decisions on the degree of fouling of certain units.

References

Ishiyama E.M., Paterson W.R. and Wilson D.I. (2009). Platform for techno-economic analysis of foulng mitigation options in refinery preheat trains. *Energy and Fuels*, 23(3) 1323 – 1337.

Ishiyama E.M., Heins A.V., Paterson, W.R., Spinelli L. and Wilson D.I. (2010). Scheduling cleaning in a crude oil preheat train subject to fouling: Incorporating desalter control. *Applied Thermal Engineering*, 30 1852-1862.

Joshi H.M., Shilpi N.B. and Agarwal A. (2009). Relate crude oil fouling research to field fouling observations, in: OCnference Proceedings of the 8th International conference of Heat Exchanger Fouling and Cleaning, Schladming, Austria.

Knudsen J.G., Lin D. and Ebert W.A. (1999). The determination of the threshold fouling curve for a crude oil, in : Bott, T.R. (editor), Understanding Heat Exchanger Fouling and Its Mitigation. Begell House Inc., Castelveccio Pascoli, Italy, pp 265-272.

(22e) Tutorial - Measuring Hydrocarbon Liquid Levels In Highly Dynamic Petroleum Processes

Steve Newton, Emerson, Sandy, UT and **Sarah Parker**, Emerson, Chanhassen, MN

Measuring Hydrocarbon Liquid Levels in Highly Dynamic Petroleum Processes

Measuring liquid levels in highly dynamic processes can be very challenging. Rapid level changes, changing liquid densities, boiling liquids, and interface fluids can combine to make it difficult to get an accurate and/or stable measurement. This presentation will discuss several of these difficult applications and the solutions found using the various measurement abilities of the guided wave radar. We will outline best practices for the mechanical installation and the level instrument and explore how to best set up the measurement in order to improve loop tuning.

In some challenging applications, initial liquid levels were changing so rapidly that the level transmitters could not track the changes in real time. Also, level measurements were not believable since liquid densities were also changing. The loops could not be properly tuned because of the volatility of the level measurements, which in turn caused liquids to boil and levels to be unstable. The units were frequently tripping due to high or low alarms, and the plant could only be operated in manual.

Using the full capabilities of the guided wave radar allowed accurate stable level measurements in a variety of highly dynamic applications. Using probe-end-projection, damping rates, and using other advanced gage settings we were able to stabilize boiling liquid level measurements, which allowed proper loop tuning to stabilize the process. The plant now operates without upsets or trips. The end result was stable process control with very little process variation.

See more of this Session: Process Control Optimization and Tutorial
See more of this Group/Topical: Topical 7: 14th Topical on Refinery Processing

(22f) Optimization of A Complex System of Crude Oil Distillation Units Using NLP and Metamodels

Diana C. Lopez[1], Luis J. Hoyos[1], Sara M. Chaparro[2] and Jes˙ s A. Castro[1], (1)Instituto Colombiano del Petróleo ICP, ECOPETROL S.A., Piedecuesta, Colombia, (2)Ambiocoop Ltda., Piedecuesta, Colombia

The objective of this paper is to show the optimisation of a complex system of Crude Oil Distillation Units (CDU) belonging to a Colombian refinery of ECOPETROL S.A. This system is composed by crude oil blending scheme and all distillation units installed in the refinery. Distillation units are made up by 5 atmospheric towers, 5 atmospheric furnaces and 4 vacuum towers; crude oil blending is integrated for 14 Colombian crude oils, which were, at first, mixed in 4 mixtures (pre-mixtures) which, at second, were used for preparing the crude oil diets.

Optimisation results were optimum feedstock composition and operating conditions for the CDU system and was carried out through Non Linear Programming (NLP) Model developed by the research institute belonging to ECOPETROL S.A. NLP model optimized the system economic profit and uses Metamodels approach to represent Atmospheric Distillation Towers (ADTs) and Assays cutting with Operating Modes to model the Vacuum Distillation Towers (VDTs). This Model has been used for short, middle and long term planning.

In NLP model, crude oil blending restrictions were mass balances, mixing rules, property boundaries like acidity, sulfur content, specific gravity. Atmospheric distillation process was modeled through Metamodels calculating volumetric flows, properties and temperatures for each ADT product and pumparound; product properties were API gravity, sulfur, neutralization number, conradson carbon, nickel, vanadium, total nitrogen. Metamodels were based on rigorous simulations of distillation units in PRO/II.

Optimization model was formulated in GAMS 23.3, was solved with CONOPT and had a user interface developed in Microsoft Office Excel 2007. NLP model found new crude oil diets and process variable points for each CDU with better economic results than those obtained with the feed compositions of normal short term planning and operation in the refinery. NLP model was able to evaluate new crude oil blending schemes such as segregation of crude oil streams in the current pre-mixtures and inclusion of new pre-mixtures for the refinery; for these new cases, the model calculated the best performance of distillation system.

See more of this Session: Process Control Optimization and Tutorial
See more of this Group/Topical: Topical 7: 14th Topical on Refinery Processing

(50a) Chemical Insights Indicate Process Opportunities for Residue Processing

Murray R. Gray[1], Ali H. Alshareef[1], Rik Tykwinski[2], Jeff Stryker[3], Alex Scherer[2] and Xiaoli Tan[1], (1) Chemical and Materials Engineering, University of Alberta, Edmonton, AB, Canada, (2)Department of Chemistry, Friedrich-Alexander-Universit%ct Erlangen-N¸rnberg, Erlangen, Germany, (3)Department of Chemistry, University of Alberta, Edmonton, AB, Canada

The vacuum residue fraction of petroleum is one of the most complex mixtures of compounds known, one which continues to defy comprehensive analysis. The combination of complexity, diversity, and aggregation of the asphaltenes has enormous implications for upgrading reactions and phase behaviour. One of the challenges is that residue molecules fall into a range of molecular weight that is intermediate between commonly available compounds and the high polymers. Extrapolation of thermodynamics and reaction chemistry from these well understood domains to describe residue processing has some merit, but also significant risks.

Our approach is to attach the problem from both directions, working downward from the whole material and building upward by chemical synthesis. We are using modern instrumental methods to quantitatively characterize the pendant groups and building blocks that comprise the residue components. The asphaltene fraction still defies satisfactory direct analysis, but analyses of the constituent building blocks are yielding quantitative results. Working from the bottom up, we are synthesizing families of components to represent key structural features of residue molecules. We then investigate the reactions of these molecules under cracking conditions, and define the molecular basis for coke formation. This research is completely revising our understanding of how cracking reactions proceed in vacuum residue materials, because when cracking occurs in the liquid phase, addition or retrograde reactions are far more important that anyone has recognized previously. This presentation will offer a synthesis of these two approaches, defining the molecular features that control residue behaviour, the main pathways for coke formation, and the implications for advanced residue processing in the future. Controlling the unwanted liquid phase reactions is the key to improving yields of distillate products from the bottom of the barrel.

See more of this Session: Heavy Oil and Bottom of the Barrel Tutorial

See more of this Group/Topical: Topical 7: 14th Topical on Refinery Processing

(50b) Petroleum Resid Conversion: Opportunities and Limitations

Irv Wiehe, Soluble Solutions, Gladstone, NJ

In the conversion of vacuum resids one has the option to form coke as a byproduct or to devise conversion processes that prevent coke formation while forming a heavy liquid byproduct. The theme of this presentation is that both options can be greatly improved over current commercial practice but there are limitations imposed by petroleum macromolecules and economics. Coking processes and indeed, the Conradson carbon residue test itself, actually form coke precursors that are not native to the vacuum resid. By inhibiting the formation of new coke precursors, one may reduce the coke yield to only that formed by the native coke precursors. As a result, the yield of coke may be less than Conradson carbon residue, rather than 1.2 to 1.6 times Conradson carbon residue, without lowering the quality of the liquid product.

The secret of converting vacuum resids without forming coke is to keep the converted asphaltenes soluble while maintaining a moderate concentration of hydrogen donors. Although hydroconversion processes clearly maintain the hydrogen donor concentration, many are limited by the solubility of converted asphaltenes. By using hydrogen gas to strip out the asphaltene nonsolvents, the volatile liquid product, hydroconversions of greater than 90% may be achieved. However, conversions in this range usually run into an economic limit caused by very slow kinetics and accelerating hydrogen consumption. Meanwhile, there are processes, such as the Eureka and the HSC processes that very effectively use steam to strip out the volatile liquids during thermal conversion, relying on only the natural hydrogen donors. As a result, conversions are in the range of coking and ebullating bed hydroconversion processes without forming coke or consuming hydrogen.

See more of this Session: Heavy Oil and Bottom of the Barrel Tutorial
See more of this Group/Topical: Topical 7: 14th Topical on Refinery Processing

REAL TIME CO_2 AND ENERGY COST REDUCTIONS: FIELD EXPERIENCES IN PETROCHEMICAL PLANTS

Carlos Ruiz, Oscar Santollani, Soteica Ideas & Technology LLC, Houston, TX

Introduction

CO_2 emissions are in most part by-products of utility systems; therefore it is only logical to model emissions along with the energy system. Any attempt to reduce the costs from the utility side (i.e.. to optimize the energy system) should be based on the appropriate utility system models (calibrated with validated and consistent set of measurements) and a compositional fuel representation to allow combustion-based Carbon calculations. Solving and optimizing the energy and Carbon balances together is key to be consistent with operational and constraints handling.

The optimization's goal should be to reduce the overall utilities cost taking into account the constraints associated with the existing equipment, fuels/power pricing and contracts. The economic optimization should be done by considering the emission limits as constraints and/or part of the objective function (for example, cost or credit of the CO_2 emission if above or below a certain quota). The energy management system models need to be executed and optimized at a scheduled frequency with online, real time data.

In addition to the optimization, Key Performance Indicators (KPIs) should also be calculated and stored in the Site Plant Information System or DCSs for Operations and Management use.

However, like in any other human activity, the technology by itself is not sufficient to guarantee the success of an energy and CO_2 reduction initiative. Over a period of more than 20 years of deploying industrial projects related to energy systems real time optimization and cost reduction in refining and petrochemical industries around the world, the following or similar issues have been found:

- Many applications of energy cost and CO_2 reduction technologies are successful when used offline by Engineers, at their offices, but only a few of them really succeed when transferred to the Operators to be routinely used in real time, tied to online field data, under unsupervised automatic execution.

- In some cases, systems that are deployed after successful, well engineered, real time projects and that are initially received enthusiastically by Engineers and Operators are only used for a few months after project completion and then fall in disuse.

- Real time systems with acceptable initial up-time figures end up degrading and finally are left offline, unused, after the Engineer in charge is promoted to a new position as a consequence of being so successful with them initially.

- Very well proven, sound technologies, fail when applied at a given Site while they succeed at a different location owned by the same Company.

- After a failure with one of those advanced technologies, the Site or the Company becomes "immunized" against similar technologies. Only after several years and with the renewal of the Engineering staff does the technology have a chance to be considered again.

This paper addresses some of the root causes for the above mentioned issues and discuss the best practices that will help to avoid project failures. Main project steps are explained and critical details to be taken into account to assure successful use and proper technology transfer are discussed. It also presents real industrial examples in which the whole utilities system cost of a production Site (i.e., steam, fuels, boiler feed water and electricity) is optimized with a real time, online, industrially well established software.

Real Time Energy Management Systems

Refineries and petrochemical plants usually operate large and complex utilities and energy systems. They burn several type of fuels, operate cogeneration units to supplement the electric power purchases or export electricity, need to provide several steam pressure levels to different types of consumers and observe emission limits.

In almost all of the cases, these complex utilities and energy systems have several degrees of freedom. Proper manipulation of the degrees of freedom with the aid of cost-based optimization software can result in significant operating cost savings. Since the electrical system represents one of the main economic trade-offs with the steam system and Site Carbon emission, electrical deregulation provides many new challenges to operate the overall combined energy system at minimum cost.

Other important aspects are that utilities systems are continuously evolving (changes are frequent) and that sometimes, there is a lack of sensors and often measurement errors need to be addressed properly.

In addition, utilities systems have several constraints that usually come from the operational and contractual sides. For example, minimum and maximum steam production capacity in the boilers, NO_x / CO_2 emission constraints, the need to maintain a steam production cushion due to safety reasons, or the fuels and electric power quotas and penalties.

Also, it is important to mention that traditionally, given the complexity of the systems, the optimization of the utilities was traditionally managed at the level of each individual Plant or Area. But the optimization of each individual Area does not necessarily give the true overall site optimum. Moreover, an attempt to optimize a whole site based on an isolated view of each Plant at a time could be worse than doing nothing.

In the dynamic operational and economical environment of a Plant, implementing the proper actions necessary for the optimum utilities system management can only be done with the aid of an on-line tool. It should execute in an unattended, automated way, providing at any time the right answers to the different users (from operators to engineers), with the man-machine interface appropriate for each one of them.

To add more complexity to the problem, the environmental constraints are becoming more and more complex and stringent. In the US, CO_2 emission reduction initiatives are a hot topic, like the State based emission reduction initiatives. For example, the California Climate Exchange Law AB-32, with a cap-and-trade scheme to be starting 2012 and the Regional Greenhouse Gas Initiative (RGGI), CO_2 Budget, whose trading program was put in effect in 2009. In the near future it is expected to have a US cap-and-trade policy for emissions reduction to be modeled on the EU scheme, that aims to cut carbon emissions to 1990 levels by 2020 and 80% by 2050.

There is a clear move towards a global carbon market. Carbon markets grew more than 3 times their worth in 2006 and it is expected to do so again by 2015, according to the World Bank. Companies have clearly identified that Carbon is money in the current and future economy while emission reporting is becoming mandatory worldwide. With all the regulations already on place and those coming on the horizon, emissions will become a significant factor in operations and planning, adding new operational constraints, increasing the operating cost, involving potential needs of capital investment and generating value from trading. In the path to compliance, organizations are facing several challenges in establishing an effective procedure to deal with the emissions under a consistent, optimum base.

And to add even more burden to the problem, control rooms worldwide are facing a shortage of people. Operators are concentrating more and more on the Units under his/her responsibility. DCS screens flourished and hundreds or thousands of tags are available per Unit and a multitude of process diagrams and trends can be projected on the walls and even the ceilings. But only a few employees are available to take care of the overall energy system operability and economics. Industrial energy systems are becoming increasingly complex and inter-related, not only between Units of the same Site but also with the power and utilities systems of neighbor facilities.

The evolution from having plant information scattered throughout many islands of automation to a unified and centralized Plant Information System was a clear enabling layer for such work. DCSs (Distributed Control Systems (DCS) and Plant information (PI) systems are the most widely available data sources in the current industrial context. PI systems usually acquire data from the DCSs and store it in a unique repository. The long term, facility wide PI system based historians constitute what is known as an *enabling technology*, because they are the cornerstone on top of which decision support systems, such as energy management systems, can be built. Centralized real-time databases provide access to massive current and historical process, laboratory and financial data.

Energy Management System Software Description and Functionality

In order to successfully address all the issues mentioned above, a software tool called Visual MESA has been extensively used. The original MESA program was developed in the early 1980's. Since the Visual MESA version was released in 1997, it has been continuously improved and adapted to be able to cope with all these changing scenarios.

Visual MESA is a computer program designed to model steam, boiler feed water (BFW), condensate, fuel, emissions and electrical systems. It is an on-line program that receives live plant data from the steam, fuel, condensate, BFW, and electrical system metering devices via a standard link to the plant information system.

It typically works uninterruptedly retrieving on-line data from the process using the OPC standard connectivity, evaluating continuously the utilities and energy system optimization alternatives and writing key performance indicators back to the plant information system.

The software includes all the necessary features to calculate and handle emissions properly: compositional description of the fuels streams, combustion properties to calculate boilers and heaters efficiencies, emission source stoichiometric calculations or estimated based on correlations for individual process point sources (for example, FCC catalyst Carbon burnt at the regenerator), heaters and boilers, etc.

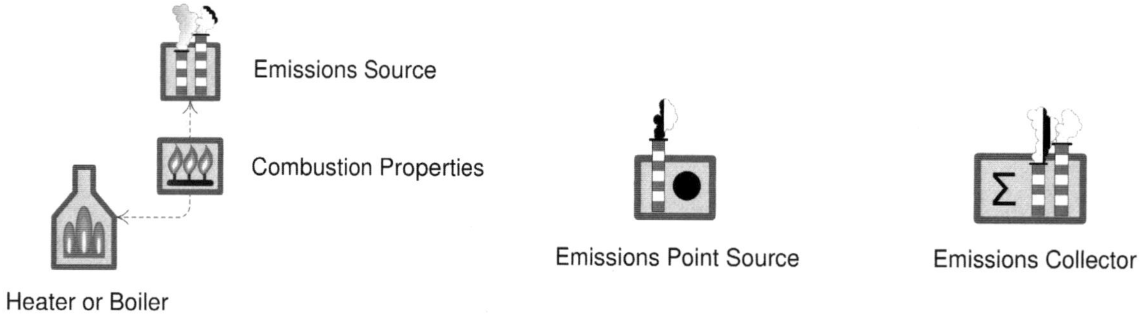

Figure 1 – Standard Visual MESA Emission Modeling blocks

Main benefit sources from where high savings can usually be obtained are:

Optimization: It helps to find the most economical way to run the utilities and energy systems, while remaining within the real operating constraints. It enables plant personnel to:

- Understand how to improve the optimum steam/fuels/electrical system operation, including emission constraints and costs.
- Quick reaction to minimize the economical impact of an operational change.
- Optimize the overall cost of fuel for steam generation and electricity, including the choice of the most convenient combination of turbines and motors.
- Make a good use of the utilities network data availability and improve its quality.

Monitoring: It provides a number of monitoring features that help access data, control data quality, and alert about changes to the system, like:

- Plant, equipments and streams information
- Trending data and calculation results

- Big changes on utilities streams and emissions regulations compliance alerts
- Data quality (balloons, graphically representing header mass balance imbalances)

Case Studies ("What If?" Planning): It allows performing and evaluating "What If?" cases that enable plant personnel to find ways to operate more efficiently and at less cost. Some examples are:

- Front-end loading on projects
- True "Plant-wide" project evaluation
- Steam, fuel, BFW, condensate, and electrical system improvements
- CO_2 emissions mitigation projects, credits availability and/or new constraints

Auditing & Accounting: It helps plant personnel to find where steam waste is occurring in the steam system, provides information to account for it properly and identifies where the imbalances are and how they behave in time.

This on-line energy management system was implemented in around 40 sites worldwide to model and optimize the overall site energy and utilities systems for a variety of industries: petrochemical, refining, chemical, sugar & alcohol, district heating and cooling, combined heat and power, etc. Several of them have been published (Refs. 4 to 17 are related to recent industrial projects). Amongst them, two particular examples will be presented and discussed.

Industrial Project Examples

The following two industrial projects are presented as examples:

- Medium size refinery complex
  - Example of daily recommendations and their impact on CO_2 emission
- Complex refinery with high conversion Units
  - Fuel gas network management and integration with the rest of the utilities system

Medium size refinery complex

This industrial example corresponds to TOTAL Feyzin refinery (Refs. 9 and 10). TOTAL is a leading refiner-marketer operating 11 refineries directly in Western Europe and one refinery in the U.S. at Port Arthur, TX. The reduction of greenhouse emissions and the enhancement of energy efficiency are among the current Company corporate challenges.

In this example, a set of manual operating recommendations given by the optimizer during a particular shift have been:

- Pump swaps (i.e., steam turbine and motors driving pumps that can be used interchangeably)
- Fuels to boilers, such as Fuel Gas (FG) and Fuel Oil (FO)

- Power production in steam turbine generators

As a result of the manual actions, adjustments to the following variables have been made by the control system:

- Steam production from boilers
- Letdown and vent rates

The following figures show the impact on steam production, fuel use and CO_2 emissions reduction:

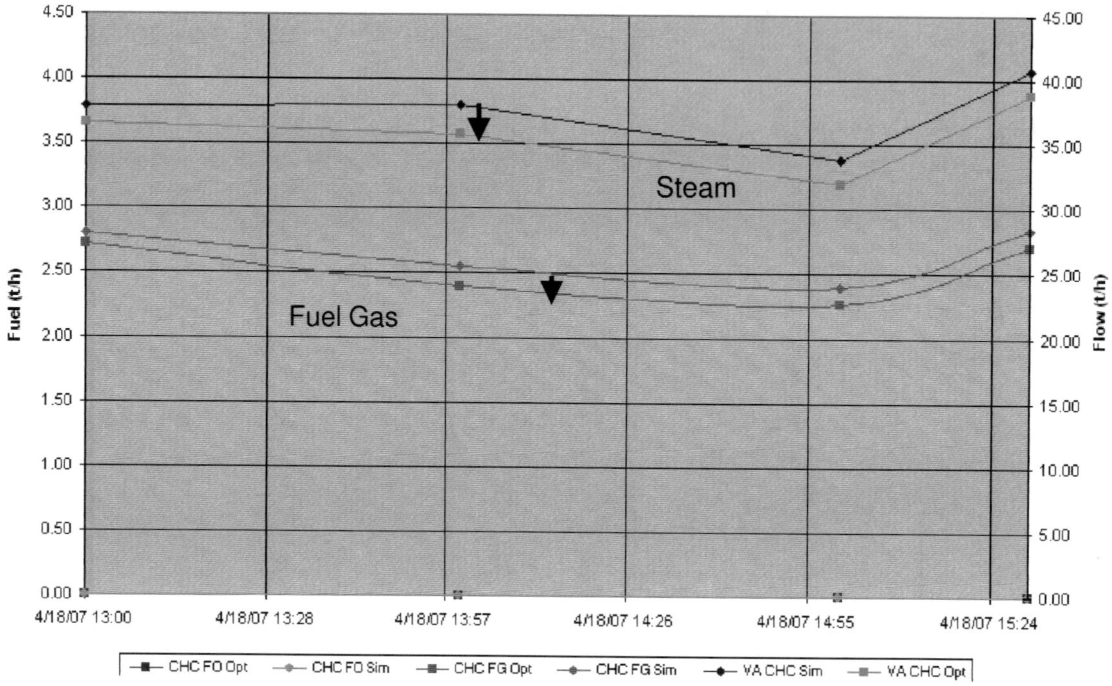

Figure 2 –Boiler C (100% Fuel Gas): 2 t/h less steam

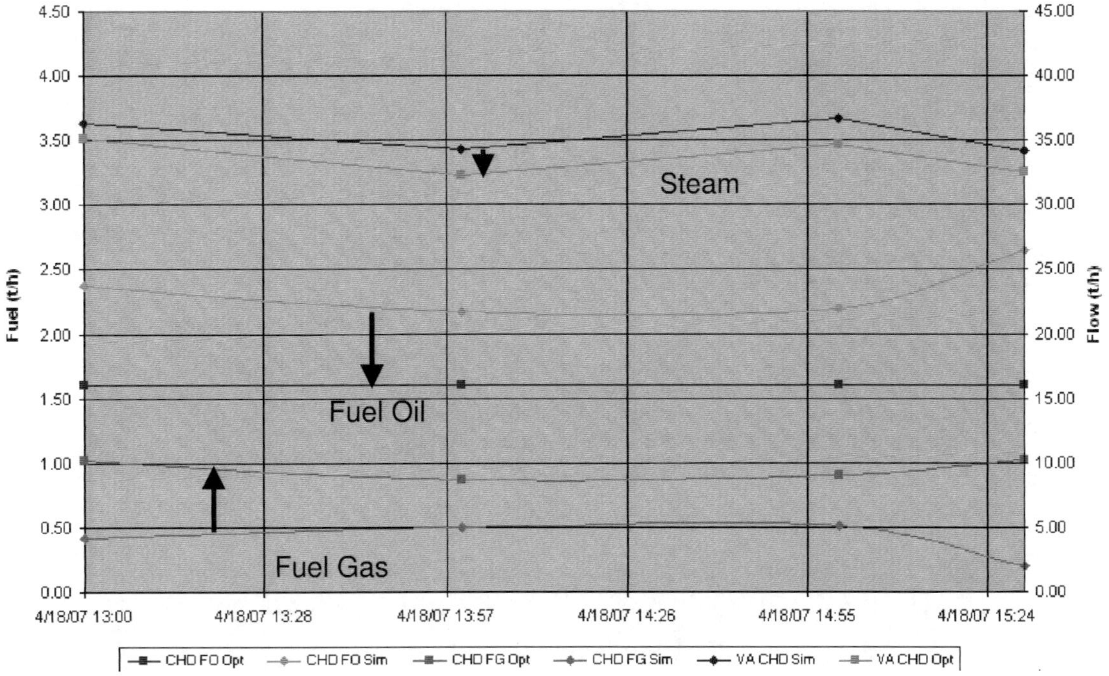

Figure 3 – Boiler D (FO and FG): 2 t/h less steam) and FO to the minimum

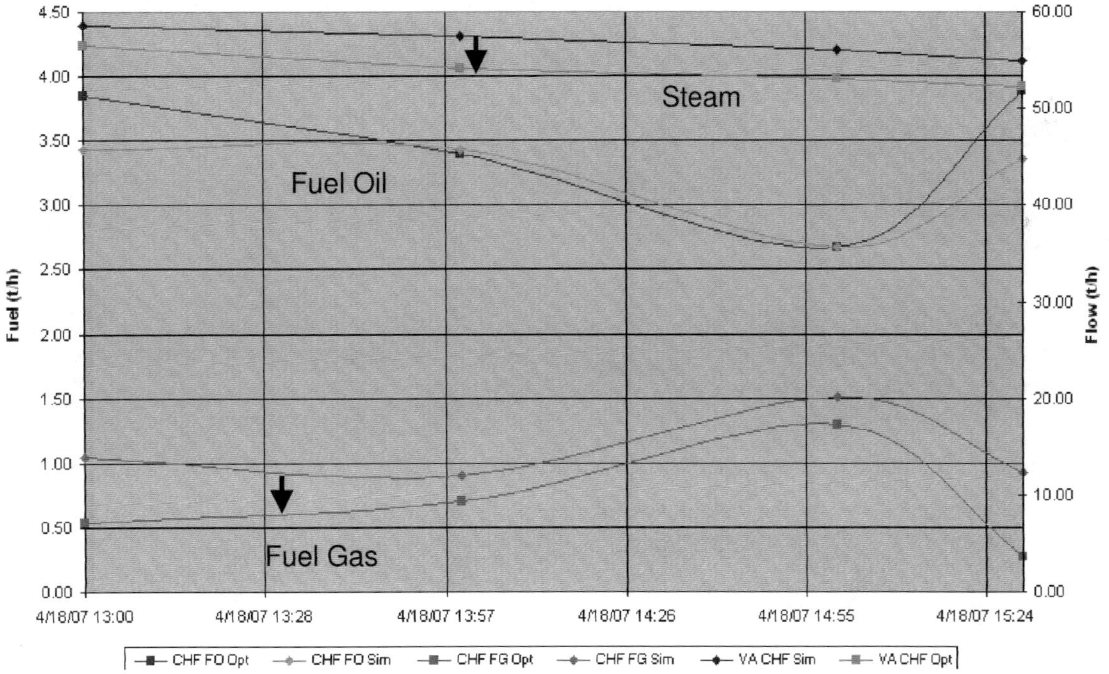

Figure 4 – Boiler F (FO and FG): more than 3 t/h less steam

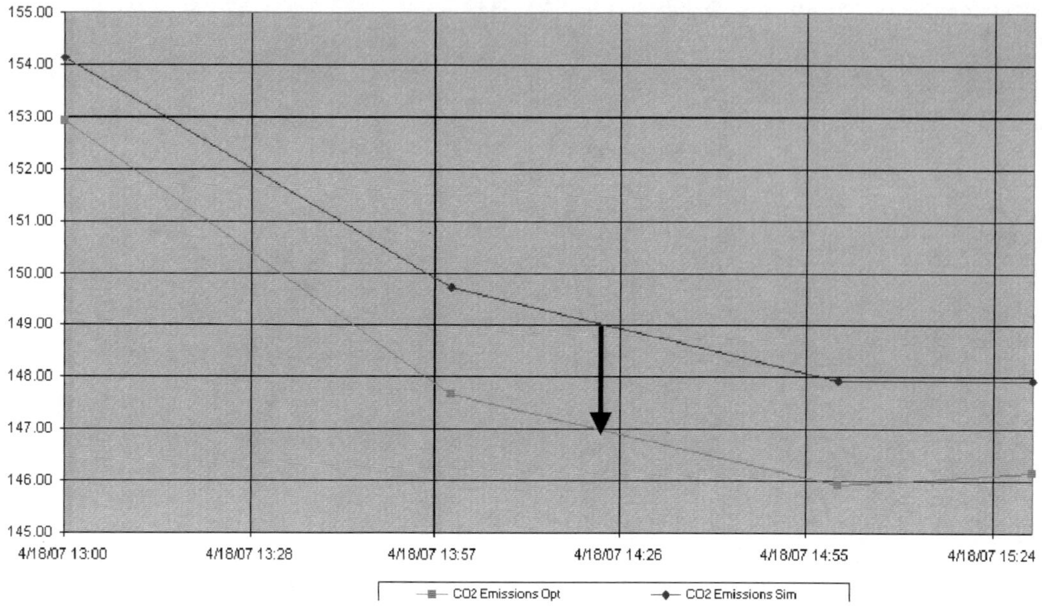

Figure 5 – CO_2 emissions reduction: 2 t/h less

In summary, the following changes were achieved while optimizing the overall cost:

- Almost 1 t/h less of FO consumed
- Approx. 7 t/h less of high pressure steam produced
- Approx. 2 t/h less of CO_2 emitted
- Approx. 200 kW more of electricity imported

When considering the CO_2 emission cost, the following is an example of the recommendation for operators:

As a result of the manual change in the boiler fuel use (more FG and less FO in overall), the Fuel Gas header pressure control system made the necessary adjustments which resulted in the need to increase the incremental Natural Gas import.

The following figure shows the fuel gas network model representation highlighting the differences between current and optimized situation (delta view, duty flows).

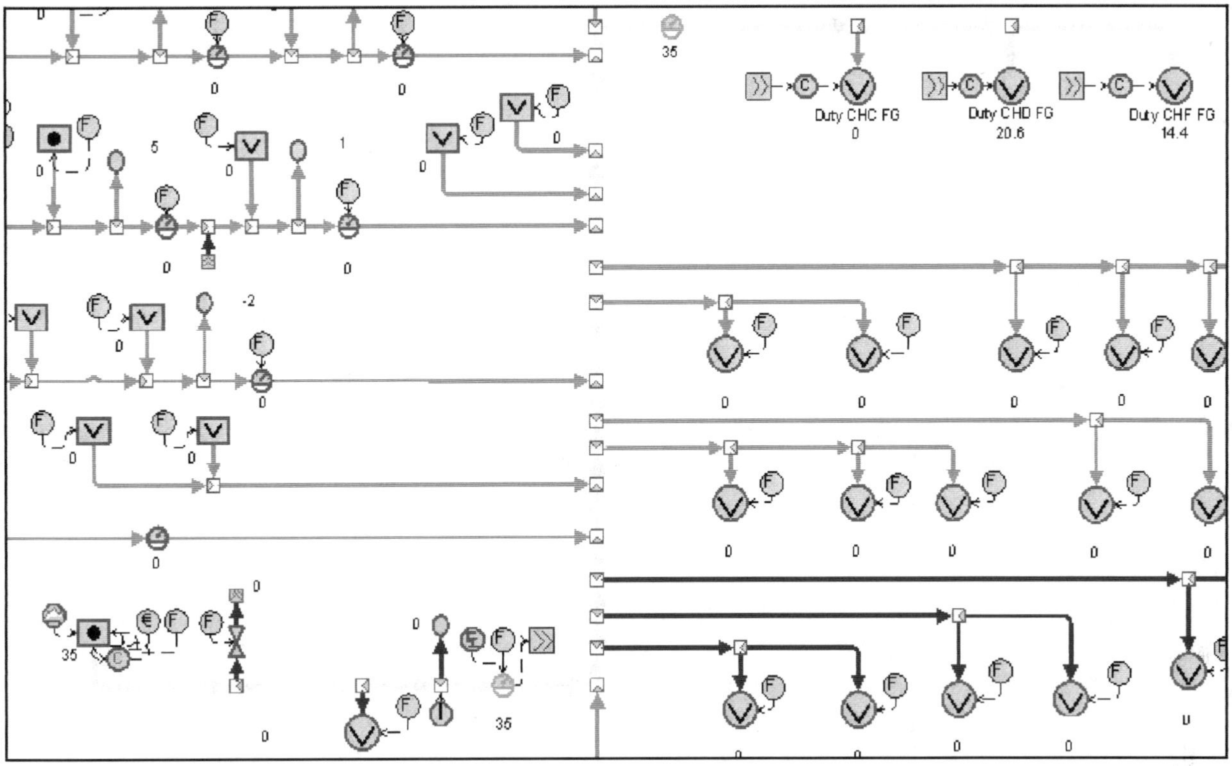

Figure 6 – Fuel Gas model showing a "delta view" (i.e., optimum – base case change)

On the left, fuel gas suppliers are represented, while on the right Fuel Gas consumers are displayed. The values indicate the corresponding changes, expressed in MW, after the application of the recommendations (zero value means no change).

As a result of replacing FO by FG (with the need to import more Natural Gas) and improving the steam efficiency usage by reducing let-downs and vents, CO_2 emissions are reduced by 4.7 t/h. This is important to consider when a trade-off between cheaper fuels that produce more CO_2 and more expensive fuels that produce less CO_2 exist. It would be an especially important consideration when the CO_2 emissions quota is expected to be exceeded by the end of the year.

Complex refinery with high conversion Units

At the Petronor site, Spain's largest refinery (ref. 1) with a processing capacity of eleven million tons of crude oil per year, a Visual MESA based on-line model for energy management has been implemented. Its energy system is large and complex; therefore, the auditing and control of energy costs are a real challenge.

The full model of the energy system includes the entire fuel, steam, boiler feed water, condensate and power systems. Steam is generated in several places, with both conventional boilers, heat recovery steam generators and a cogeneration unit with steam injection. Electricity and fuel supply contract details have been included in the model. In particular, the purchased Natural Gas daily quota was rigorously modeled and the constraints respected during the routine optimization.

All the constraints have been included and the model is continually populated with live, validated data. Performance monitoring is performed and includes the follow up of calculated equipment efficiencies using the validated data mentioned before as well as the site Carbon emissions.

By auditing the energy system, imbalances are detected and diagnosed. As a result, the data can be relied on for evaluating the value of energy production and usage, and waste can be eliminated. Planning for a better operation of the energy system by performing case studies is usually done by using the validated model in off-line mode. Finally, it is important to mention that the optimization of operating conditions is performed on a continuous, daily basis since the system was commissioned in 2005. As a result of the project, substantial savings in energy costs have been achieved.

In order to illustrate the obtained benefits, optimization results are shown in Figure 7. Results correspond to one day of operation using Visual MESA. Savings are indicated as a percentage of total energy costs. After implementing the optimization recommendations, operation is closer to the minimum global energy cost, as it can be seen in the plot, after 9:36 AM. When the operator saw in the report potential savings higher than 1% of the overall utilities system cost, he implemented the optimization recommendations (in this case, a combination of changes in boiler loads and steam flows among production areas).

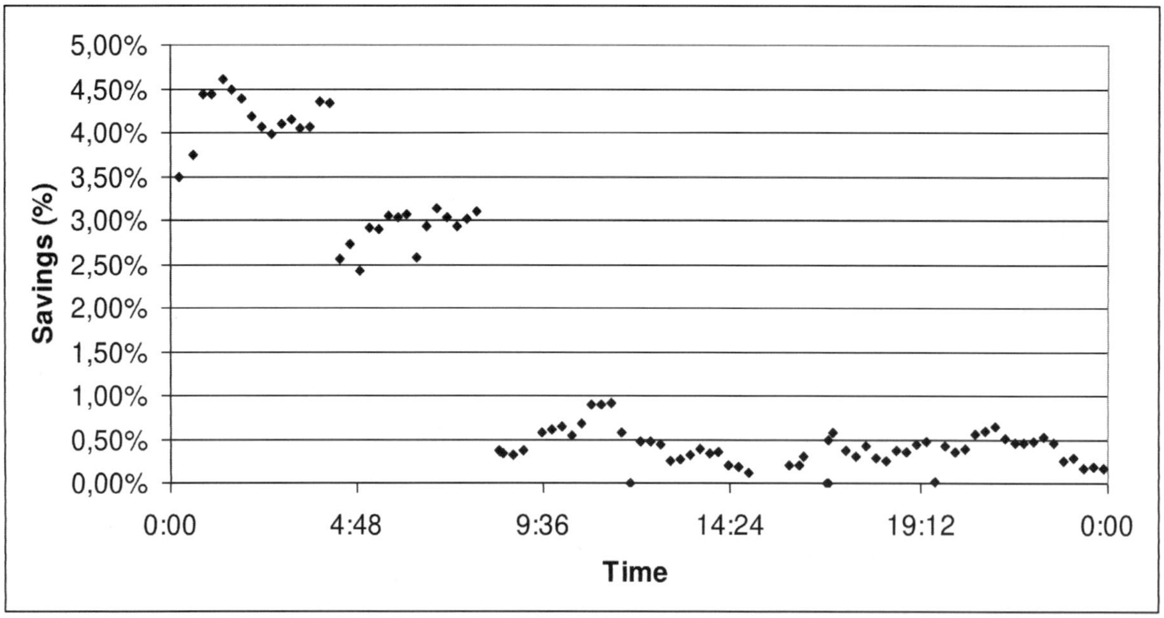

Figure 7 – Energy Costs Reduction Performance: One Day of Operation

Real Time Energy Management System Projects Key Issues

To successfully implement, deploy and sustain a real time energy management system energy cost and CO_2 emission reduction project, there are several issues that need to be addressed and recommended practices we have found useful to be taken into account.

Proper Technology Selection: Select the technology that fits with the problem solution rather than force the problem to be solved with an a priori selected technology. For example, if a simple linear regression with two input variables is sufficient enough to infer the NO_x emission, do not force the use of a several inputs neural network. But always high rank systems based on sound process engineering principles (i.e., mass and energy balances and rigorous non-linear equipment modeling) than those based on correlations or non-physical sense solutions (Friedman, 1999).

Avoid "Computerization": To solve a given problem, engineers need to apply the *EBC* approach suggested by Richalet, 2000, rather than the straight use of *C*. Follow the *EBC* procedure implies:

- First, use your ***E***ye to inspect trends, data, process diagrams, DCS screens...
- Second, use your ***B***rain to carefully understand and design the real time system and select or design the proper software tool.
- Finally, only at the end, use the ***C***omputer to solve the problem.

The enthusiastic but inexperienced engineers tend to go straight to the ***C***omputer, forgetting or minimizing to analyze and perfectly understand the problem to be solved, their real handles and constraints.

Balance Model Complexity against Expected Results: Energy system model complexity must be balanced against expected benefits. Remember the golden rule: "the simplest is generally the better". Engineering judgment and experience plays a key role in this task. Operators and Engineers will appreciate any efforts towards simplification, especially for future re-training and system maintenance.

Real Time Data Seldom Can be Used "as is": Carefully designed and engineered "safety envelops" must be built around real time on-line data, in order to validate all the signals, in and out of the applications. Minimum and maximum limits for each variable, controller windup identification and associated valve openings and controllers statuses must be combined to properly identify and isolate measurement problems. Steady state detection procedures must also be considered to warn Operators or directly skip an optimization run if the Process is unstable. What should a system do if bad data is detected? If the bad data feeds "critical" variables of the system, then its execution should be suspended and Operators warned. If bad data is related to "noncritical" variables then some kind of backup or default values could be used but such an action needs to be defined on each individual variable. The implementation of data validation procedures is a key activity for every real time energy management system project.

Pricing Information Must be Properly Updated: As economics is the main driver to apply real time energy management systems, prices used for the optimization or objective function calculations must be diligently updated. Manual data entering should be avoided.

Automatic pricing information must be implemented acquiring the prices directly from process information systems or web pages (some examples we have used in the past include the web sites of ERCOT in Texas, US, OMEL in Spain, both for electricity market spot prices and ICE in the US for energy prices).

Have Operators Always in Mind: From the very beginning try to involve Operators, inform them about the project objectives and solve their concerns. A common question is about being replaced by the new system and fired after completion. Second in rank is their worries regarding if the new system will increase their burden. First question is usually outside our control but we never have seen a single Operator being fired due to our projects, all of them successful, by the way. With respect to the second concern, an especial effort needs to be done to not increase Operator's burden, implementing automatic data updates or procedures to minimize his/her manual intervention. Clear and easy to understand operating interfaces need also to be built. Operators must be properly trained and assisted right by the DCS console after being left alone with the new system. It is always a good idea to spend many hours at the DCS console with the Operators to discuss the proposed solutions and technology, explaining to them what is needed so they will accept the introduction of the new technology. Properly designed Operating manuals must be provided and updated as frequently as needed.

Do not "Prototype" but Deploy Well Engineered Systems: Do not expose Operators nor final users to "interim or prototyping versions" but only to finished, very well built models, including the Operating interface and true handles limits and constraints. Operator's general attitude is to closely adhere to their first impression. There is a single "silver bullet" to use with them: they will never forget if early disappointed or frustrated when trying to use the system for the first time. Moreover, many of them would became "vaccinated" against the energy management technology and refuse to use it, subtle or frontally.

Sustain the Application to be Maintained "Forever Green": Engineering documentation must be provided and updated as soon as changes are implemented in the model or to the Operating interface. The application needs to be maintained "forever green", reflecting continuously the current Process handles and constraints. Changes on the energy system or process related side must be immediately reflected in the model. Local resources need to be properly trained and technology providers need to assist them and remotely maintain the applications. As a side note, it is important to mention the huge impact of remote access capabilities on systems performance. Only a few years ago just a minor model adjustment would take a lot of time for the new information to be exchanged back and forth. Nowadays, VPN or Webex like secure remote access can be easily implemented and solutions can be provided in minutes, instead of hours or even days.

Conclusions

CO_2 emissions and energy cost reductions have been obtained taking advantage of an energy management system software functionalities. It demonstrated to be a robust optimizer and very well suited to be used on-line, providing optimization recommendations to the operators on a routine basis.

The authors, working tightly with clients and Soteica engineering team members, following the generic outlined project rules, have participated in many successful implementations which resulted in significant economic benefits for the refining and petrochemical industries involved.

The success ratio of the industrial real time project deployments was highly increased by following a set of simple rules and ensuring a proper knowledge transfer to both, Operators and Engineers. Coordination among plant areas in order to implement the proposed optimization recommendations, as well as management involvement, were critical issues. The robustness of the tool helped Operators to gain confidence on the overall optimization system.

Application sustainability, which includes both maintenance and support, helps to sustain the economic benefits after commissioning allowing them to be always in good shape and still used and appreciated by the owners many years after handing the project over to them.

References

1. OPC Foundation, http://www.opcfoundation.org/, 2010
2. "Avoid Advanced Control Project Mistakes", Y. Z. Friedman, Chapter of the book Advanced Process Control and Information Systems for the Process Industries, L. Kane (editor) , ISBN 0-88415-239-1, Gulf Publishing Co., Houston, Texas, Pages 42-48, 1999
3. Personal communication, J. Richalet, 2000
4. "Auditing and control of energy costs in a large refinery by using an on line tool", D. Ruiz, C. Ruiz, J. Mamprin, Depto. de Energías y Efluentes Petronor, (2005), ERTC Asset Maximisation Conference organized by Global Technology Forum, Budapest, May 23-25, 2005
5. "Reducing refinery energy costs", Ruiz D., Ruiz C., Nelson D., Roseme G., Lázaro M. and Sartaguda M., (2006), Petroleum Technology Quarterly, pages 103-105, Q1 2006
6. "Online energy management", S. Benedicto, B. Garrote, D. Ruiz, J. Mamprin and C. Ruiz, Petroleum Technology Quarterly, pages 131-138, Q1 2007
7. "The Use of an On-line model for Energy Site-Wide Costs Minimisation", García Casas, J.M., Kihn, M., Ruiz D. and Ruiz C., (2007), ERTC Asset Maximisation Conference organized by Global Technology Forum, May 21-23, Rome, 2007
8. "Energy System Real Time Optimization", D. Uztürk, H. D. Franklin, J. M. Righi, A. T. Georgiou, NPRA Plant Automation and Decision Support Conference, Phoenix, USA, 2006

9. "Site-wide Energy Costs Reduction at TOTAL Feyzin Refinery", Département Procédés - Energie, Logistique, Utilités, TOTAL - Raffinerie de Feyzin, D. Ruiz, J. Mamprin and C. Ruiz, ERTC 12th Annual Meeting organized by Global Technology Forum, 19-21 November, Barcelona, 2007
10. "A Watchdog System for Energy Efficiency and CO2 Emissions Reduction", Diego Ruiz, Carlos A. Ruiz, European Refining Technology Conference (ERTC) Sustainable Refining, Brussels, Belgium, April 2008
11. "Online energy costs optimizer at petrochemical plant", Marcos Kihn, Diego Ruiz, Carlos Ruiz, Antonio García Nogales, Hydrocarbon Engineering, pages 119-123, May 2008
12. "Finding Benefits by Modeling and Optimizing Steam and Power Systems", M. Reid, C. Harper, C. Hayes, Industrial Energy Technology Conference (IETC), New Orleans, May 2008
13. "Energy Real-Time Optimizer Baytown Refining/Chemical Complex", M. Lorenz, Industrial Energy Technology Conference (IETC), New Orleans, May 2009
14. "Real Time Energy Management at Repsol Cartagena Refinery", Fernán Serralunga, Diego Ruiz, Carlos Ruiz, Process Department REPSOL Cartagena Refinery, European Refining Technology Conference (ERTC) Asset Maximisation, Prague, May 2009
15. "Industrial Experience on the Implementation of Real Time On Line Energy Management Systems in Sugar and Alcohol Industry", Douglas C. Mariani, Marcos A. Kihn, Carlos A. Ruiz, 10th International Symposium on Process Systems Engineering - PSE2009, Salvador de Bahia, Brazil, Elsevier, August 2009
16. "Emissions and Energy: An Integral Approach Using an On-Line Energy Management and Optimization Mode"l, Diego Ruiz, Fernán Serralunga, Carlos Ruiz, European Refining Technology Conference (ERTC) Annual Meeting Energy Workshop, Berlin, Germany, November 2009
17. "Online real-time optimisation helps identify energy gaps", Marcelo Galleguillos, Michel Maffet, Marcos Kihn, Ruben Monje and Carlos Ruiz. Petroleum Technology Quarterly (PTQ), Q4, 101-109, October 2010

(51b) Reduction of CO_2 Emissions In a Refinery

Jason Stalhman, UOP, Des Plaines, IL

The CO2 Emissions from a refinery may be reduced in a variety of ways. This presentation will focus on methods to optimize Utilities and Offsites related equipment; both inside and outside the process unit battery limits. By reducing fuel fired in Utility Boilers, and CO2 production in other areas of the refinery, the total CO2 emission footprint will be reduced. Specific topics that will be discussed: optimization of an MP steam header pressure, steam generation, multiple BFW headers, HRSG supplemental firing, and tank vapor recovery.

See more of this Session: CO_2 Reduction

See more of this Group/Topical: Topical 7: 14th Topical on Refinery Processing

(51c) Reducing CO_2 Emissions In FCC Process

X. X. Frank Zhu, UOP LLC, Des Plaines, IL and **Saadet Ulas Acikgoz**, R&D Adsorption and Gas Processing, UOP LLC, Des Plaines, IL

The FCC process unit has been a major work horse in the petroleum industry since the 1940's. The attention for improving FCC performance in the past has been on enhancing the reaction conditions and catalysts to deliver value through yield improvements. However, in the current high energy cost environment and forthcoming CO_2 regulations, reducing CO_2 emissions in FCC unit started drawing much more attention. Refineries contribute to ~4% of total CO_2 emissions in the United States. FCC units account for 15-20% of refinery energy consumption and CO_2 emissions. In this project, novel and major improvement opportunities for FCC have been identified which can create a step change in reducing CO_2 emissions in FCC units.

In summary, improvement opportunities are identified based on viewing FCC as ISBL (Inside Battery Limit) and integrating FCC with the refinery as whole, i.e. OSBL (Outside Battery Limit). The opportunities include optimization of fractionation/separation flowsheet and conditions, integration of advanced technologies for fractionation/separation, maximizing heat and power recovery within FCC and integration of generated heat and power in ISBL with OSBL balances. Collectively, these improvements could help establish the image of FCC as a utility hub, delivering high value energy utilities to the refinery and lessen the CO_2 emissions from the refinery.

The new concepts developed as a result of this project include:

- *New flue gas cooler design:* Integrating a raw feed preheating system into the flue gas cooler to reduce flue gas cooler discharge temperature and maximize energy recovery.

- *Dividing wall column for naphtha splitting:* Using dividing wall columns for naphtha splitting and column sequencing to minimize utility usage and capital costs

- *Main fractionator optimization:* Improving the design of FCC Main Fractionator to recover more energy

- *Steam turbine integration:* Extraction HP (High Pressure) steam turbine in power recovery section to provide all the MP (Medium Pressure) and LP (Low Pressure) internal steam requirements in FCC unit and maximize power recovery

(51d) Profitable Emission Reduction In Petroleum Refineries

Jian Zhang and Qiang Xu, Dan F. Smith Department of Chemical Engineering, Lamar University, Beaumont, TX

Emission reduction is becoming increasingly important for petroleum refineries, as more environmental concerns and stringent regulations have been posed. Air emission reduction in refineries is also a very challenging task. It involves a large-scale complex manufacturing system. In such system, air emissions could spatially scatter at many reactors, furnaces, and flaring facilities; and temporally occurs at any time. Thus, cost-effective emission-reduction strategies should neither depend on end-of-pipe approaches, nor focus on solutions for single units or subsystems. The major thrusts should focus on the comprehensively study of material, energy, and information exchanges within the entire manufacturing system, so as to identify the best solutions to the entire plant. From the systems theory point of view, it's unnecessary to enforce emission sources from each individual facility to reach the minimum. As long as the generated emission sources from some facilities could be appropriately reused by others facilities to the maximum extent, the entire system emissions may still have chances to approach the minimum. Therefore, to optimally balance emission source generations and utilizations among the entire system might be the smartest way to reduce refinery emissions.

A greedy concept of profitable emission reduction (PER) is proposed in this paper, which has merits of economically attractive, environmentally benign, and technologically viable for air emission reductions in petroleum refineries. To identify PER strategies, a methodology framework and a new general plant-wide scheduling model have been developed, which couples generic production activities and the characterization of major air emissions from refineries, such as CO_2, VOC (volatile organic compounds), NO_X (nitrogen oxides), and PM (particulate matters). A case study has demonstrated the efficacy of the PER concept and the developed methodologies.

See more of this Session: CO_2 Reduction

See more of this Group/Topical: Topical 7: 14th Topical on Refinery Processing

Synthesis and Characterization of Trimetallic Ni(Co)-Mo-W HDS Catalysts Supported on Al-HMS Substrates

R. Huirache-Acuña[1], B. Pawelec[2], E.M. Rivera-Muñoz[3], R. Nava[4]
B. Torres[5], G. Alonso-Núñez[6], J.A. Cortés[1] and J.L.G. Fierro[2]

[1]Facultad de Ingeniería Química, Universidad Michoacana de San Nicolás de Hidalgo, Ciudad Universitaria, 58060, Morelia, Michoacán, Mexico.
[2]Instituto de Catálisis y Petroleoquímica, CSIC, c/ Marie Curie, 2, Cantoblanco, 28049 Madrid, Spain.
[3]Centro de Física Aplicada y Tecnología Avanzada, Universidad Nacional Autónoma de México, A.P. 1-1010, 76000. Querétaro, Qro., Mexico.
[4]Facultad de Ingeniería, Universidad Autónoma de Querétaro, Centro Universitario, Cerro de las campanas, 76010, Querétaro, Mexico.
[5]University of Texas at El Paso, Materials Research Technology Institute, El Paso, TX ,79968, USA.
[6]Centro de Nanociencias y Nanotecnología, UNAM, Ensenada, B. C., 22800, Mexico.

E-mail: rafael_huirache@yahoo.it

Abstract

In order to clarify the effect of promoter (Co vs Ni) on the HDS activity of ternary systems (Co(Ni), Mo, W), Al-HMS supported catalysts were synthesized by simultaneous impregnation. Prior to the catalytic test, the catalytic materials were sulfided *ex situ* at atmospheric pressure in a tubular furnace heated up 673 K under a flow of H_2S in hydrogen (15%) and kept under those conditions for 4 h. The sulfided catalysts were tested in the HDS of dibenzothiophene (DBT) performed in a batch reactor at 623 K and total pressure of 3.1 MPa. The calcined and fresh sulfide catalysts were characterized by a variety of techniques (N_2 adsorption–desorption isotherms, XRD, DRS UV–vis spectroscopy and HRTEM).
Keywords: HDS, Catalysts, Al-HMS

1. Introduction

Traditionally, the removal of sulfur from petroleum-derived feedstocks has been achieved by a hydrodesulfurization (HDS) process using alumina-supported Mo or W catalysts promoted by Co or Ni. However, to satisfy new stringent environmental restrictions on the sulfur content of fuels [1], more active HDS catalysts are being requested by the industry. The new generations of catalysts, such as NEBULA®, which are based on totally different concept of bulk-like materials [2] are more expensive than alumina-supported ones. Thus, the challenge is to design novel cost-effective and active supported catalysts.
Many different approaches have been taken to achieve this goal, including varying the catalyst support and catalyst synthesis methods. Considering the new supports, recently, the use of mesoporous silica materials for supporting hydrotreating catalysts has attracted interest of the research community [3-11]. In particular, the use of hexagonal mesoporous molecular sieves such as HMS and SBA-15 are intensively studied [3-9,11,12] because of their hexagonal array of uniform mesopores, a narrow pore size distribution, high surface area, high sorption capacity and thermal stability. In addition, the former could be improved by support modification with different cations such as Al, Ti, Zr [11].

In particular, a comparison of HDS activities of WS_2/Al-HMS [4] and MoS_2/Al-HMS [5] catalysts with γ-Al_2O_3-supported ones indicated that the former exhibit outstanding activities. The results on MoS_2/Al-HMS catalysts indicated that HMS mesoporous structure has a unique ability to impart higher activities in the supported MoS_2 phase through metal-support interaction [5]. Effect of Al content in HMS material on the catalytic activity of NiMo and CoMo hydrotreating catalysts in the HDS of dibenzothiophene was studied by Zepeda et al. [6]. It was found that all supported NiMo and CoMo catalysts on Al-HMS substrates showed higher catalytic activity than their Al-free counterparts. Moreover, it was found two correlations between the structure and chemical coordination of the supported oxide precursors and catalytic activity. The differences observed in catalyst performance were attributed to the structure and specific electronic properties of the supported active species [6].

For technical and economic reasons, the impregnation method is the one most frequently used for preparing heterogeneous catalysts with a high surface area in order to improve the accessibility of active sites. As compared with the impregnation method, supercritical methanol–carbon dioxide deposition synthesis is more complicate and expensive method. However, the results indicated that supercritical impregnation of Co(Ni) and Mo into support has a positive effect on the dibenzothiophene hydrodesulfurization activity of CoMo/Al-HMS and NiMo/Al-HMS nanocatalysts [7-9]. Furthermore, it was found that addition of aromatics in the feed showed less inhibition effect of CoMo/Al-HMS nanocatalyst with respect to the commercial catalyst [7]. The higher catalyst performance was attributed by authors to higher and more uniform deposition of the active phases on the support with increased active surface area.

To the best our knowledge there are no references in the literature about the use of the Al-HMS material for supporting trimetallic Co(Ni)MoW catalysts. All investigations are on the monometallic [4, 5] or bimetallic catalysts [3, 6-9]. In order to clarify the effect of promoter (Co vs Ni) on the HDS activity of ternary systems Al-HMS supported catalysts were synthesized and studied.

2. Experimental

The Al-HMS material was synthesized following a procedure similar to Zhang and coworkers [13], via neutral templating pathway using dodecylamine as the surfactant. In a typical synthesis of Al-HMS material, 0.25 mol of dodecylamine were dissolved in 10 mol of ethanol and 130 mol of deionized water. Subsequently, 1.125 mol of Mesitylene were added and the solution was stirred for 20 min. After this, a solution of the required amount of tetraethyl orthosilicate (TEOS, 98%, Aldrich) and aluminum isopropoxide (Si/Al = 40) was added to the surfactants solution, which was stirred at room temperature for 20 h. After this, the solid obtained was filtered, washed thoroughly with deionized water, dried first in air at room temperature and then to 110 °C for 18 h and calcined at 540 °C for 4 h in order to remove the organic template.

Ni(Co)MoW catalysts were prepared using the Al-HMS substrate. The support was loaded with fixed amounts of molybdenum (8.53 wt% as MoO_3), tungsten (13.75 wt% as WO_3) and cobalt/nickel (3.81 wt% as CoO or NiO), by simultaneous impregnation via incipient wetness method. For this, an aqueous solution containing ammonium heptamolybdate tetrahydrate ($(NH_4)_6Mo_7O_{24}.4H_2O$, Aldrich), ammonium metatungstate

(($(NH_4)_6H_2W_{12}O_{40}.xH_2O$, Aldrich) and cobalt nitrate hexahydrate ($Co(NO_3)_2.6H_2O$, Aldrich, 98%) or nickel nitrate hexahydrate ($Ni(NO_3)_2.6H_2O$, Aldrich, 98%) was prepared. The concentrations were calculated to achieve a Mo/W atomic ratio of 0.5 and a Ni(Co)/Mo atomic ratio of 0.3. The impregnates were dried at 85 °C for 16 h and then calcined at 500 °C for 4 h. Prior to the catalytic test, the catalysts were sulfided *ex-situ* at atmospheric pressure in a tubular furnace heated up 400 ºC (4 ºC/min) under a flow of H_2S in hydrogen (15%) and kept under those conditions for 4 h. Then, the sample was cooled down to room temperature under a nitrogen flow.

3. Results and discussion

3.1 Textural properties

The textural properties of the pure Al-HMS support and the calcined Co(Ni)MoW/Al-HMS catalysts were studied by N_2 adsorption-desorption isotherms at 77 K (Fig. 1). According to the IUPAC classification [14], the isotherms of all samples are type IV with a sharp step at intermediate relative pressures.

The pure support shows the sharp inflection of the adsorption branch, which implies the uniformity in the pore size distribution. The shape of the isotherms of both calcined CoMoW/Al-HMS and NiMoW/Al-HMS catalysts is similar to those of the pure Al-HMS material indicating that the porous characteristics of the support have not been damaged after metal loading following by calcination.

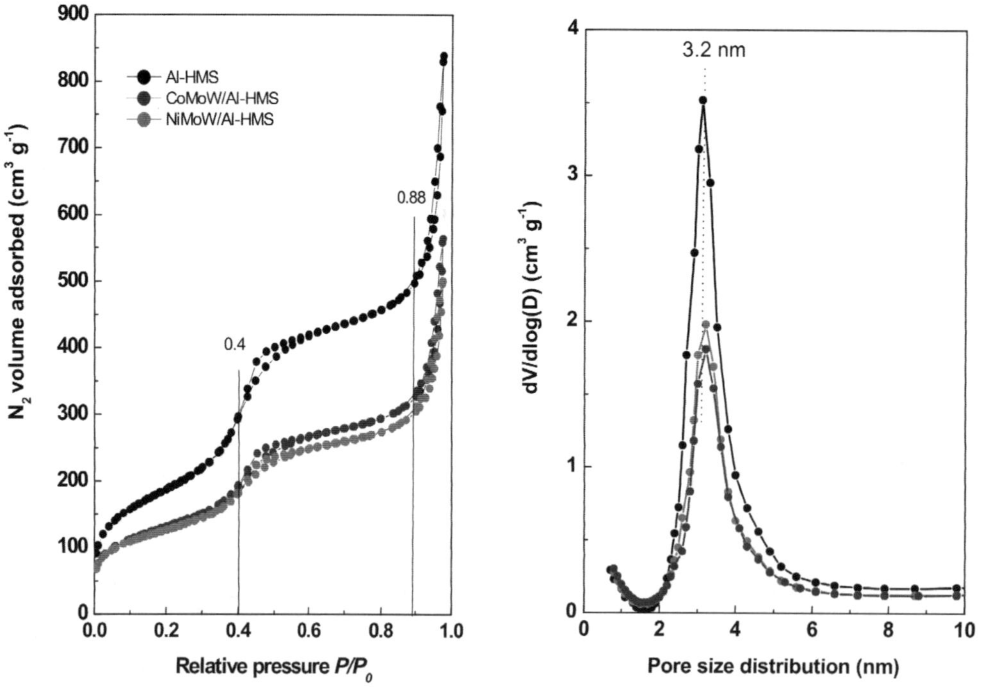

Figure 1. Nitrogen adsorption-desorption isotherms and Pore size distribution determined from adsorption branch of the N_2 adsorption data using the BJH method of pure Al-HMS support and calcined Co(Ni)MoW/Al-HMS catalysts

The N_2 isotherms of all samples have two capillary condensation steps: (i) the first hysteresis loop for these materials starts at a partial pressure of about 0.40, indicating the presence of framework mesoporosity and (ii) the second hysteresis loop starting at a relative pressure of about 0.88 might to be attributed to either to macropores or to textural inter-particle mesoporosity, which takes place when the particle diameters are very small and closely packed [15].

The pore size distributions calculated from the adsorption branch of the N_2 isotherm by using the Barret-Joyner-Halenda model are shown in Fig. 2. As seen in this figure, all samples show a uniform, narrow pore size distribution centred at 3.2 nm. The specific area (S_{BET}) and the pore size/volume of the samples derived from the N_2 isotherms are given in Table 1. A comparison of the specific surface area values of the pure support and calcined catalysts indicates that S_{BET} decreases upon the incorporation of metal oxides.

Simultaneously, the total pore volume decreases after metal oxides incorporation during catalyst preparation. Since the NiMoW/Al-HMS catalyst show only a slightly lower S_{BET} than its CoMoW/Al-HMS counterpart (471 vs 451 $m^2\ g^{-1}$), similar metal oxides dispersion on the support surface could be expected for both catalysts.

Considering the experimental error values, the CoMoW/Al-HMS catalyst recorded similar average pore diameter value (7.4 nm) and pore volume (0.87 $cm^3\ g^{-1}$) in comparison with the NiMoW/Al-HMS sample (6.9 nm and 0.77 $cm^3\ g^{-1}$, respectively).

Table 1. Textural properties of the pure support and calcined catalysts as determined by N_2 sorption and XRD data

Support/Catalyst	S_{BET} ($cm^3\ g^{-1}$)	V_{total} ($cm^3\ g^{-1}$)	Average pore diameter (nm)
Al-HMS	689	1.30	7.5
CoMoW/Al-HMS	471	0.87	7.4
NiMoW/Al-HMS	451	0.77	6.9

3.2. Low- and wide-angle X-ray diffraction

The low-angle X-ray diffractograms of the pure support and both Co(Ni)MoW/Al-HMS catalysts are shown in Figure 2(A). Pure Al-HMS material shows an intense single *d*100 reflection accompanied by more or less pronunciated diffuse scaterring centered at 1.73°, which indicates hexagonal symmetry. Higher order Bragg reflections are not resolved in the patterns of this material. However, it was demonstrated that similar "single-reflection" materials can still exhibit local hexagonal symmetry, as observed by selected area electron diffraction and TEM [16]. After metal loading, the intensity of this peak decreases very much and the peak is shifted to higher angles (from 1.73 to 1.8)

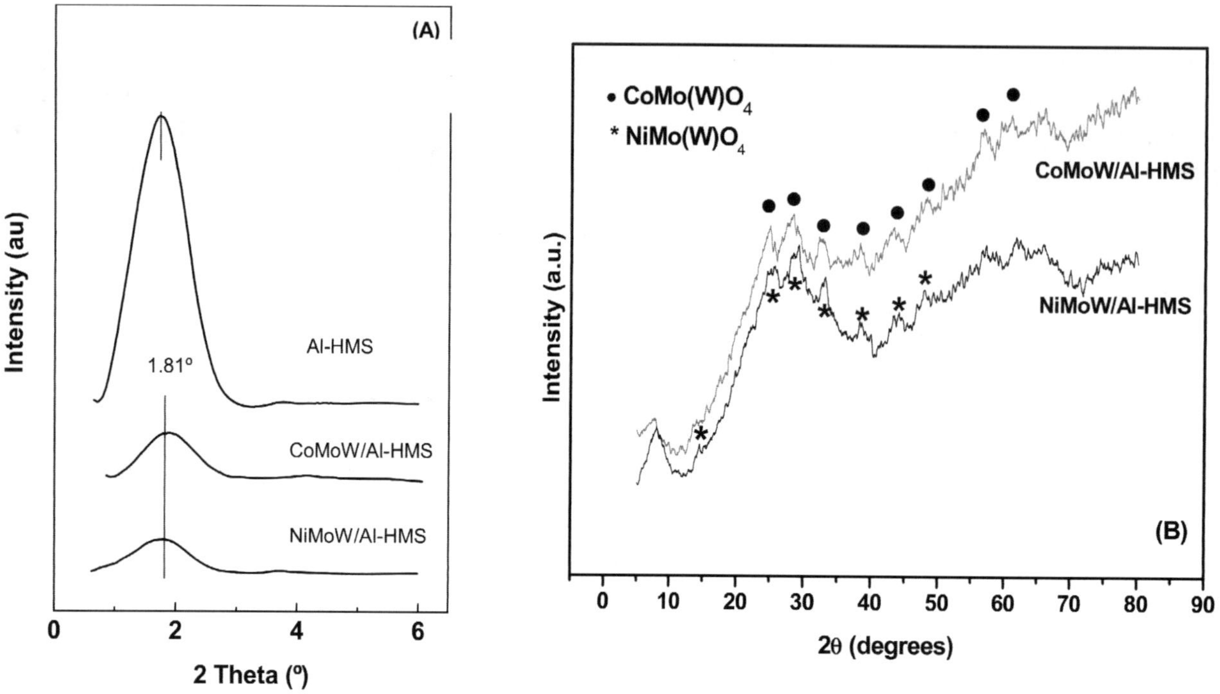

Figure 2. Low and wide angle XRD patterns for Ni(Co)MoW/Al-HMS catalysts

Wide angle XRD profiles were recorded with the aim to investigate the presence of any crystallite species in the catalysts. Figure 2(B) illustrates the diffraction profiles of the calcined CoMoW/Al-HMS and NiMoW/Al-HMS catalysts. All oxide catalysts reveal typical diffraction patterns, with a very broad peak at around $2\theta = 24°$, typical of amorphous silica [17]. Moreover, the XRD patterns contain low peak intensities at $2\theta = 25$, 28, 32.5, 38, 43, 48, 57 y 59.5° corresponding to β-$CoMoO_4$ phase (JCPDS-ICDD 21-868). On the other hand, peak intensities corresponding to $NiMo(W)O_4$ was observed for NiMoW/Al-HMS catalysts (JCPDS-ICDD 33-0948).

3.3. UV-Vis diffuse reflectance spectra (DRS)

The coordination environment of $Ni(Co)^{2+}$ and $Mo^{6+}(W^{6+})$ ions in the calcined catalysts was studied by UV-Vis diffuse reflectance spectroscopy. The DRS spectra of the calcined catalysts are plotted in Fig. 3 as a Kubelka-Munk-Schuster function. Regardless of promoter, all catalysts show two strong bands at about 210-270 nm and 320-350 nm. The first one imply the presence of a ligand-to-metal charge transfer involving isolated transition metal sites which is ascribed in literature to tungsten in tetrahedral coordination, like that in the WO_4^{2-} group [18], but differed form that in the WO_3 crystal [19]. The second one is assigned in literature to O^{2-}-Mo^{6+} (W^{6+}) ligand-to-metal charge transfer transition (LMCT) of molybdenum and tungsten ions having octahedral coordination [20, 21]. A displacement observed for the mentioned bands could be attributed to the influence of tungsten on the electronic properties of this kind of catalysts. It is noteworthy that both catalysts showed very similar amount of $Mo^{6+}(W^{6+})$ ions in octahedral coordination and a larger amount of those ions in tetrahedral coordination was observed for CoMoW/Al-HMS.

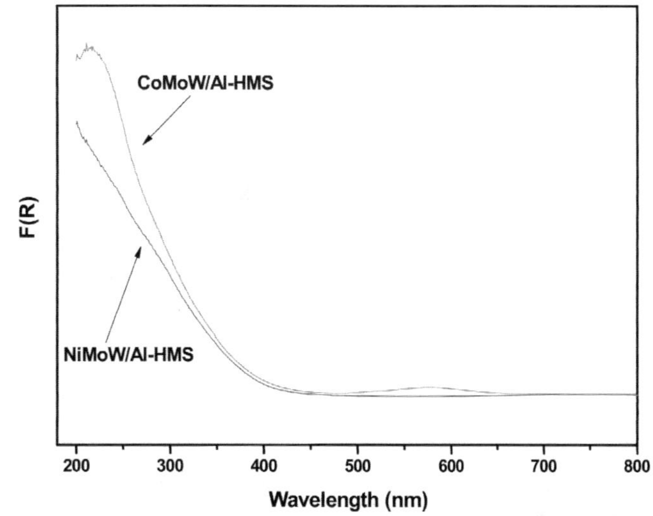

Figure 3. DRS UV-vis spectra of the calcined Co(Ni)MoW/Al-HMS catalysts

This behavior could be related with a larger pore diameter of the CoMoW/Al-HMS with respect to NiMoW/Al-HMS (7.4 vs. 6.9 nm) this is because the location of species within the inner porous structure of the former catalysts which did not inhibited a strong metal-support interaction. As a consequence, for the Co-supported catalyst the formation of the species having Mo^{6+}/W^{6+} ions in tetrahedral coordination is favored.
In addition, the visible spectra of the SBA-16-based catalysts exhibit bands in the 450-700 nm region which are ascribed to *d–d* transitions of high spin octahedral Co(Ni) complexes [22, 23, 6]. Additionally, the DR spectra provide strong evidence that well-dispersed octahedral cobalt and nickel species are present on the surface of Al-HMS substrates. The octaedral species are easy reduced and sulfided, and therefore are susceptible to develop the HDS active sites [23, 24].

3.4. HDS of DBT

In this work, the HDS of DBT was carried out in a Parr model 4575 high-pressure batch reactor. In order to minimize internal diffusion limitations, all catalysts were thoroughly ground in a mortar to a fine powder. All catalysts (including the reference sample) were sulfided by the same procedure. The sulphided catalyst (ca. 0.5 g) was then introduced into the batch reactor containing a solution of 5% vol of DBT in decaline ($[DBT]_0$= 0.239 mol/L]. The reactor was then pressurized to 3.1 MPa with hydrogen and heated up to 350 °C at a rate of 10ºC/min. The stirring rate of the reaction mixture was sufficiently intense (600 rpm) to exclude external diffusion limitations. After the working temperature was reached, sampling for chromatographic analysis was performed during the course of each run to determine conversion versus time dependence. The reaction time was averaged to 5 h. The resulting products were analyzed using a HP 4890 gas chromatograph provided with a 10-ft packed column containing 3% OV-17 as separating phase on Chromosorb WAW 80/100. The main reaction products from the HDS of DBT are biphenyl (BP), cyclohexylbenzene (CHB) and tetrahydrodibenzothiophene (THDBT). The total DBT conversion was calculated as DBT disappearance, and direct desulfurization (DDS) and hydrogenation (HYD) selectivities were defined as:

DDS: (BP) x 100% / (CHB+BP+THDBT) (1)

HYD: (CHB+THDBT) x 100% / (CHB+BP+THDBT) (2)

HDS kinetic rate constants (*k*) were calculated assuming pseudo-zero order kinetics referred to initial DBT concentration, DBT conversion (*XDBT*) and reaction time (*t*), taking into account that the excess of hydrogen was employed, thus, its partial pressure could be considered as constant. The equation employed for *k* values calculation was:

$$[DBT]_0(1-X_{DBT}) = kt \quad (3)$$

The activities of the Co and Ni promoted MoW/Al-HMS samples in hydrodesulfurization reaction were compared and were found to fall in the order: NiMoW/Al-HMS ($17x10^{-7}$ $mol_{DBT}/g \cdot s$) > CoMoW/Al-HMS = CoMo/Al_2O_3 (commercial) ($12x10^{-7}$ $mol_{DBT}/g \cdot s$). This behavior might be related with a larger amount of $Mo^{6+}(W^{6+})$ ions in tetrahedral coordination for CoMoW/Al-HMS catalyst compared with that observed for their nickel-promoted counterpart. The tetrahedral species are difficult to be reduced and sulfided, and therefore are not susceptible to develop the HDS active sites.

4. Conclusions

The effect of promoter (Co vs Ni) on the HDS activity of ternary systems Al-HMS supported catalysts was studied. The highest catalytic activity was observed for NiMoW/Al-HMS catalyst. This behaviour could be related with a larger amount of Mo^{6+} (W^{6+}) ions in octahedral coordination which are susceptible to develop the HDS active sites.

References

[1] Final Report by the European Commission on the Revision of Directive 98/70/EC.
[2] F.L. Plantenga, R.G. Leliveld, Appl. Catal. A: Gen. 248 (2003) 1.
[3] T. Halachev, J.A. de los Reyes, C. Araujo, G. Cordoba, L. Dimitrov, in B. Delmon, G.F. Froment, P. Grange (Eds.), Hydrotreatment and Hydrocracking of Oil Fractions, Stud. Surf. Sci. Catal. 127 (1999) 401.
[4] T. Chiranjeevi, P. Kumar, M.S. Rana, S.K. Maity, G. Murali Dhar, T.S.R. Prasada Rao, Microporous Mesoporous Mater. 44-45 (2001) 547-556.
[5] T. Chiranjeevi, P. Kumar, M.S. Rana, G. Murali Dhar, T.S.R. Prasada Rao, J. Mol. Catal. A: Chem. 181 (2002) 109-117.
[6] T.A. Zepeda, B. Pawelec, J.L.G. Fierro, A. Olivas, S. Fuentes, T. Halachev, Microporous Mesoporous Mater. 111 (2008) 157-170.
[7] M. Alibouri, S.M. Ghoreishi, H.R. Aghabozorg, Chem. Eng. Transactions 17 (2009) 1005-1010.
[8] M. Alibouri, S.M. Ghoreishi, H.R. Aghabozorg, J. Supercritical Fluids 49 (2009) 239-248.

[9] M. Alibouri, S.M. Ghoreishi, H.R. Aghabozorg, Ind. Eng. Chem. Research 48 (2009) 4283-4292.

[10] J. Ramírez, R. Conteras, P. Castillo, T. Klimova, R. Zárate, R. Luna, Appl. Catal. A: Gen. 197 (2000) 69-78.

[11] B. Pawelec, S. Damyanova, R. Mariscal, J.L.G. Fierro, I. Sobrados, J. Sanz, L. Petrov, J. Catal. 223 (2004) 86-97.

[12] R. Huirache-Acuña, B. Pawelec, E. Rivera-Muñoz, R. Nava, J. Espino and J.L.G. Fierro, Applied Catalysis B: Environmental, Vol. 92 Issues 1-2 (2009) 168-184.

[13] W. Zhang, T.R. Pauly, T.J. Pinnavaia, *Chem Mater* 9 (1997) 2491-2498.

[14] A.Y. Khodakov, A. Griboval-Constant, R. Bechara, F.J. Villain, Phys. Chem. B 105 (2001) 9805

[15] J.A.Hernandez, S. Gómez, B. Pawelec, T.A. Zepeda, Appl. Catal. B: Environ. 89 (2009) 128-136.

[16] P.T. Tannev, M. Chibwe, T.J. Pinnavaia, Nature, 368 (1994) 321.

[17] R. Nava, R. A. Ortega, G. Alonso, C. Ornelas, B. Pawelec and J.L.G. Fierro, Catalysis Today 127 (2007) 70.

[18] J.G. Graselli, B.J. Bulkin, Analytical Raman Spectroscopy, Wiley, New Cork, 1991, p. 352.

[19] Z. Zhang, J. Suo, X. Zhang, S. Li, Appl. Catal. A: Gen. 179 (1999) 11.

[20] R.A. Schoonheydt, Diffuse reflectance spectroscopy, in: F. Delannay (Ed.), Characterization of Heterogeneous Catalysts, Marcel Dekker, New York, 1984, p. 125.

[21] D.K. Lee, H.T. Lee, I.Ch. Lee, S.K. Park, S.Y. Bae, Ch. H. Kim, S.I. Woo, J. Catal. 159 (1996) 219.

[22] J.E. Herrera, D.E. Resasco, J. Catal. 221 (2004) 354.

[23] H.K. Matralis, Ch. Papadopoulou, A. Lycourghiotis, Appl. Catal. A: Gen.116 (1994) 221.

[24] Ch. Papadopoulou, J. Vakros, H.K. Matralis, G.A. Voyiatzis, Ch. Kordulis, J. Colloid. Interface Sci. 274 (2004) 159.

(52b) Intrinsic Vs. Environmental Factors In Ultra-Deep Hydrodesulfurization

Teh C. Ho, Corporate Strategic Research Laboratories, ExxonMobil Research and Engineering Co., Annandale, NJ and Jonathan M. McConnachie, ExxonMobil Research and Engineering Co., Annandale, NJ

A hydrogenation index (HI), measured in the hydrodesulfurization (HDS) of dibenzothiophene (DBT), is used to estimate the intrinsic hydrogenation selectivities of MoS_2, $Co_{0.1}MoS_2$, and two supported HDS catalysts. The HI and catalyst activity for desulfurizing 4,6-diethyl-DBT follow the same trend: MoS_2 >> $Co_{0.1}MoS_2$ >> supported catalysts. For desulfurizing a petroleum fraction rich in 4,6-alkyl-DBTs and 4-alkyl-DBTs, the activity decreases as follows: $Co_{0.1}MoS_2$ > supported catalysts >> MoS_2. These results introduce an apparent conundrum: If MoS_2 has such a high hydrogenation power and activity for desulfurizing 4,6-diethyl-DBT, why is that it perform poorly in real-feed tests? This conundrum is resolved by showing that an ultra-deep HDS catalyst requires an optimum balance between an intrinsic factor (hydrogenation function) and an environmental factor (tolerance of organonitrogen). Incorporating Co into MoS_2 lowers the hydrogenation function of MoS_2 and hence improves tolerance of organonitrogen. This conclusion corroborates the prediction of an early modeling study.

See more of this Session: Advances In Hydroprocessing
See more of this Group/Topical: Topical 7: 14th Topical on Refinery Processing

Predicting Operating Regimes in Heavy Gas Oil Hydrotreating Reactors

Jinwen Chen, Mugurel Catalin Munteanu

CanmetENERGY, Natural Resources Canada, One Oil Patch Drive, Devon, AB, T9G 1A8, Canada

Abstract

In this study, flash experiments were conducted in a continuous-flow unit to obtain experimental data on vapor-liquid equilibrium (VLE) using a heavy gas oil feedstock. The VLE data were then used to estimate the interaction coefficients between hydrogen and hydrocarbon pseudo-components that are needed for flash calculation, which predicts the flow rates of the vapor and liquid phases as well as their physical properties. This information was used to predict fluid flow regimes in a trickle-bed pilot plant hydrotreater and to generate maps of operating conditions and reactor configurations that meet plug flow and complete catalyst wetting criteria.

1. Introduction

Trickle-bed reactors are widely used in the petroleum refining, heavy oil upgrading, chemical/petrochemical, and other industries for three-phase reactions. During heavy oil upgrading and petroleum refining, trickle-bed reactors are used to eradicate sulfur, nitrogen, oxygen, and metals; to saturate olefins and aromatic rings; or to crack heavy hydrocarbons. These processes referred to as hydrotreating or hydrocracking, are normally operated in the presence of catalyst and hydrogen under high temperatures and pressures (320–420°C and 30–120 bars). Under such conditions, the reaction system (hydrogen and hydrocarbons) is at vapor-liquid equilibrium (VLE), with both the vapor and liquid phases containing hydrogen and hydrocarbons. In a trickle-bed reactor, under conditions of complete catalyst wetting, only the reacting components present in the liquid phase can come in contact with the catalyst surface and experience reactions. With increases in temperature and hydrogen flow rates, and decreases in pressure, more hydrocarbons vaporize into the vapor phase, which can considerably change the fluid flow rates, hydrodynamics, and the concentrations of various reacting species in both phases. Determination of VLE is especially challenging in pilot plant studies, where a relatively small reactor is used, and any discrepancies from the desired ideal operating regime (plug flow, full catalyst wetting, and absence of reactor wall effects) could result in defective and unrepresentative data for commercial scale-up and studies of kinetics. Therefore, it is essential to investigate, understand, and correctly predict the VLE in hydroprocessing reactors and, additionally, to predict the operating regimes under which pilot plant studies can generate reliable, repeatable, and representative results.

Few studies have been published considering VLE in 3-phase reactors,[1-5] and even fewer accounting for VLE effects in petroleum hydroprocessing.[6,7] Experimental data on VLE of hydrogen–petroleum fraction systems obtained under commercial hydroprocessing conditions are rarely published in the open literature. Instead, systems consisting of hydrogen and a single hydrocarbon or a simulated hydrocarbon mixture are often studied.[8-12] The dearth of research on VLE during petroleum hydroprocessing is probably due to the significant difficulty of experimental measurements (or simulations) and the poor definition of the problem due to the complexity of the system. In addition, high temperature and pressure and the required sample characterization and analysis add to the difficulty of the experiments.

In this research, a calibrated flash calculation program was used to predict VLE for a typical heavy distillate feed in pilot plant hydrotreaters under various operating conditions. The calculated VLE results were further used to predict flow dynamics in the hydrotreaters to determine whether they were operating under the desired operating regime (plug flow, full catalyst wetting, and absence of reactor wall effects). The hydrotreater operating conditions investigated are discussed in detail.

2. Mapping of Hydroprocessing Trickle-Bed Reactor Operating Regimes

2.1 Calculation Basis

In this study the calculation and determination of operating regimes in hydrotreating trickle bed reactors centered on an pilot plant hydrotreater. Units such as this are frequently used in hydrotreating catalyst development, evaluation, and studies of activity and stability to generate reliable data for commercial scale-up. The specifications of the pilot plant hydrotreater used were:

Reactor ID: 2.54 cm
Catalyst volume: 100 to 200 mL
Catalyst particle size (average): 1.3x5 mm
Diluent particle size: 0.20 to 0.4 mm
Catalyst-to-diluent volumetric ratio: 1:1

The operating conditions were:

Temperature: 350, 360, 370, 380, 390, 400, 410 and 420°C
Pressure: 50, 60, 70, 80, and 90 bars
Liquid hourly space velocity (LHSV): 1.5/h
Gas/oil ratio: 500, 800, and 1000 NL/h

A single heavy vacuum gas oil (HVGO) feed was used with the properties given in Table 1.

Table 1. Physical properties and chemical composition of the HVGO feed

Density (15.6°C), g/cm³ (ASTM D4052)		0.9609
Paraffins, wt%		30.94
Cyclo-paraffins, wt%		7.86
Aromatics, wt%		61.2
Boiling points, °C:	IBP	201.8
	10 wt%	292.6
	30 wt%	350.2
	50 wt%	386.4
	70 wt%	445.4
	90 wt%	529.6
	FBP	702.6

2.2 Criteria for Operating Regime Mapping

In order to generate consistent and reliable experimental data in trickle-bed pilot plant hydrotreating studies, it is critical to maintain the reactor operation in the desired regime by satisfying the following criteria:

2.2.1. Plug Flow Criterion

In a pilot plant trickle-bed reactor, attention should always be paid to axial dispersion, or backmixing, in the liquid phase, especially in a short catalyst bed with a high conversion. Otherwise, the catalyst evaluation results and subsequent kinetics analysis, normally based on the assumption of plug flow, could cause discrepancies during reactor scale-up. Mears[9] derived a criterion from the perturbation solution of an axial dispersion model with N-series mixers, which was based on the assumption that the minimum bed length to ensure freedom from axial diffusion should deviate less than 5% from the bed length required for plug flow. In this equation, an implicit assumption that conversion is less than 90% was applied by Mears.[9] In our computation, an even more relaxed criterion of 15% deviation from plug flow was used:

$$\frac{L_B}{d_{pe}} > \frac{\sqrt{20}n}{Pe_L}\ln\frac{1}{(1-x)} \tag{1}$$

It should be pointed out that the above equation contains the Peclet number P_{eL} based on catalyst particle diameter. In this work, we used the Hochman and Effron correlation:[11]

$$Pe_L = 0.034\,\mathrm{Re}_L^{0.53} \tag{2}$$

in which the particle Reynolds number is defined as

$$\mathrm{Re}_L = \frac{d_{pe}u_L\rho_L}{\mu_L} \tag{3}$$

2.2.2. Full catalyst wetting criterion

In a trickle-bed reactor, it is desirable to have all the catalyst particles in the reactor completely covered by the flowing liquid to ensure maximum catalyst utilization. Any partial catalyst wetting could result in bypassing of the feed and, thus, reduced catalyst performance. If the liquid superficial velocities are very low, which is common in a pilot plant reactor due to the relatively small volume of catalyst loading, partial catalyst wetting may occur. In this work, the criterion for wetting efficiency proposed by Gierman[10] was used:

$$\frac{\mu_L u_L}{\rho_L d_{pe}^2 g} > 5.0\times10^{-6} \quad (4)$$

2.2.3. Absence of Reactor Wall Effects Criterion

Normally reactor wall effects on reactor performance are not a concern for a commercial hydrotreating reactor, since the reactor diameter is much larger than the catalyst particle size. However, in a pilot plant reactor, such a concern arises when a small-diameter reactor and relatively large catalyst particles are used. To keep wall effects minimum, a trickle bed reactor should meet the following criterion:[12]

$$\frac{D_R}{d_{pe}} > 25 \quad (5)$$

In our study, the calculated D_R/d_{pe}=49.7>>25. Therefore, reactor wall effects were negligible and are not discussed further.

2.3 Calculation Results and Discussion

2.3.1. Plug Flow Operating Regime Mapping

The calculated operating regimes that meet the plug flow criterion are shown in Figure 1. Each curve in the figure represents, under certain gas/oil ratios, the boundary of pressure and temperature at which plug flow is ensured. The zone above the curve represents the operating temperatures and pressures that meet the plug flow criterion and the zone below the curve represents the operating temperatures and pressures that cannot meet the plug flow criterion. Normally, for a fixed gas/oil ratio, increased reactor temperature requires increased pressure in order to ensure plug flow of the liquid phase. At a given temperature, increased gas/oil ratio requires increased pressure to maintain liquid plug flow.

For some pilot plant studies, the operating conditions, catalyst particle size, and diluent particle size are fixed according to test requirements. Therefore, to maintain plug flow of the liquid inside the reactor, the catalyst bed has to be a certain length under a particular set of conditions. Figure 2 shows the requirement for catalyst bed length at different reactor temperatures and pressures for a fixed LHSV of 1.5/h and gas/oil ratio of 800 NL/kg. Each curve in the figure represents, for a given pressure, the boundary of temperature and catalyst bed length at which plug flow is ensured. As seen in this figure, increased temperature requires an increased catalyst bed length in order to ensure plug flow of the liquid. At higher temperature and lower pressure, the required catalyst bed length increases sharply.

Figure 3 presents the plug flow operating regime criterion in terms of the conversion. Although conversion was considered as 90% for the rest of the calculation, conversion in real HVGO hydrotreaters can vary between 80% and 95%. As the figure shows, higher conversion requires higher pressure and/or lower temperature in order to meet the plug flow criterion. At a certain temperature the required pressure does not increease much when the conversion increases from 80% to 90%; but increasing conversion from 90% to 95% leads to a significant increase in the pressure at which the plug flow criterion is met.

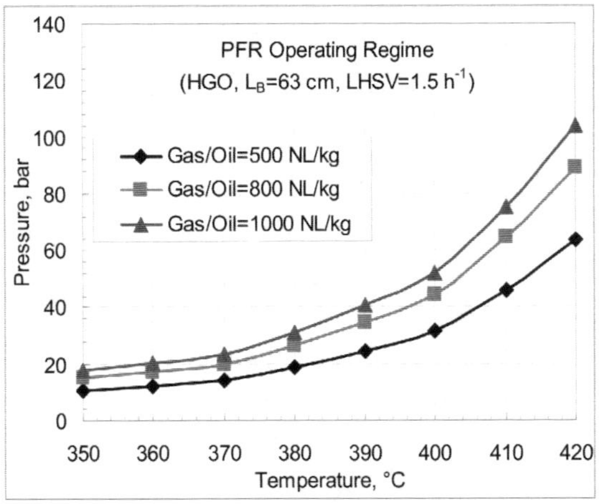

Figure 1. Plug flow operating regime mapping, temperature-pressure-gas/oil ratio

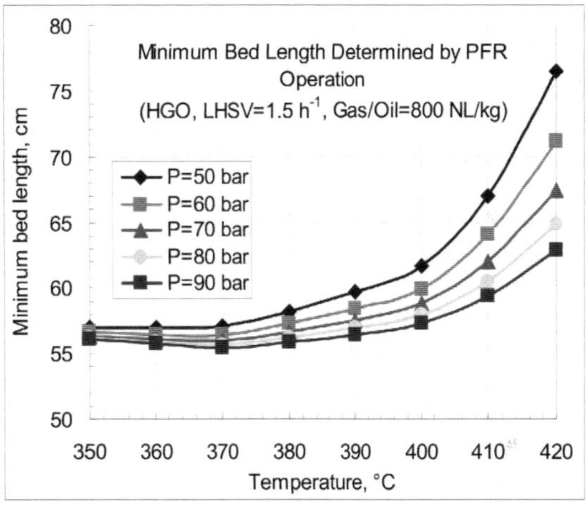

Figure 2. Plug flow operating regime mapping, temperature-pressure-catalyst bed length

Another set of calculations were performed in order to evaluate the minimum LHSV needed to ensure plug flow. Figure 4 depicts the calculation results. At the lower temperatures (350–370°C) there is no significant difference between the LHSVs at which the criterion is met. This is no longer the case when the temperature is above 400°C. If the reactor is operated at higher pressures and higher temperatures, the LHSV needed to maintain plug flow in the reactor is lower than when the reactor is operated at lower pressures and higher temperatures. At a given temperature, lower pressure requires higher LHSV to meet the criterion.

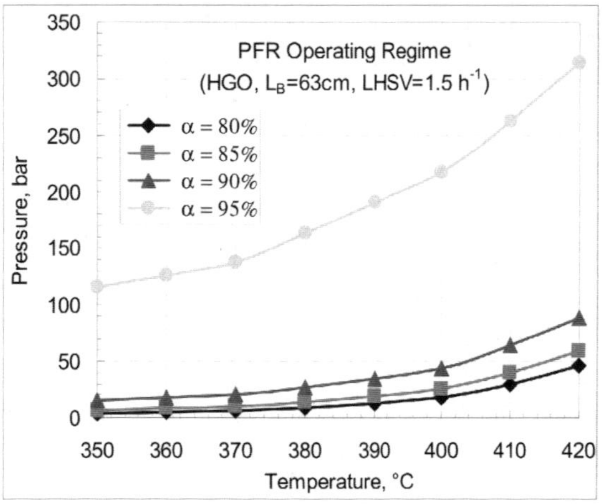

Figure 3. Plug flow operating regime mapping, temperature-conversion-pressure

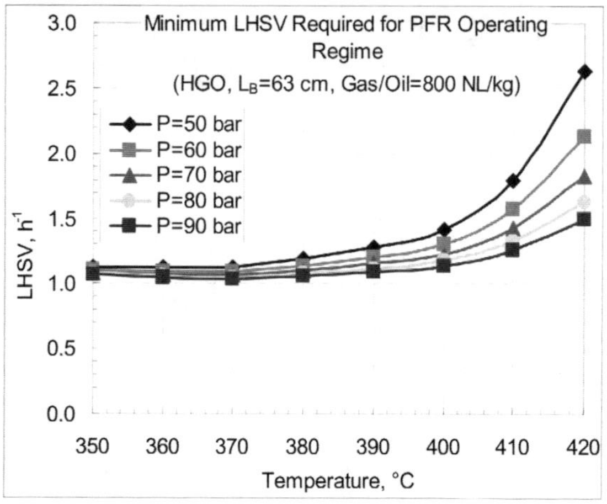

Figure 4. Plug flow operating regime mapping, temperature-LHSV-pressure

2.3.2. Full Catalyst Wetting Mapping

The liquid-phase flow rate in a pilot plant hydrotreater can be significantly affected by partial vaporization of the oil into vapor phase. Such changes in liquid flow rate can result in incomplete catalyst wetting, leading to bypassing of the feed and, thus, reduced catalyst performance. It is therefore important to maintain full/complete catalyst wetting in a pilot plant hydrotreater in order to generate reliable and reproducible data. Unlike the plug flow criterion, the full catalyst wetting criterion is not related to the catalyst bed length. It is only related to the operating conditions and the catalyst and diluent particle sizes. Figure 5 shows the mapping of full catalyst wetting for the HVGO feed. The zone above the curve represents the operating temperatures and pressures that meet the full catalyst wetting criterion and the zone below the curve represents the operating temperatures and pressures that cannot achieve full catalyst wetting. Normally, at fixed gas/oil ratios, increased reactor temperature requires increased pressure in order to ensure full catalyst wetting. At a given temperature, increased gas/oil ratio also requires increased pressure to maintain full catalyst wetting. It is interesting to note that at the same temperature and gas/oil ratio, the pressure required for achieving complete catalyst wetting is lower that that required for plug flow if one compares Figures 1 and Figure 5. For this particular study the full catalyst wetting criterion is easier to meet, or more relaxed, than the plug flow criterion. Therefore, as long as the plug flow criterion is met, full catalyst wetting will be automatically be achieved for the system discussed in this study. However this trend can be reversed for a commercial scale hydrotreater in which the plug flow criterion is easier to meet than the full catalyst wetting criterion.

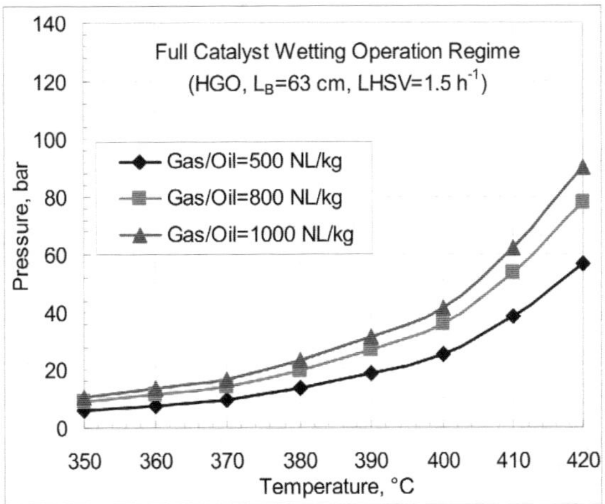

Figure 5. Full catalyst wetting operating regime mapping, temperature-pressure-gas/oil ratio

3. Conclusions

In this research VLE calculations were performed to predict the flow dynamics in a pilot plant hydrotreater using a single HVGO feed. Further calculations were performed to provide maps of operating conditions under which the desired operating regime (plug flow, full catalyst wetting, absence of reactor wall effects) can be maintained. The calculations of pilot plant hydrotreater operating regimes showed that higher pressure, lower temperature, lower gas/oil ratio, longer catalyst bed, lower conversion, and/or higher LHSV are required in order to maintain plug flow of the liquid in a HVGO trickle-bed hydrotreater. At the same time, as long as the plug flow criterion is met, full catalyst wetting will be achieved automatically for the system discussed in this study.

Nomenclature

BP = boiling point (°C)
$C_{Aromatics\%}$ = aromatic content of feed (wt%)
d_{pe} = effective particle size (cm)
D_R = diameter of reactor bed (cm)
g = acceleration of gravity (981 cm/s^2)
L_B = reactor bed length (cm)
$LHSV$ = liquid hourly space velocity (h^{-1})
n = reaction order
P = pressure (bars)
Pe_L = particle Peclet number
Re_L = particle Reynolds number
T = temperature (°C or K)
u_L = liquid superficial velocity (cm/s)
x = conversion of component
Suffixes
L = liquid phase
Greek symbols

μ_L = liquid phase viscosity (cP)
$_L$ = liquid phase density (g/cm^3)

Acknowledgements

The authors are grateful to Dennis Carson for conducting the VLE experiments and Conrad Gietz for editing this paper. The authors are grateful to the staff from CanmetENERGY pilot plant and analytical lab for their support and help. Partial funding has been provided by the Canadian Program for Energy Research and Development (PERD).

Literature Cited

(1) Akgerman, A.; Collins, G. M.; Hook, B. D. Effect of Feed Volatility on Conversion in Trickle-bed reactors. Ind. Eng. Chem. Fundam. **1985**, 24, 398-401.

(2) Smith, C. M.; Satterfield, C. N. Some Effects of Vapour-Liquid Flow Ratio on Performance of a Trickle-bed reactor. Chem. Engng. Sci. **1986**, 41, 839-843.

(3) LaVopa, V.; Satterfield, C. N. Some Effects of Vapour-Liquid Equilibria on Performance of a Trickle-Bed Reactor. Chem. Engng. Sci. **1988**, 43, 2175-2180.

(4) de Jong, K. P. Effect of Vapour-Liquid Equilibrium on Coke Deposition in Trickle-Bed Reactors during Heavy Oil Processing. 2. Modeling. Ind. Engng. Chem. Res. **1994**, 33, 3141-3145.

(5) Tesser, R.; Di Serio, M.; Santacesaria, E. Influence of the Vapour-Liquid Equilibrium (VLE) on the Kinetics in Gas-Liquid and Gas-Liquid-Solid Systems. Catal. Today. **2003**, 79&80, 323-331.

(6) Kocis, G. R.; Ho, T. C. Effects of Liquid Evaporation on the Performance of Trickle-Bed Reactors. Chem. Engng. Res. Des. **1986**, 64, 288-291.

(7) Bellos, G. D.; Papayannakos, N. G. The Use of a Three Phase Microreactor to Investigate HDS Kinetics. Catal. Today. **2003**, 79&80, 349-355.

(8) Carberry, J. J.; M. M. Wendel. A Computer of the Fixed Bed Catalytic Reactor: The Adiabatic and Quasi-adiabatic Cases. AIChE J. **1963**, 9, 129.

(9) Mears, D. E. The Role of Axial Dispersion in Trickle-Flow Laboratory Reactors. Chem. Engng. Sci. **1971**, 26, 1361.

(10) Gierman, H. Design of Laboratory Hydrotreating Reactors Scaling Down of Trickle-flow Reactors. Appl. Catal. **1988**, 43, 277.

(11) Hochman J.; Effron E. Two-phase Cocurrent Downflow in Packed Beds. Ind. Engng. Chem. Funds. **1969**, 8, 63.

(12) Chu, C. F.; K. M. Ng. Flow in Packed Tubes with a Small Tube to Particle Diameter Ratio. AIChE J. **1989**, 35, 148.

(52d) Single Event Kinetics Based Simulation of Vacuum Gas Oil Hydrocracking Using Deterministic Molecular Reconstruction

Jorge Martinis, Development, Bryan Research & Engineering Inc, Bryan, TX and Gilbert F. Froment, Artie McFerrin Department of Chemical Engineering, Texas A&M University, College Station, TX

The usual characterization of vacuum gasoil practiced today is inadequate for the simulation of hydrocracking based upon a modern fundamental and realistic kinetic model. Significant progress has been achieved in this field by Molecular Reconstruction techniques. In the present work, a deterministic approach to Molecular Reconstruction of vacuum gasoil has been combined with Single-Event Kinetics in the simulation of a Hydrocracking Process Unit targeting maximum Diesel production. Three different process flow configurations combining pre-treatment, reaction, fractionation and recycle have been investigated. At the core of the simulation model, a computer-generated reaction network consisting of 126 million elementary steps and 18 million species describes the fundamental chemistry of hydrocracking. Subsequent identification of the rate-determining steps and application of equilibrium constraints to fast elementary steps permits to solve for intermediate species, leaving about 180,000 reaction rates and 1267 species to solve for in the formulation. Rate parameters of the elementary steps are modeled by means of the Single-Event concept, thus reducing the number of model parameters to be determined from plant data to a realistic level. In addition, a rigorous model for trickle-flow in packed-bed reactors has been validated for extreme limiting conditions such as local hydrogen depletion and transition to complete vaporization.

See more of this Session: Advances In Hydroprocessing

See more of this Group/Topical: Topical 7: 14th Topical on Refinery Processing

Accurate Estimation of Si Accumulation in Coker Naphtha Hydrotreaters Using an Improved Direct Injection ICP (ICPDIN) Analysis and Hot Loop Feed Sampling

Thienan D. Tran*, Patrick J. Gripka, & Lawrence (Larry) S. Kraus

Criterion Catalysts & Technologies L.P.

Houston, Texas

***Corresponding Author: thienan.tran@cri-criterion.com**

Abstract

Delayed coking processes typically use silicon containing oils, polydimethylsiloxane (PDMS), to suppress foaming in the coker drums. Inside the coker, these high molecular weight oils crack into fragments (e.g. dimers and trimers of the dimethylsiloxane). The majority of these fragments boil in the naphtha range and are processed in the coker naphtha hydrotreater where they absorb onto the catalyst surface causing irreversible loss of catalytic activity.

The cycle life of a coker naphtha hydrotreater is normally dictated by: 1) the silicon capacity of the selected catalyst system and 2) the amount of silicon entering with the coker naphtha feed. The silicon capacity of a selected catalyst system is known. However, the amount of silicon accumulated on a catalyst system cannot usually be accurately determined from feed properties because: 1) feed samples are not collected frequently enough to represent the feed and 2) the standard Inductively Coupled Plasma (ICP) test used by the industry does not accurately measure the silicon concentration in the coker naphtha feed. This makes the cycle life of coker naphtha hydrotreaters difficult to predict and therefore often results in refiners changing out coker naphtha hydrotreater catalyst based on fixed cycle length or silicon slippage. This either underutilizes the silicon capacity of the catalyst system or results in an unplanned shutdown.

This paper focuses on a case study in which a hot loop feed sampling station is used in combination with an improved silicon analysis applying Direct Injection ICP (ICPDIN) to estimate the amount of silicon accumulated in a naphtha hydrotreater as the cycle progressed. At the end of the cycle, spent catalysts from the unit were analyzed to determine the actual accumulated silicon. Results demonstrate that the silicon deposition estimated using hot loop sampling and ICPDIN was within 10% of the silicon deposition determined from spent catalyst analysis. The improved accuracy of the estimate would have allowed the catalyst cycle life of the unit to be extended up to 4.5 months beyond the scheduled 12 month cycle length.

Background

Silicon (Si) poisoning is a major concern in coker naphtha hydrotreaters. The source of Si in coker naphtha can be traced back to the delayed coking process which typically uses Si-containing oils, polydimethylsiloxane (PDMS), to suppress foaming in the coker drums. At the elevated temperature inside the coker drums, these high molecular weight Si-containing oils crack to form lighter silica fragments (e.g. dimers and trimers of the dimethylsiloxane). The majority of these silica fragments boil in naphtha range and therefore are routed to the downstream naphtha hydrotreaters together with the coker naphtha.

Under the operating conditions of the naphtha hydrotreaters, the silica fragments present in the feed transform to modified silica gels and absorb onto the catalyst surface(1). As Si accumulates on the catalyst surface, it covers active sites and restricts catalyst pores; the latter process eventually blocks access to active sites. Once Si is bound to the catalyst surface it cannot be removed and results in an irreversible lost of catalyst activity. Without Si in the feed, the typical cycle length of a naphtha hydrotreater is three or more years. When processing coker naphtha, the cycle length is typically 12 months. In extreme cases, the cycle length can be six months or less.

The cycle life of a coker naphtha hydrotreater is dictated by the silicon capacity of the selected catalyst system and the Si accumulating rate. The silicon capacity of a selected catalyst system is known. However, the Si accumulating rate often cannot be determined due to lack of accurate feed characterization data.

Refiners normally collect feed sample for Si analysis on a frequent basis. However, because of the transient nature of the delayed coking process, the frequency of the feed sampling is often not sufficient to determine the actual amount of Si being fed to the unit. In addition to the unrepresentative feed sampling issue, the commonly used ICP test method does not accurately measure the silicon species present in the coker naphtha feed. These issues cause the calculated Si accumulating rate to be unreliable and therefore the cycle life of coker naphtha hydrotreaters often cannot be predicted. This results in refiners changing out coker naphtha hydrotreater catalyst based on fixed cycle length or Si slippage. This either underutilizes the Si capacity of the catalyst system or results in an unplanned shutdown.

In this case study, a hot loop feed sampling station was installed on a commercial coker naphtha hydrotreater to obtain composite samples. The weekly composited samples were tested for Si using a Shell proprietary Direct Injection ICP (ICPDIN) analysis. The results were used to estimate the amount of silicon accumulated on the catalyst as the cycle progressed. At the end of the cycle, spent catalysts from the unit were analyzed to determine the amount of Si accumulated on the catalyst. Results indicated the Si deposition estimated using hot loop sampling and ICPDIN was within 10% of the Si deposition determined from spent catalyst analysis. Due to the accuracy of the estimate, the cycle life of the unit could have been extended up to 4.5 months beyond the scheduled 12 month cycle length if the unit was not shut down due to furnace fouling.

Case study

The coker naphtha hydrotreater in this case study consists of a guard reactor followed by two main reactors which are in parallel. The guard reactor contains OptiTrap grading materials and DN-200. The primary function of the guard reactor is to saturate diolefins. The main reactors contain 45% MaxTrap[Si], 20% DN-140, and 35% DN-3531. All of these materials are Criterion catalyst grades. MaxTrap[Si] is a silicon trap catalyst. DN-140 is a dual functional NiMo catalyst which was used to provide both significant silicon uptake capacity and hydrotreating activity for the load. DN-3531 is a high activity NiMo hydrotreating catalyst which provides the majority of the HDS and HDN activity requirements for the unit. The catalyst system had a silicon capacity of 8,750 pounds and was scheduled for a 12 month run length. The unit process conditions and feed properties are shown in Table 1.

Table 1 - Unit Description

Feed	
Composition	18 vol% Coker Naphtha
	82 vol% SR Naphtha
Sulfur, wppm	580
Nitrogen, wppm	18
Operating Conditions	
WABT, ^{o}F	> 600
LHSV in the Main Reactors	4.1
Product Properties	
Sulfur, ppmw	0.2
Nitrogen, ppmw	< 0.2

Feed Sampling

The majority of coker naphtha hydrotreaters sample feed once per week for Si analysis. Fluctuations in the amount of coker naphtha in the feed and the amount of foam suppressing oils used in the coker drums will affect the Si content in the feed. Recognizing the fact that frequent feed sampling is important to capture the changes in the Si content, a hot loop feed sampling system was installed. This sampling system was designed to collect a defined amount of feed sample on a daily basis and combine seven daily samples to obtain a weekly composite sample.

Silicon Analysis of Coker Naphtha Feed

The Si present in the coker naphtha is in the form of dimers and trimers of the dimethylsiloxane. These dimethylsiloxane molecules are volatile. Due to their volatility, the standard Inductively Coupled Plasma (ICP) method, which is used by most refiners, can not accurately measure the Si content in the coker naphtha. Depending on the form of the Si compound and the dynamics of the sample introduction system of the ICP method being used, the Si concentration determined can be off by a factor of 10 to 20[(2)].

To obtain reliable Si content data, the weekly composite samples were sent to the Shell Global Solutions Westhollow Technology Center for Si analysis. A Shell proprietary ICP analysis was used to measure the Si present in the weekly composite sample. This analysis uses a custom design direct injection nebulizer (DIN) to introduce the sample into the ICP. Based on results of extensive research, the error margin of the ICPDIN test is typically less than 10%.

Spent Catalyst Analysis

After 10 months online, the unit was shut down due to furnace fouling issues. The decision was made to change out the catalyst while the unit was down. The spent catalyst was unloaded by vacuuming, which allowed an accurate Si deposition profile to be constructed. Samples of the spent catalyst were collected to determine the Si uptake capacity of each type of catalysts and the total amount of Si accumulated on the catalyst system.

Results and discussion

Results of the weekly Si analysis indicated the Si content in the feed varied from 0.26 to 6.5 wppm with 62% of the data population in the 1.5 and 2.5 wppm range (Figure 1). Without the composite feed sampling program, the periods with high or low Si content could have been unaccounted for. This could have significantly affected the estimated average feed Si content. This shows that a good feed sampling program is required to capture the changes in the Si content of the feed and accurately estimate the total Si fed to a naphtha hydrotreater unit.

Figure 1 - **Weekly Average Si Content in Feed**

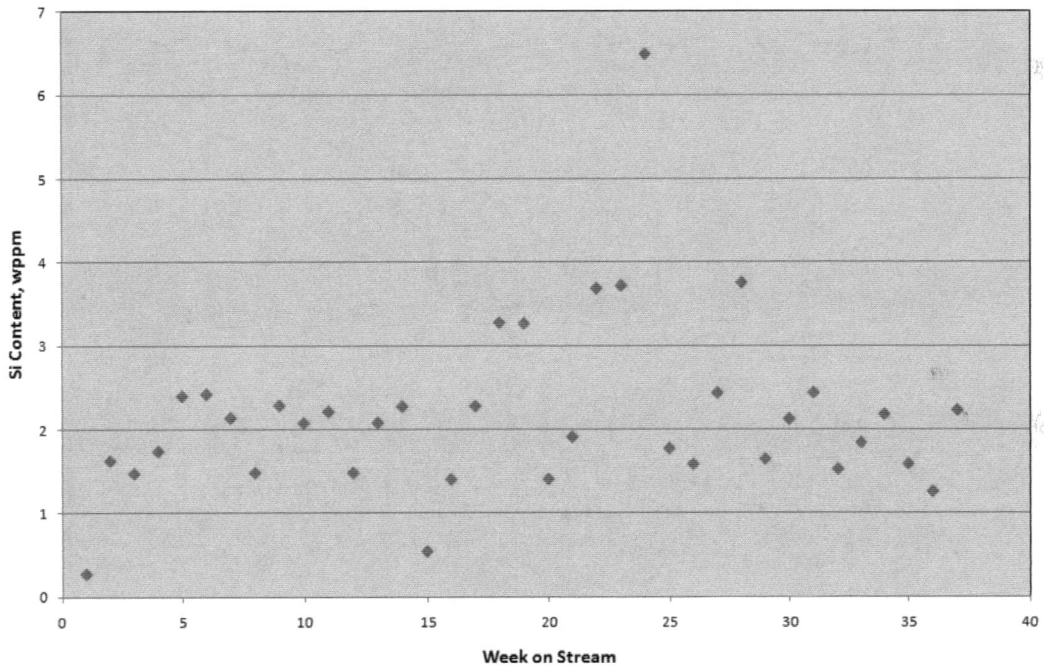

Based on the weekly test results of the Si content in the feed, the Si accumulation was calculated as the cycle progressed. At the time the unit was changed out, the total Si accumulation was estimated to be 4,770 pounds (~ 55% of the total catalyst system Si capacity).

Results of the spent catalyst analyses indicated the actual amount of Si accumulated on the catalyst was 5,220 pounds which equated to 9% higher compared to the estimated number. Results of the spent catalysts also confirmed that the ICPDIN test method used in this case study can accurately measure the Si content in the coker naphtha. If the unit was not shut down due to furnace fouling, the cycle length could have been extended up to 4.5 months beyond the scheduled 12 month cycle length (Figure 2).

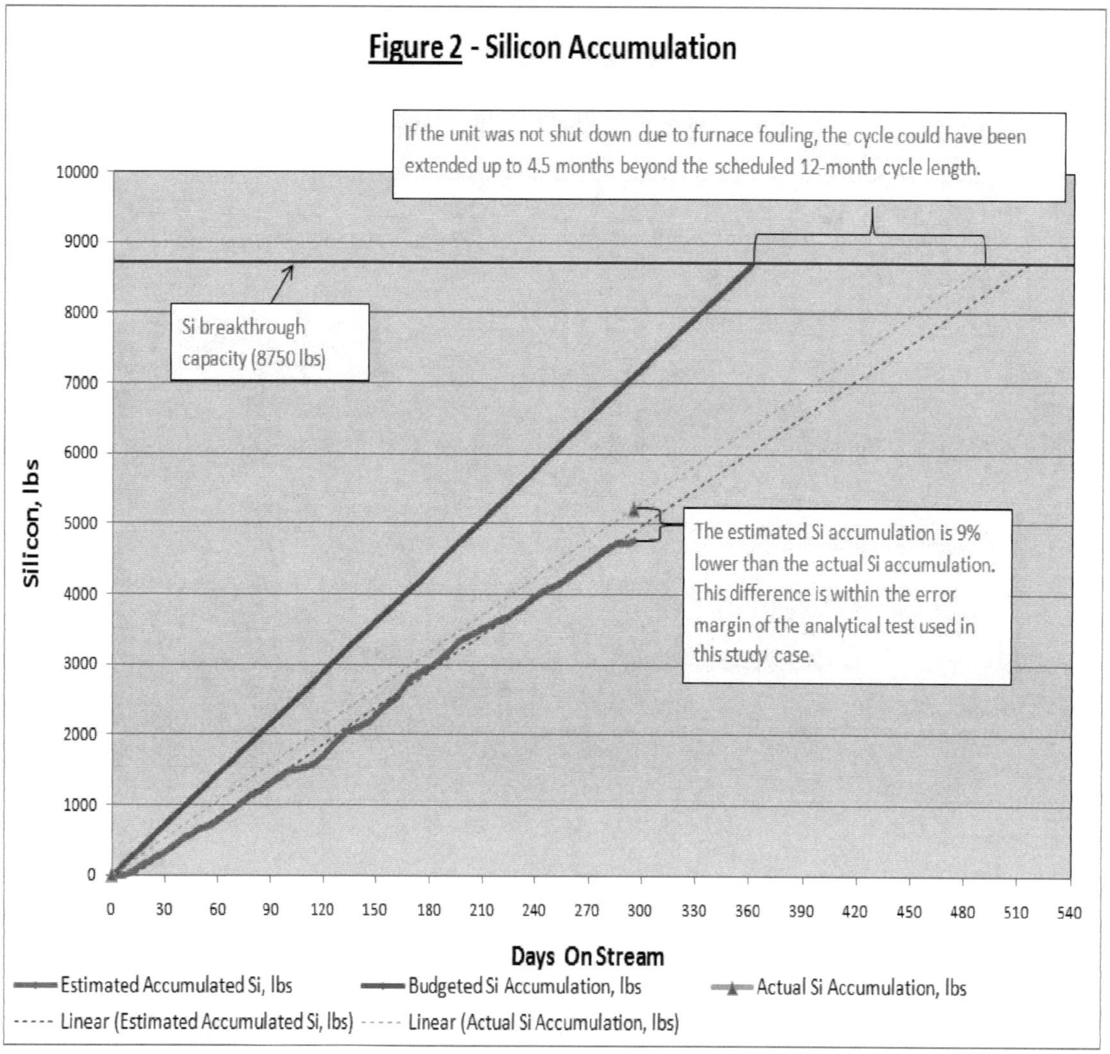

In addition, results of the spent catalyst analyses also showed that the guard reactor picked up a minimal amount of Si (Figure 3). This was expected due to the low operating temperature (<450°F) in the guard reactor. The MaxTrap[Si] in the main reactors picked up as much as 21 wt% Si. Almost no Si deposited in the DN-140 and DN-3531 layer due to the premature shut down of the unit.

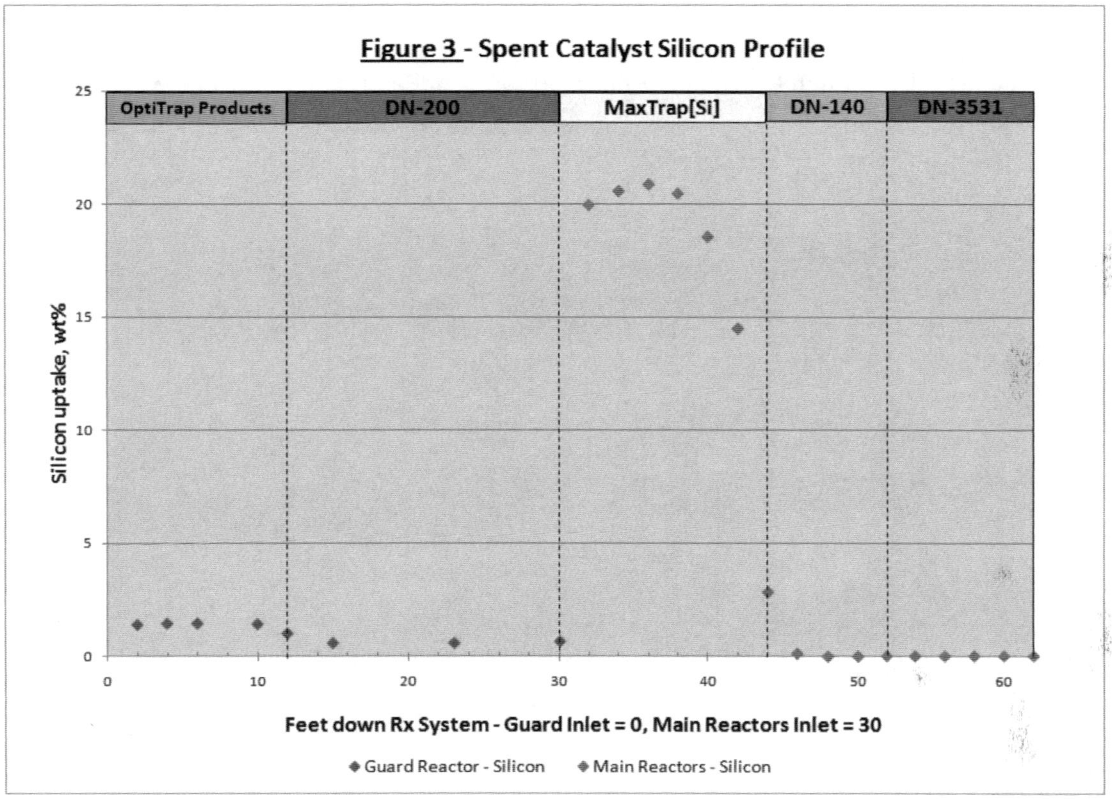

Conclusions

In coker naphtha hydrotreaters, the quality of the estimates of silicon deposition on a catalyst system depends on the quality of the feed sampling program and the accuracy of the analytical method used to determine the Si content in the feed samples. A hot loop sampling system can be used to obtain composite feed samples which capture the changes in the Si content of the feed. The direct injection ICP analysis used in this case study accurately measured the Si content in coker naphtha feed. The combination of the good feed sampling program and the accurate Si analysis resulted in a more reliable estimate of silicon deposition and could have extended the cycle length up to 4.5 months beyond the scheduled 12 month cycle length.

Acknowledgment

The expertise provided by Michael Shepherd, Research Chemist in Elemental Analysis department at Shell Global Solutions Westhollow Technology Center, is gratefully acknowledged.

References

1) Kellberg L., Zeuthen P. and Jakobsen H.J., Journal of Catalysis, (1993) 143, p 45-51
2) Raquel Sanchez, Jose-Luis Todoli, Charles-Philippe and Jean-Michel Mernet, *J. Anal. At. Spectrom.*, 2009, 24, 391-401

(61a) Performance Investigation for Constrained Model Predictive Control of the Shell Fractionater Problem

Peyton C. Richmond and M.A.K. Rasel, Chemical Engineering Department, Lamar University, Beaumont, TX

Model predictive control (MPC) is one of the most profitable present day control strategies because of features like constraint handling, integrated multivariable control, optimal and safe performance. The model used to predict the future outputs in MPC plays an important part in the control calculations, thus the performance of MPC depends on the accuracy of the model used.

In this work, performance degradation is evaluated for different plant-model mismatch cases using performance assessment techniques described in the literature. Closed loop simulation in Matlab was used. A constrained MPC was formulated using a state-space model without state augmentation. The controller was used for the plant-model mismatch cases described in the Shell Oil fractionator problem using random walk signals as input disturbances. A performance index for the constrained MPC system was calculated and the results for the different mismatch cases in the Shell Oil fractionator problem were explored.

See more of this Session: Advanced Control and Applications I
See more of this Group/Topical: Topical 7: 14th Topical on Refinery Processing

(61b) Quick Hit Financial Benefits From Pre-Engineered APC Packages

Pete Sharpe, Emerson Process Management, Glen Allen, VA, **Thomas Novotny**, CITGO Lemont Refinery, Lemont, IL and **Gary Hawkins**, Emerson Process Management LLLP, Hinsdale, IL

The refining industry is faced with a challenging economic environment which has led to a high emphasis on cost reduction and cash flow management. With a renewed focus on energy efficiency, green house gas reduction and yields, many refineries have embarked on a number of strategic initiatives to achieve their financial goals, but these investments don't all generate positive cash flow in the near term. Advanced Process Control (APC) projects have historically demonstrated high payback opportunities, but often take 6-9 months to complete and another 9-12 months to recoup the benefits. A large, midwest US refinery tried a different approach. Using the embedded APC tools in their new DCS and a pre-engineered application supplied by the DCS vendor, they implemented an APC application on one of their distillation columns in 2 weeks, start-to-finish, that paid for itself in less than 40 days.

See more of this Session: Advanced Control and Applications I

See more of this Group/Topical: Topical 7: 14th Topical on Refinery Processing

(61c) Multi-Model MPC for a Complex pH Neutralization Process

Weiting Tang, Chemical Engineering, Texas Tech University, Lubbock, TX and M.Nazmul Karim, Chemical Engineering Department, Texas Tech University, Lubbock, TX

Control of feed pH control is an important task in chemical industries, wastewater treatment, polymerization reactions, fatty acid production and biochemical processes. However, this is a challenging problem due to *(i)* presence of large and varying time delays and *(ii)* complex time varying nonlinear relation exhibited by the titration curves. Several advanced control algorithms like multi-model predictive control and neuro-controllers have been used to address the pH control problem. However, the corresponding drawbacks in these approaches are: *(i)* model-plant mismatch (plant dynamics constantly varies) and *(ii)* computational complexity involved in designing of a single neuro-controller. In this work, a novel genetic algorithm based neural multi-model predictive controller is developed to control the highly nonlinear and challenging pH control problem. The advantages of the proposed approach are *(i)* multi-model predictive control approach reduces the computational burden involved in designing a single nonlinear controller and *(ii)* neural network model can provide better approximation of the process over various regions compared to the linear models and *(iii)* genetic algorithm, a global optimization approach used to obtain the number of hiden nodes and connecting weights of the multiple neural models helps in reducing the model-plant mismatch that could occur in various regimes. The proposed approach is demonstrated on a wastewater neutralization process and a comparison of results with other commonly used approaches is provided. Simulations studies illustrate the practicality and utility of the proposed scheme to control nonlinear processes.

See more of this Session: Advanced Control and Applications I

See more of this Group/Topical: Topical 7: 14th Topical on Refinery Processing

(61d) Experiment Design for Identification of Integral-Controllable Dynamic Multivariable Models

Parag Kulkarni, Electrical & Computer Engineering, University of Houston, Houston, TX, Mark Darby, CMiD Solutions, Houston, TX and Michael Nikolaou, Chemical and Biomolecular Engineering, University of Houston, Houston, TX

Integral controllability is a fundamental property of multivariable models to be used in decoupling model-based control that maintains closed-loop stability when detuned. It has been shown that experiments for identification of integral-controllable models cannot rely on standard PRBS (pseudo-random binary sequence) inputs, particularly when the process to be identified is ill-conditioned. A number of investigators (MacGregor et al., Braatz et al., Darby and Nikolaou) have shown that experiment design based on rotated inputs can produce better results, namely the integral controllability condition can be satisfied by models resulting from experiments of shorter duration. All prior results rely on direct identification of the steady-state gain matrix of the process to be controlled, because the integral controllability condition involves the steady-state gain matrix of the controlled process. However, a full dynamic model must be identified for use in model-based control. Consequently, experiment design methods are needed for efficient identification of dynamic models that satisfy the integral controllability condition. In this work, we present an approach to designing such experiments. Specifically, we extend prior results that transform the integral controllability condition into a manageable inequality that can be used directly in numerical optimization. Optimal inputs result from this optimization. The approach proposed in this work assumes that a reasonable frequency range for all inputs is approximately the same, whereas the correlation between input amplitudes is determined through numerical optimization. A numerical simulation on an ill-conditioned distillation column is used to illustrate the proposed approach.

See more of this Session: Advanced Control and Applications I
See more of this Group/Topical: Topical 7: 14th Topical on Refinery Processing

(61e) Using a Continuous MPC Algorithm to Control An Unsteady Batch Fermentation Process

William Stewart and Michael Tay, Rockwell Software, Austin, TX

Classical MPC has been leveraged to improve control on a nonlinear unsteady-state system, batch alcohol fermentation. The controller uses a hybrid model with fundamental reaction equations along with empirical process data-based modeling. More than thirty lines of batch quality control have been successfully commissioned. These applications include property estimators, optimal batch trajectory management and classical sense and respond MPC with lab feedback within control algorithms that were designed to use dynamics on stable processes.

The algorithmic stretch occurs because of the instability. With classical MPC algorithms, step tests produce a stable response within an operating condition. However, in an unstable batch fermentation, by the time one observes a process response, a new operating condition already has been established. The solution is to leverage hybrid modeling using known fundamental fermentation equations, tuned through process historic data and dynamic testing to match the process performance. This hybrid modeling system runs against the process and targets a dynamically optimal batch quality trajectory. Periodic lab analyses of the fermenter contents provide feedback. Yield improvements have been documented.

See more of this Session: Advanced Control and Applications I

See more of this Group/Topical: Topical 7: 14th Topical on Refinery Processing

(63a) Process Integration for Dimethyl Ether Synthesis Via Catalytic Dehydration of Methanol

Mansour Emtir, Research and development, Libyan petroleum institute, Triploi, Libya

Process Integration for Dimethyl Ether Synthesis via Catalytic Dehydration of Methanol

Mansour Emtir and Aymen Elbabour

Academy of Graduate Studies, Chemical Engineering Department, *P.O. Box 6431 Tripoli, Libya* email: memtir@yahoo.com

Abstract

Distillation is the most widely used separation technique in the process industry. Its main disadvantage is the high consumption of thermal energy: the heat requirements of the distillation columns often represent most part of the total energy cost of the plant. The energy saving in distillation is then an essential field of the chemical engineering research since the early 1970's; appropriate integration of the distillation columns with the overall process often results in a significant reduction of the energy consumption in many cases.

In this work the concept of the integration with the overall process has been applied by two integrated design approaches:

- Heat-integrated design; where the distillation system of the process will be heat-integrated with the background process (i.e. integration between the distillation columns and the heat exchangers).
- Reactive distillation design; where the reaction and distillation will be accomplished in the same unit operation.

The process of dimethyl ether (DME) synthesis by the catalytic dehydration of methanol has been taken as a case study.

Heat-integrated design of DME process showed a significant total annual cost saving of 41% compared to the conventional process, and the reactive DME distillation column design showed highest saving in total annual cost by 81%.

In addition to the cost saving of the two proposed designs, also the flue gas emissions have been reduced in large amounts using these designs, where 51% of the total emissions has been reduced by the heat-integrated design and 86% by the reactive distillation column design.

See more of this Session: Energy Conservation
See more of this Group/Topical: Topical 7: 14th Topical on Refinery Processing

(63b) Exploitation of Low-Grade Heat In Site Utility Systems

Ankur Kapil, The Mill, Centre for Process Integration,, Manchester, United Kingdom, Jin-Kuk Kim, Ceas, University of Manchester, Manchester, United Kingdom and Robin Smith, University of Manchester, Centre for Process Integration, Manchester

The site utility system provides the heating, cooling load along with power production for the processes on the total site. Low-grade heat is available in large amount across the process industry from temperatures of 30°C to 250°C as gases (e.g. flue gas) and/or liquids (e.g. cooling water). Various technologies are available for generating, distributing, utilizing and disposing of low grade energy. Low grade heat can be recovered by integration of heat pumps, Organic Rankine cycle, energy recovery from gas turbine exhaust, absorption refrigeration, and boiler feed water heating etc with the total site. It is not straightforward to identify the most appropriate technology along with the optimal design option to be implemented.

In this work the integration of heat pump, absorption refrigeration, absorption heat pump, organic Rankine cycle (ORC) and boiler feed water (BFW) heating to the total site has been studied with the help of a case study. HYSYS models have been developed to estimate the coefficient of performance (COP) of the low grade heat upgrade technologies. Linear correlations are generated to estimate the capital cost of the various design options. It has been shown that the integration of heat pump and absorption heat pump with the total site is not economically viable. However, integration of absorption refrigeration, ORC and BFW heating has shown to reduce the total annual cost for the site utility system.

See more of this Session: Energy Conservation
See more of this Group/Topical: Topical 7: 14th Topical on Refinery Processing

ENERGY CONSERVATON ON MULTICOMPONENT NON-IDEAL DISTILLATION USING HEAT PUMP AND SPLIT TOWER TECHNIQUE

Nada Bahjat Nakkash, Nada Dhiaa
Al-Nahrain University, Baghdad - Iraq.

Abstract

This study deals with reducing energy requirement in multi-component distillation using heat pump with and without split tower technique for two ternary feed systems Methanol-Ethanol-Water (a) and Acetone-methanol-Water (b), and one quaternary feed system n-Hexane – MCP – Ethanol and Benzene (c). The separation was carried out using one configuration for each feed system according to their non-ideality. Many variables have studied such as operating pressure of each column in configuration in the range of 0.5-3.0atm, feed composition and fractional recovery of 0.9, 0.925, 0.95, and 0.99 for feed systems a and b and 0.9 for feed system c. Computer programmers written in FORTRAN 90 were developed for the design of multi-component distillation with and without heat pump system with and without split tower technique. The results show that the percentage reduction in energy requirements using heat pump was ranging from 12.14% to 76.02% when the same operating pressure was used in each column in the configuration. While with split tower technique the percent reduction ranging from 12.02 % to 78.00 %. Also this technique gives a percentage saving in total annual cost ranging from 10.00% to 77.36% for all cases studied. The present work was compared with previous work using heat integration for the same systems and the same variables. The results show that when using heat pump technique there is an average of 20% saving in total energy consumption over that with heat integration.

Keywords: *Energy Conservation and Efficiency, Heat Pump, Distillation*

Introduction

One alternative to recover the waste heat in the condenser is the use of heat pump. Freshwater [1] proposed several arrangements where the heat pump would extract heat from somewhere between the feed entrance and the condenser at the top.

Wolf et al and Weiler [2] showed that heat pumps optimized with the union carbide multiple down comer (MD) trays and high flux tubing could reduce equipment costs by 30% and energy cost by half, when compared to heat pumps having conventional multipass valve trays and bare tube reboilers.

Null [3] proposed a simple system where the heat pump could be used to transfer heat from the vapor at the top of an adiabatic column to the reboiler at the bottom i.e. between the extreme points of the column. This was to be carried out by an external working fluid, where direct refrigeration or chiller water was needed to condense the overhead.

Patterson and wells [4] proposed an arrangement where the vapor at the top is compressed and later condensed in the reboiler and the bottom product is flashed through the expansion valve and vaporized in the condenser. The first arrangement eliminates the condenser and has the advantage of not being restricted by the cooling medium. The second eliminates the reboiler.

Danziger [5], described process of vapor recompression in a pilot plant to measure the energy saving. The plant was running at total reflux with the mixture cis-/trans – declain, the boil up rate was 360 kg/hr, at top pressure of 175mbar. The turbo –blower have a speed of 3600 rev/min and it compressed vapor from 175mbar to 314mbar. This pressure corresponds to a saturated vapor temperature of 143C, the bottom temperature was 135.5C consequently the temperature difference in the reboiler /condenser was 7.5C. The energy saved through the application of heat pump for an adiabatic column is over 80% compared with that of conventional distillation apparatus.

Quadri [6] analyzed the propane – propylene (p-p) splitter heat pump using generalized equations for the process modeling of the complete system , stresses the importance of proper selection of the compressor and reboiler – condenser.

Wisz et al [7] discussed the use of high efficiency components in the (p-p) separation and the design and operation of heat pumps with both single and two stages compressors. Systems with two stages of overhead compression, offer an additional degree of energy saving over the single stage since only a small fraction (10% or less) of the compressed vapor need to be elevated to a pressure high enough to reject the heat of compression for cooling water or air.

Fredrick and Alexandra [8] considered recovering energy by mechanical vapor recompression. An example was given for extraction of xylene in a refinery plant using (MVRS) give saving nearly 47.3% in total annual cost.

Neill et al [9] discussed the use of high efficiency components with the operation of heat pumps in separations that involved close boiling components mixtures, like (p-p) separation or iso –n-butane or mixed butanes. The result of using heat pump system gives a large reduction in cooling water or air and elimination of steam or other heating media for the reboiler. The heat of compression and other sensible heat affected in the system must be balanced by trim cooler.

Lynd and Grethelin [10] presented a new approach for using heat pump; distillation with intermediate heat pumps and optimal side stream return (IHOSR distillation). The (IHOSR) strategy appears to be most attractive in cases with either a dilute feed or a large column temperature drop.

Meili and Guxens [11] showed that vapor recompression technique was extremely economical solution to save energy, this was especially true whenever the distillation column were equipped with regular packing which have a low pressure drop three examples; C3 – splitter, ethylene styrene separation and EDC – purification, and the percentage saving in energy for the three examples were 86%, 83%, 81% respectively. They presented a simulation program for distillation columns with vapor recompression together with an optimization algorithm based on the box's complex algorithm, and using this program for both high and low volatility systems.

Nakkash and Al Ramadhani [12] studied heat pumps in multicomponent distillation of nearly ideal hydrocarbon systems. Heat pump system gave a saving in total annual cost from 5% to 77% over the conventional system and the percentage reduction in exergy losses with heat pump was 27% to 97% over the conventional system.

Annakou and Mizsey [13] investigated a propylene-propane splitter assisted by a heat pump where single and double compressor schemes were compared to with conventional column; they found that heat pump is 37% cheaper than the conventional column.

Oliveira et al [14] simulate an ethanol-water distillation column assisted by an external a vapor compression heat pump determining how much energy could possibly be saved.

Ozcelik [15] study consist of modeling and optimization of binary and multi-component distillation columns related heat pumping and integration using the simulated annealing method and aim it to the total cost.

Present work

The aim of this work is to investigate the possibility of reducing energy requirements in non-ideal multi-component distillation using heat pump with and without split tower technique and their effects on energy saving. The analysis emphasizes to investigate these possibilities when two ternary and one quaternary system feed stock mixtures are separated into a relatively pure products, and how these possibilities vary when the feed composition, degrees of recovery and operating pressure vary. The separation was carried out using different configuration for each feed system depending upon its non-ideality [16].

The material and energy balances were carried out for each configuration using short-cut method (empirical method) and rigorous method. Computer programmers were developed written in FORTRAN 90 to design multi-component non-ideal distillation with and without heat pump and with and without split tower technique. The two programmers were modified in order to calculate percentage saving in total energy consumption and total annual cost when split tower and heat pump techniques have been considered.

The flow rates, composition and temperature of all process streams within the configuration are determined together with the process design of all columns, number of plates, actual reflux ratio, feed plate location, heat load for the condensers and reboilers and total annual cost using the short-cut method. The composition, flow rates and temperature profiles were obtained using rigorous method.

Many variables have been studied, two ternary and one quaternary non-ideal feed systems with different feed compositions Table (1, 2), four different recoveries of 0.9, 0.925, 0.95, and 0.99, for feed systems a and b and 0.9 for feed system c for their conventional configuration, Figure (1a, b, c) and for heat pump technique when considered for the three feed systems, Figure (2a, b, c) with and without split tower and when the column was operated at same pressure and/or at different pressure. All feed systems are assumed to be liquids at their boiling points deviated from ideality in liquid phase only. Total condensation is considered to give a maximum heat removed at the top of the column. The operating pressure of each column in configuration was changed from 0.5 – 3.0atm in order to minimize the total energy consumption.

The number of possibilities studied for heat pump system using ternary feed system were (1 configuration * 4 feed composition * 4 degrees of recovery *36 possible operating pressure), and for quaternary system were (1 configuration * 5 feed composition * 1 degree of recovery * 27 possibility of operating pressure).

Theoretical Aspects

The design calculations were made using the two developed computer programmes for empirical and rigorous methods.

The empirical method calculations based on Fenske-Underwood-Gilliland equations, the design of each column was made in the sequence as it occurred in the configuration, so that the input streams to the next column in the configuration could be specified. Underwood [17] equation was used to determine the minimum reflux ratio:

$$\sum_{i=1}^{N} \frac{\alpha_i x_{i,d}}{\alpha_i - \phi} = R_m + 1 \qquad \text{... (1)}$$

$$\sum_{i=1}^{N} \left(\frac{\alpha_i x_{if}}{\alpha_i - \phi} \right) = 1 - q \qquad \text{... (2)}$$

The value of ϕ is calculated by trial and error.

Gilliand [18], Van Winkle and Todd [19] relationships were used to calculate the theoretical number of plates:

For $(R - R_m)/(R + 1) < 0.125$

$$\frac{N - N_m}{N + 1} = 0.5039 + 0.5068 \left(\frac{R - R_m}{R + 1} \right) - 0.0908 \log \left(\frac{R - R_m}{R + 1} \right) \qquad \text{... (3)}$$

For $(R - R_m)/(R + 1) > 0.125$

$$\frac{N - N_m}{N + 1} = 0.6257 - 0.9868 \left(\frac{R - R_m}{R + 1} \right) + 0.516 \left(\frac{R - R_m}{R + 1} \right)^2 - 0.1738 \left(\frac{R - R_m}{R + 1} \right)^3 \qquad \text{... (4)}$$

For plate-by-plate calculation (rigorous model) the material and energy balances and equilibrium relation comprise the set of MESH equations to be solved for column of N-equilibrium stages including a condenser and reboiler, the stages numbered from top to bottom with the condenser as first stage and the reboiler as the N-th stage, assumed that one feed stream, one vapor side stream, one liquid side stream and one intercooler or interheater exist at each stage except for the condenser and reboiler.

- Material balance equation (M – equation)

For $1 \le i \le m$ and $1 \le j \le N$

$$M_{i,j}(X_{i,j}, V_j, T_j) = L_{j-1} X_{i,j-1} - (V_j + W_j) y_{i,j} - (L_j + U_j) X_{i,j} + V_{j+1} y_{i,j} + F_j Z_{i,j} = 0 \qquad \text{... (5)}$$

- Vapor – liquid equilibrium (E – equation)

$$E_j(X_{i,j}, V_j, T_j) = y_{i,j} - K_{i,j} X_{i,j} = 0 \qquad \text{... (6)}$$

Where

$$K_{i,j} = \gamma_{i,j}(x_j T_j) p_{i,j} / p \qquad \text{... (7)}$$

- Summation equation (S – equation)

$$S_j(X_{i,j}, V_j, T_j) = \sum_{i=1}^{m} y_{i,j} - 1.0 = 0 \qquad \text{... (8)}$$

$$S_j(X_{i,j}, V_j, T_j) = \sum_{i=1}^{m} x_{i,j} - 1.0 = 0 \qquad \text{... (9)}$$

- Heat balance equation (H – equation)

$$H_j(X_{i,j}, V_j, T_j) = L_{j-1} h_{j-1} - (V_j + W_j) H_j \qquad \text{... (10)}$$

Equation (5) and (6) were combined and the liquid flow rate (L's) is expressed in terms of vapor flow rate (V's) by an overall material balance of all stages from the condenser through j-th stage.

$$L_j = V_{j+1} + \sum_{k=2}^{j} (F_k - W_k - V_k) - D \quad , 2 \le j \le n-1 \qquad \text{...(11)}$$

$$D = V_1 + U_1 \qquad \text{....(12)}$$

There are N (2m+2) equations and N (2m+3) unknown variables, the solution of these equations is by transforming the material balance equation (11) to a tridiagonal matrix [16].

Antoine equation is used to calculate the vapor pressure, the feed systems are deviated from ideality in liquid phase, and Wilson equation was used to calculate the activity coefficient γ.

Heat pump made a large reduction in cooling water or airflow and elimination of steam or other heating media for the reboiler. The heat of compression and other sensible heat effects in the sequence must be balanced by a trim cooler. Trim cooler heat load is usually a small fraction of the reboiler load so this load can be neglected. After the trim cooling the overhead vapors are totally condensed in the reboiler and the resulting stream split into reflux and product. Sub-cooling the condensate in sub-cooler, this in turn reduces flash vapor recycle when the reflux was returned to the column. The only change from conventional system is the elimination of the condenser.

The ratio of heat pumped to the reboiler Q_R to compression work (W) required, can be:

$$\frac{Q_R}{W} = (\varepsilon T_2)/(T_2 - T_1) \quad \text{... (13)}$$

Where ε is compressor efficiency, T_2 is the temperature of compression vapor and it must be equal to the reboiler temperature plus the temperature difference in the reboiler (i.e. $T_2 = T_{reb} + \Delta T_{reb}$), and T_1 is the overhead vapors temperature. For any heat pump system there is a limited range of ΔT_{reb} which economics will be favorable. In this study, the value of ε was set to be 0.7. In heat pump system, the important equipment is the compressor and its cost.

In order to investigate the effect of using heat pump system in multi-component distillation many variables must be calculated for the condenser and reboiler:

$$Q_C = V_{ac}(H_{VI} - h_{LI}) \quad \text{... (14)}$$

$$Q_R = D \cdot h_D + W \cdot h_w + Q_c - F \cdot H_{iF} \quad \text{... (15)}$$

Where Q_c and Q_R are the heat loads for condenser and reboiler.

$$W_{comp} = V_{ac}(H_{VI} - H_{VO}) \quad \text{... (16)}$$

Where, W_{comp} is the work of the compressor in and H_{VO} is the enthalpy of vapor stream out of the compressor. The heat exchanged by the sub-cooler was given by:

$$Q_{SC} = V_{AC}(h_{LO} - h_D) \quad \text{... (17)}$$

Results and Discussion

The effect of changing feed system, operating pressure, feed composition and fractional recovery on total reboiler load, percentage saving in total energy consumption and in total annual cost have been studied for two ternary and one quaternary feed system with and without heat pump and with and without split tower technique.

Effect of Pressure - Using split tower technique six operating pressures have been taken 0.5, 1.0 1.5, 2.0, 2.5 and 3.0 for feed system a and b. Column 1 assumed to be operated at 0.5atm and column 2 was operated at different operating pressure for six times from 0.5 to 3.0atm. Then the first column was assumed to be operated at 1atm and the second changed from 0.5 to 3.0atm until column one reached the operating pressure 3.0atm, this means that there are 36 possibilities for feed systems a and b at different feed composition and recovery.

Figures 3 and 4 show the change in total reboiler load as a function of changing the operating pressure of column 1 and 2 for feed system a and b. For feed system c there are 216 possibilities if the three columns in Figure (1c) operated at different pressure for six times. Therefore, only three different operating pressures of 1, 2, and 3 were used and the possibilities reduced to 27. The results show that as the operating pressure increases the total reboiler load increases for all feed systems and feed compositions when the columns operate at same pressure or when using split tower technique.

Using heat pump technique the results show that increasing the operating pressure is not proportional to the % saving in total energy consumption where when operating the columns at the same pressure the % saving in total energy consumption increases while when using split tower technique the % saving in total energy consumption not proportional to pressure change, Figures 5 and 6. Increasing the operating pressure will decrease the %saving in total annual cost, Figures (7, 8).

Effect of feed composition – Using split tower technique without heat pump the total reboiler load as a function of feed composition and fractional recovery shown in form of histogram for feed system a and b Figures 9 and 10 and for feed system c Figure 11. The results show that for feed systems a and b the maximum total reboiler load is in feed composition 3, this is due to the high reflux ratio in which will increase the reboiler load and will decrease as the feed composition changes from 2,1, and 4, respectively. For feed system c there are 27 possibilities and the results show that the maximum total reboiler load is for feed composition 2 this decreases as the feed composition changes to 3, 4, 1 and 5, respectively.

Figures 12, 13 and 14 give the % saving in total energy consumption with heat pump for feed systems a, b and c. The results show for feed system a, there is a maximum % saving in total energy consumption in feed composition 2 and this decrease when the feed composition change to 1, 3 and 4 respectively. While, for feed system b, there is a maximum % saving in total energy consumption in feed composition 3 and this decrease when the feed composition change to 4, 2 and 1 respectively.

For feed system c, feed composition 5 gives a maximum %saving in total energy consumption in possibility 1, 3, 7, 8, 10, 13, 18, 24 and 27, for feed compositions 4 and 3 the maximum % saving in total energy consumption is in possibility 2, 5, 6, 14, 19, 21, 25 and 4, 9, 11, 12, 15, 16, 17, 20, respectively. While for feed composition 1 the maximum % saving is in possibility 22 and 23 and for feed composition 2 is in possibility 26, Table (3).

For feed systems a and b feed composition 3 and 1 give a maximum saving in total annual cost. For system c the maximum % saving in total annual cost is in possibility 1, 7, 8 and 27 for feed composition1, possibility 13, 15, 17, 18, 19, 21, 23 and 25 for feed composition 2 and possibility 3, 6, 9, 10, 11, 12, 14, 16, 20, 26 for feed composition 3, Figures(15,16 17).

Effect of fractional recovery – For all feed systems and compositions the results show that increasing the fractional recovery will increase the total reboiler load for all cases, Figures (9, 10, and 11) and as the recovery fraction increases the % saving in total energy consumption decreases this is due to the increase in temperature of streams which require more energy input to the system in order to get high purity Figures (12, 13 and 14). Increasing the fractional recovery the percentage saving in total annual cost for all feed systems decreases.

Tables (4, 5) give the regions of optimality with respect to energy requirements for feed systems a and b depending upon operating pressure, feed composition and fractional recovery.

The results show that the % saving in total energy consumption for feed systems a, b and c ranging from 22.55%-76.02%, 12.14%-76.00% and 19.00%-73.25%, respectively when the towers operated at the same pressure with heat pump while when using split tower technique with heat pump the% saving ranging from 15.99%-71.26% for feed system a, 12.91%-78.00% for feed system and 12.02%-77.36% for feed system c. The previous work [20] on the same feed systems using split tower technique with heat integration gives % saving in total energy consumption for feed system a ranging from 2.9%-57.8%, for feed system b 2.6%-65.8%

and for feed c 2.3%-65.8%. This means there is 20% increase in % saving in total energy consumption when heat pump is considered.

Conclusion

Increasing the pressure for any column in all conventional configurations will increase the total reboiler load and decrease the % saving in total energy consumption and % saving in total annual cost. When using split tower technique with heat pump the % saving in total energy consumption increases. For feed systems a and b feed composition 4 requires less total reboiler load of $0.83x10^6$ kcal/hr and $0.42x10^6$ kcal, respectively. While for feed system c composition 5 require less total reboiler load of $4.55x10^6$kcal. The maximum saving in total energy consumption is 45.63%, 50% and 75.23% for feed composition 2 feed system a, feed composition 3 feed system b and feed composition 1 feed system c, respectively and the maximum % saving in total annual cost is 75%, 60% and 77% for feed composition 3 feed system a, feed composition 1 feed system b and feed composition 2 feed system c. In all cases increasing the recovery fraction will increase the total reboiler load and decrease the % saving in total energy consumption and total annual cost.

References

1. Freshwater, D. C., (1961) Brit. Chem. Eng. Vol. 6, June, p. 388-391.
2. Wolf, C.W., and Weiber, D.W., (1975) the Oil and Gas Journal, Sept.
3. Null, H. R., (1976) Chem. Eng. Prog., Vol. 72, No 7, July, p.58.
4. Patterson, W.C., and Wells, T.A., (1977) Chem. Eng., September 26, p.78.
5. Danziger, R. (1979) Chem. Eng. Prog., Vol. 75, (9), Sept., p. 58.
6. Quadri, G. P., (1979) Hydrocarbon Processing, 60, 2, Feb.
7. Wisz, M.W., Antonelli R., and Ragi, E.G. (1981) Industrial Energy conservation technology, Conference, Houston Apr.
8. Frederic, E. and Alexandra, I., (1985) Chem. Eng. Prog., July, p. 45.
9. Neill et al, (1985) Chem. Eng. Prog. July, p.57.
10. Lynd, L. R., and H. E. Grethlein, (1986) AIChE.J, Vol.32, No.8 August, p.1347.
11. Meilli, A. and Guxens, I. (1987) Chem. E. Symposium series. No. 104.
12. Nakkash, N. B., and Ramadhani, S.A.A, (1992) Energy conservation in multi-component distillation using Heat pump system, International Renewable Energy Conference, Amman, Jordan.
13. Annakou, O. and Mizsey, P. (1995) Rigorous Investigation of Heat Pump assisted Distillation, Heat Recovery Systems & CHP, 15, (3), p 241-247.
14. Oliveira, S.B.M., Parise, J.A.R. and Marques R. P. (2002) Modeling of an Ethanol-Water Distillation Column Assisted by an External Heat Pump, Int. J. Energy Res., 26, p 1055-1072.
15. Ozcelik, Y. and Ozcelik, Z. (2004) Utility Optimization for Distillation Columns Via Simulated Annealing, International Journal of Exergy, Vol.1, No.3, p.334-349.
16. Al-Mo'otasim, N.D., (2005) "Heat Pump in Multicomponent Non-Ideal Distillation", M.Sc. Thesis, Al-Nahrain University, Baghdad-Iraq.
17. Underwood, A.J.V. (1948) Chem.Eng. Prog., Vol. 44, Aug., p. 603-614.
18. Gilliland, E. R. (1940) Ind. Eng. Chem., Vol. 32, p.1220-1223.
19. Van Winkle, M., and Todd, W.A. (1971) Chem. Eng., Vol. 133, Sept., p.20-28.
20. Nakkash, N.B., and Hababa, N.F. (1991) energy conservation of multicomponent distillation column for non-ideal mixture with heat integration, Al-Muhandis Journal, p.108.

Nomenclature

D	Distillate rate	kg mole/ hr
F	Feed rate	kg mole/ hr
H	Enthalpy of vapor	kJ/ kg mole
h	Enthalpy of liquid	kJ/kg mole
K	Equilibrium ratio y/x	
k	Iteration number	
L	Liquid flow rate	kg mole/hr
$M_{i,j}$	Material balance function	
m	Number of component	
N	Number of theoretical plates	
N_m	Minimum number of plates	
P	Total pressure	atm
p_i	Vapor pressure	atm
Q	Heat load	kJ/hr
q	Heat required to vaporize the Feed/latent heat of vaporization	
R	Reflux ratio	
R_m	Minimum reflux ratio	
S_j	Summation function	
T	Temperature	C°
V	Vapor flow rate	kg mole / hr
W	Bottom product flow rate	kg mole / hr
x_i	Liquid mole fraction of component i	
y_i	Vapor mole fraction of component i	
Z_i	Mole fraction of component i in the feed	

Greek

α_{ij}	Relative volatility of component i to j
ϕ	Parameter in Underwood equation for calculating minimum reflux ratio
ε	Compressor efficiency
γ_i	Activity coefficient.

Subscripts

c	Condenser
D	Distillate
f	Feed
i	Component
j	Stage number
L	Liquid
m	Minimum
O	Out
R	Reboiler
SC	Sub-Cooler
T	Top
V	Vapor
w	Bottom product

Table 1. Feed systems

Components	Ternary feed system	
	a	b
A	Methanol	Acetone
B	Ethanol	Methanol
C	Water	Water

Components	Quaternary feed system c
D	n-Hexane
E	MC
F	Ethanol
G	Benzene

Table 2. Ternary and Quaternary feed systems composition

Components	Ternary feed system composition			
	1	2	3	4
A	0.333	0.8	0.1	0.1
B	0.333	0.1	0.8	0.1
C	0.333	0.1	0.1	0.8

Components	Quaternary feed system composition				
	1	2	3	4	5
D	0.25	0.7	0.1	0.1	0.1
E	0.25	0.1	0.7	0.1	0.1
F	0.25	0.1	0.1	0.7	0.1
G	0.25	0.1	0.1	0.1	0.1

Table 3. Effect of % saving in total energy consumption with change in feed composition for (27 possibilities) of feed system c.

Possibility No.	Feed composition				
	1	2	3	4	5
1	19.00	41.25	20.03	19.36	58.00*
2	17.05	18.01	42.15	54.26*	42.36
3	15.20	46.36	40.25	42.13	71.00*
4	18.79	25.16	46.36 *	33.21	28.36
5	30.12	42.36	52.12	63.22*	13.15
6	19.12	45.10	13.20	64.13*	52.33
7	35.36	19.66	42.12	29.12	43.00*
8	15.00	42.36	53.41	25.52	55.00*
9	46.01	36.12	60.12 *	41.10	41.32
10	22.00	34.13	18.36	32.02	44.00*
11	20.23	41.26	62.36 *	56.04	56.02
12	27.15	28.21	67.21 *	56.22	42.36
13	16.00	32.01	50.13	41.00	74.00*
14	30.36	47.12	55.28	66.32*	53.01
15	41.02	28.20	50.11 *	46.20	33.36
16	42.06	23.14	46.25 *	16.02	31.02
17	33.26	13.36	74.32 *	18.42	42.36
18	42.36	42.01	19.25	45.25	63.23*
19	22.15	18.04	46.23	73.12*	42.00
20	36.01	12.02	71.55 *	69.23	56.36
21	25.13	41.32	52.10	68.40*	46.25
22	75.13*	18.36	55.13	62.22	40.20
23	52.14 *	26.36	28.16	35.02	30.05
24	46.39	19.82	25.33	41.25	64.23 *
25	52.36	71.12	66.12	75.23 *	41.25
26	45.00	52.36 *	17.33	45.36	48.02
27	41.12	50.00	59.23	70.00	73.25 *

Table 4. Optimum operating pressures with respect to percentage saving total energy consumption for feed system a (atm.)

Feed comp-osition	R_c=0.9		R_c=0.925		R_c=0.95		R_c=0.99	
	Column		Column		column		Column	
	1	**2**	**1**	**2**	**1**	**2**	**1**	**2**
1	3.0	2.0	3.0	2.0	3.0	2.0	3.0	2.0
2	3.0	2.0	3.0	2.0	3.0	2.0	3.0	2.0
3	3.0	2.0	3.0	2.0	3.0	2.0	3.0	2.0
4	3.0	2.0	3.0	2.0	3.0	2.0	3.0	2.0

Table 5. Optimum operating pressures with respect to percentage saving total energy consumption for feed system b (atm.).

Feed comp-osition	R_c=0.9		R_c=0.925		R_c=0.95		R_c=0.99	
	Column		Column		column		Column	
	1	**2**	**1**	**2**	**1**	**2**	**1**	**2**
1	3.0	1.5	3.0	1.5	3.0	1.5	3.0	1.5
2	3.0	1.5	3.0	1.5	3.0	1.5	3.0	1.5
3	3.0	1.5	3.0	1.5	3.0	1.5	3.0	1.5
4	3.0	1.5	3.0	1.5	3.0	1.5	3.0	1.5

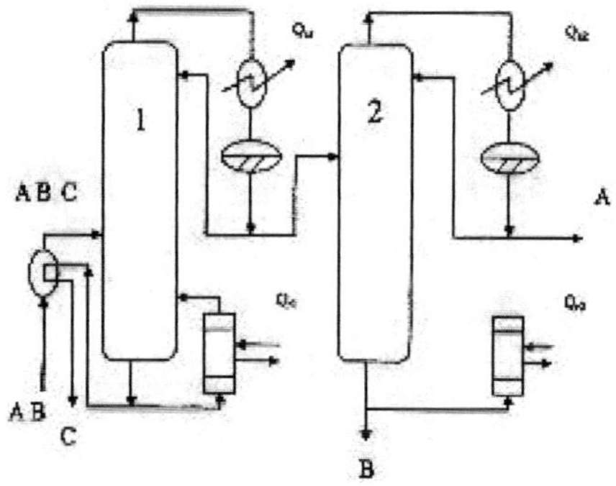

Feed System a

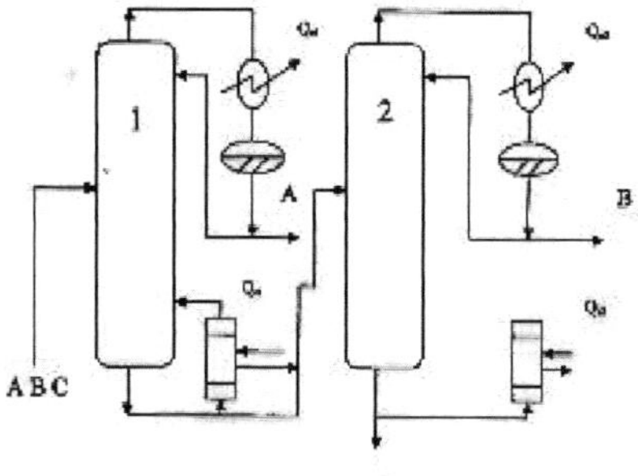

Feed System b

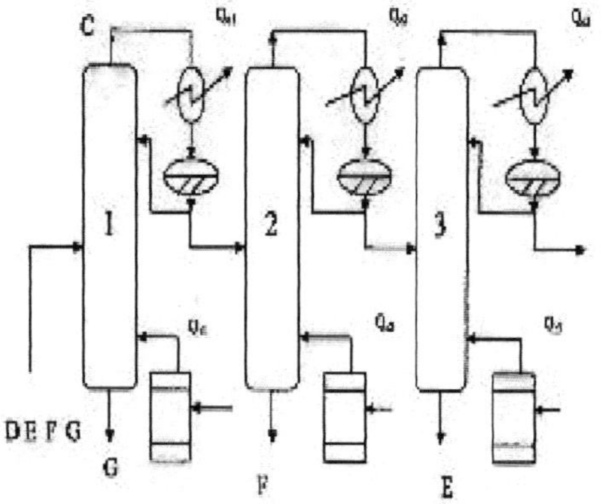

Feed System c

Figure 1. Conventional configurations for feed sysyem a,b,c

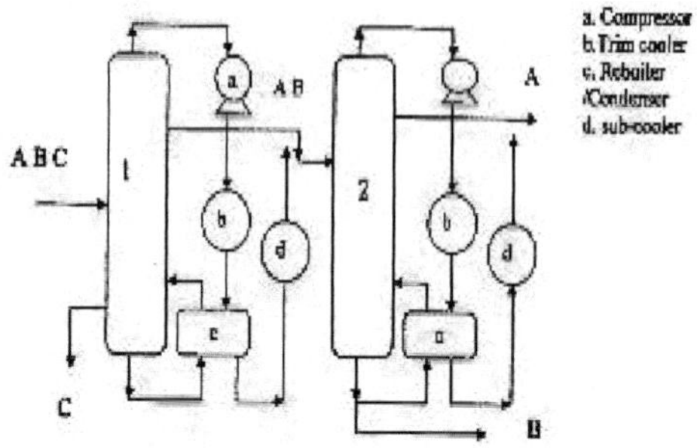

Feed System a

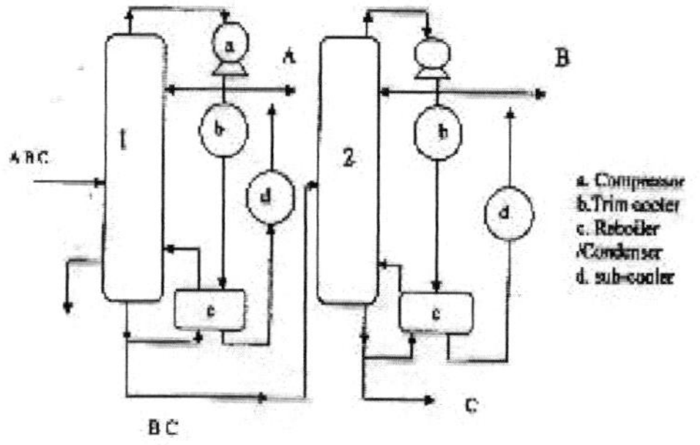

Feed System b

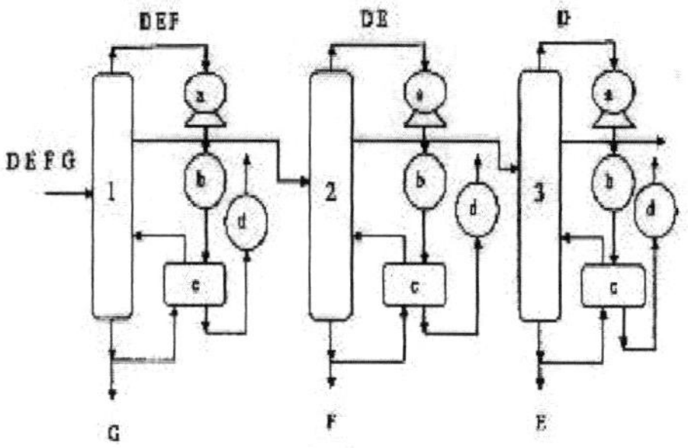

Feed System c

Figure 2. Heat Pumps for feed sysyem a,b,c

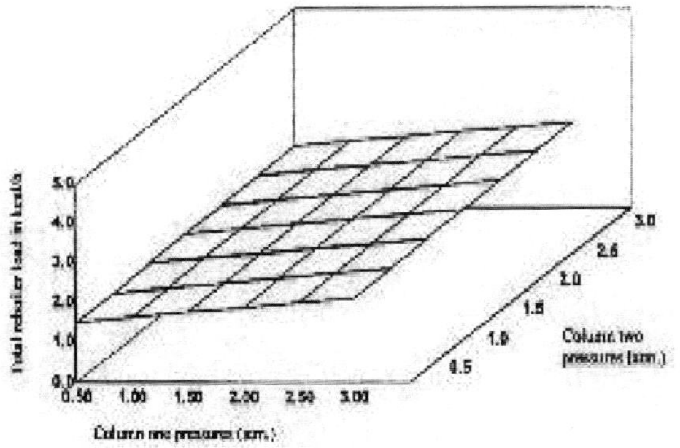

Figure 3. The effect of pressure change of column 1and 2 on total reboiler load for feed system a (feed composition 1 and recovery 0.9)

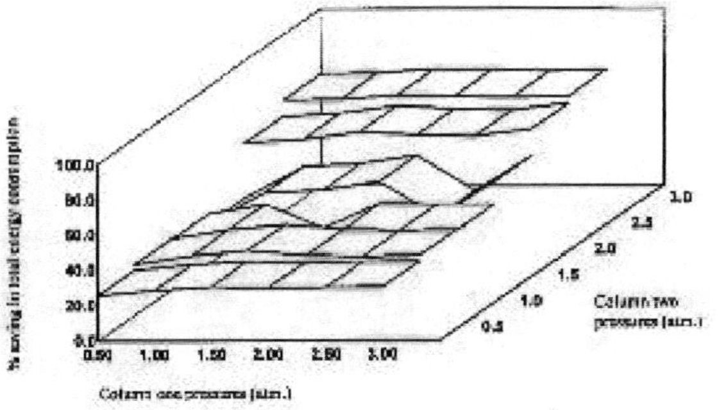

Figure 5. The effect of pressure change of column 1and 2 on % total energy consumption for feed system a (feed composition 1 and recovery 0.9)

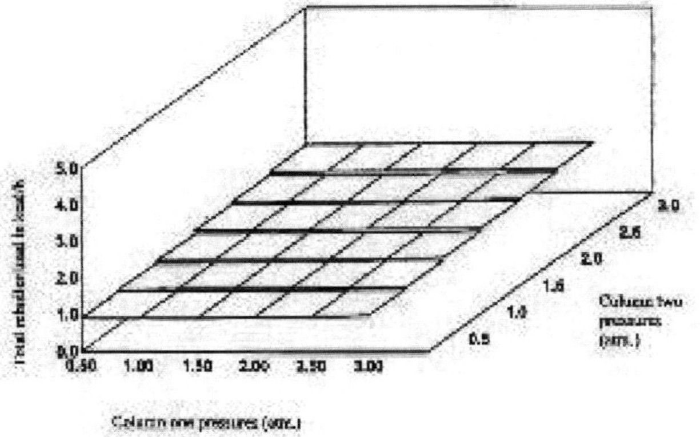

Figure 4. The effect of pressure change of column 1 and 2 on total reboiler load for feed system b (feed composition 1 and recovery 0.9)

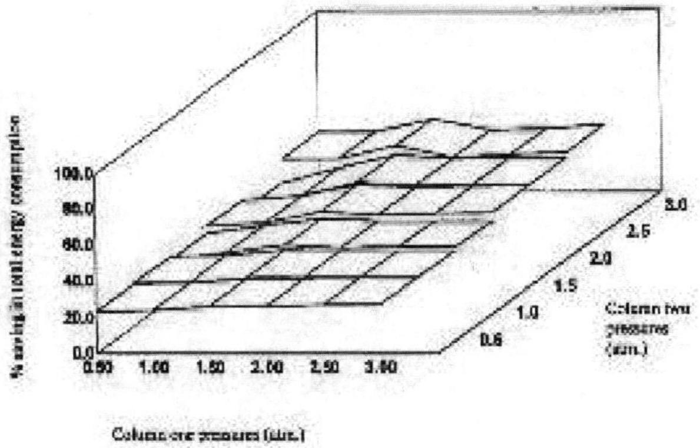

Figure 6. The effect of pressure change of column 1 and 2 on % total energy consumption for feed system b (feed composition 1 and recovery 0.9)

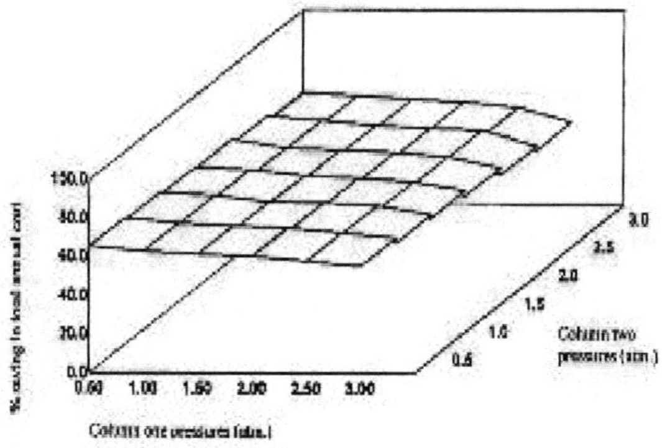

Figure 7. The effect of pressure change of column 1and 2 on % total annual cost for feed system a (feed composition 1 and recovery 0.9)

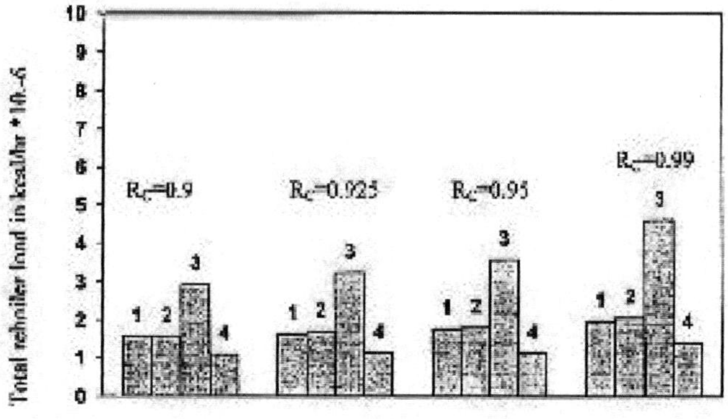

Figure 9. The effect of feed composition change on total reboiler loads for feed system a (column one at 3 atm and column two at 0.5atm).

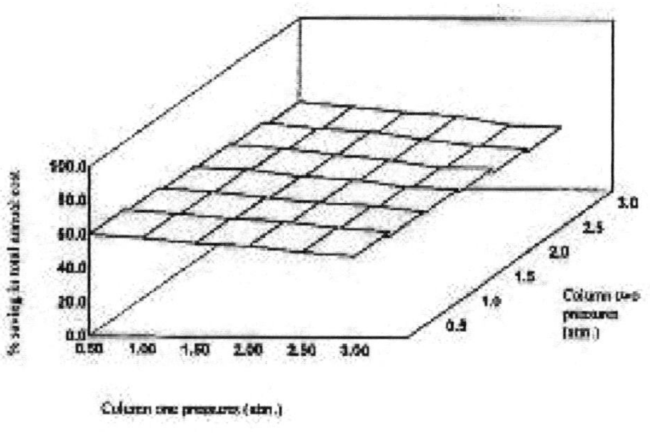

Figure 8. The effect of pressure change of column 1and 2 on % total annual cost for feed system b (feed composition 1 and recovery 0.9)

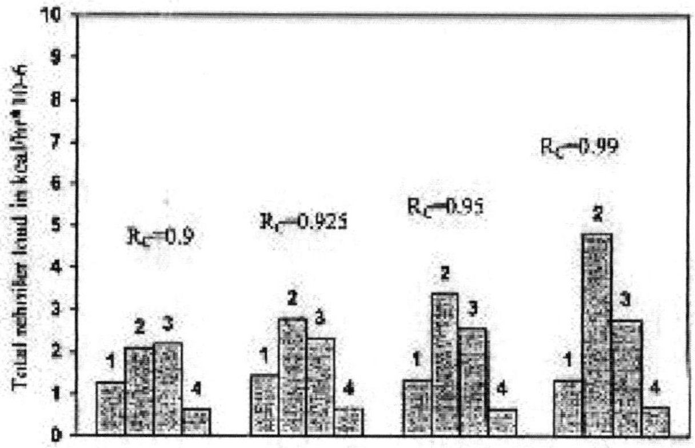

Figure 10. The effect of feed composition change on total reboiler load for feed system b (column one at 3 atm and column two at 0.5atm).

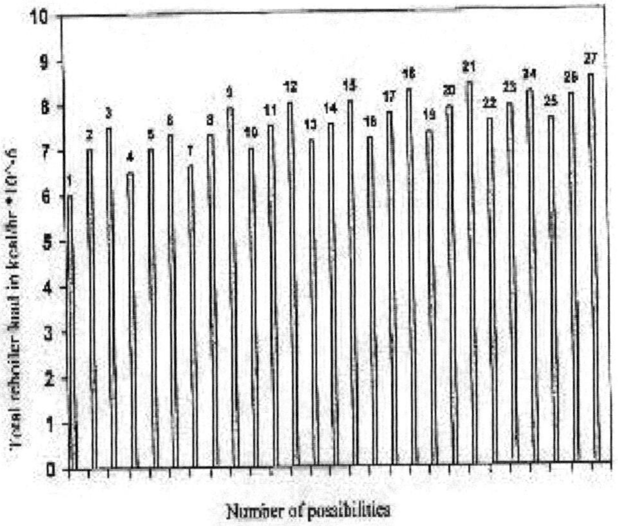

Figure 11. The effect of all possibilities of the pressure change for the three columns on total reboiler load for feed system c (feed composition one).

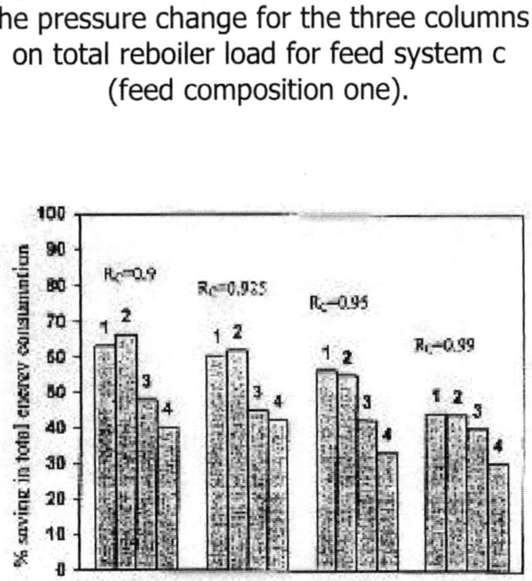

Figure 12. The effect of feed composition change on % saving in total energy consumption for feed system a (column one at 3 atm and column two at 0.5atm).

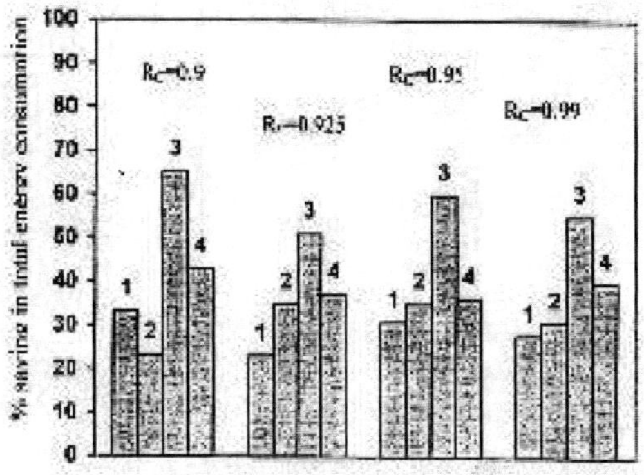

Figure 13. The effect of feed composition change on % saving in total energy consumption for feed system a (column one at 3 atm and column two at 0.5atm).

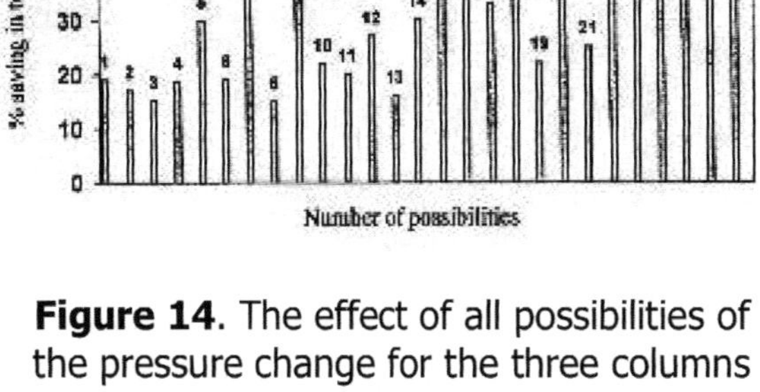

Figure 14. The effect of all possibilities of the pressure change for the three columns on % saving in total energy consumption for feed system c (feed composition one).

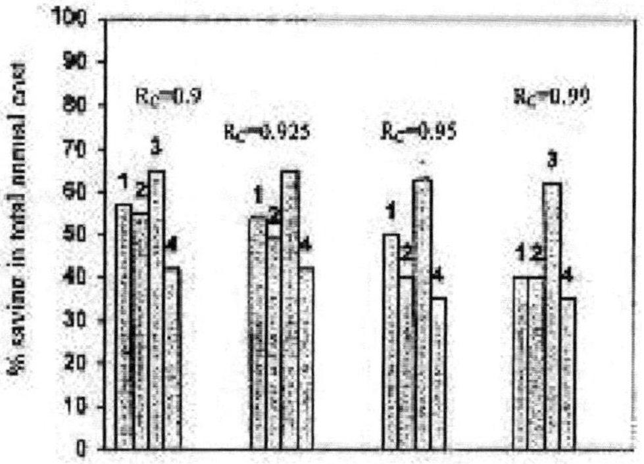

Figure 15. The effect of changing feed composition on % saving in total annual cost for feed system a (column one at 3 atm and column two at 0.5atm).

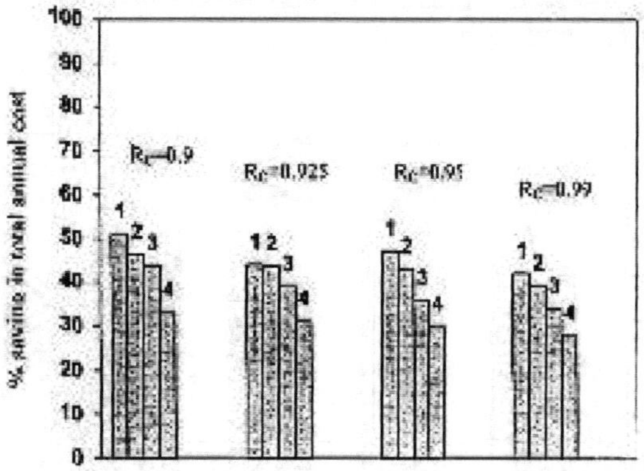

Figure 16. The effect of changing feed Composition one % saving in total annual cost for feed system b(column one at 3 atm and column two at 0.5atm).

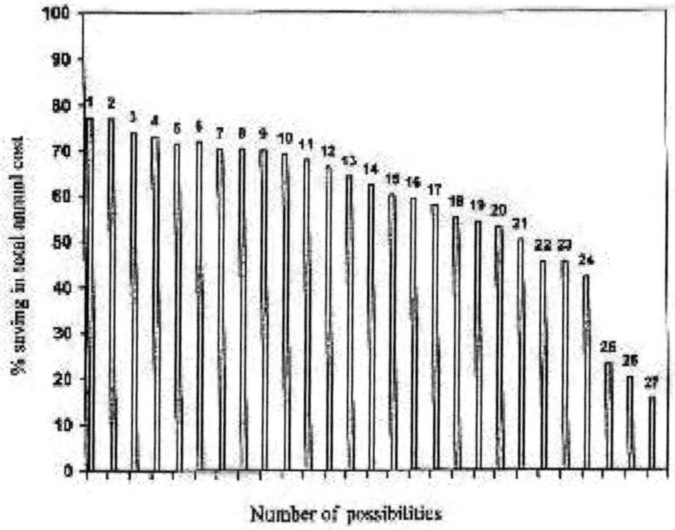

Figure 17. The effect of all possibilities of the pressure change of column 1 and 2 on % saving in total annual cost for feed system c (feed composition one).

Steam Utility Systems are not 'Business as Usual' for Chemical Process Simulators

J. Currie * D.I. Wilson ** N. Depree *** B.R. Young ****
S. Zainal Abidin † L. Karim †

* *Electrical & Electronic Engineering, AUT University, Auckland, New Zealand (e-mail:* jonathan.currie@aut.ac.nz*).*
** *Industrial Information & Control Centre, AUT University, Auckland, New Zealand*
*** *Light Metals Research Centre, University of Auckland, Auckland, New Zealand*
**** *Industrial Information & Control Centre, University of Auckland, Auckland, New Zealand*
† *Group Technology Solutions, PETRONAS, Kuala Lumpur, Malaysia*

Abstract: The operating costs of a typical refinery utility system are so significant that small operational improvements can result in large annual savings. This paper introduces a new utility modelling package designed for PETRONAS to allow in-house modelling and optimization of their refinery and processing plant utility systems. Using this package, engineers can construct a rigorous utility system model which can calculate the actual cost of their utility system, and quantify how operational changes can reduce the bottom line. This paper also contrasts the fundamental differences in approach between utility systems and a steady-state chemical process simulator and provides methods to resolve these differences.

Keywords: Utility, Steam, Modelling, Simulation

1. INTRODUCTION

Industrial utility systems such as steam or electricity form an integral part of most processing plants, supplying the process demands for heat and power, Varbanov et al. (2004a). Utility energy is the largest managed operating cost for the hydrocarbon process industry according to Fernandez-Polanco and Richard (2004), yet Eastwood (2002) argues from both a control and operation point of view, it typically takes a back seat to production.

Within the utility system, cogeneration is a typical modern feature, where steam turbines and/or gas turbines can be used to provide both mechanical and electrical power, either for the plant itself, or if economically viable, to sell energy back to the national grid. However the cost of fuel to generate steam or run a gas turbine can be significant, with large plants costing hundreds of millions of dollars per year to run, as is purchasing external power from the local grid, Varbanov et al. (2004b).

Searching for the most efficient way to operate a utility system given varying electricity and fuel prices, changes in process demands, and equipment availability is a complex problem, Fernandez-Polanco et al. (2004), Eastwood and Bealing (2004), where the optimal selection of running equipment is not always directly obvious. This provides an excellent opportunity for a rigorous utility system model to be built, which can not only provide run time information, but also forms the basis for optimization.

Utility system modelling and optimization has been explored by multiple authors and methods, with perhaps the most well known being Pinch, Linnhoff and Hindmarsh (1983). Using graphical curves, networks of heat exchangers (HENs) can be optimized based on the'pinching' of two composite curves, created from the hot and cold sides of the network. An alternative method is the R-curve proposed by Kenney (1984), which was further explored by Kimura and Zhu (2000) building on the co-generation targeting methodology developed by Mavromatis and Kokossis (1998). This method focused on the redistribution of steam flows through the system's steam turbines to maximise the steam boiler fuel efficiency.

Introducing constraints on maximum or minimum flows, site-wide power requirements or a refinery fuel balance and the methods described so far will fail to establish the optimum, nor will they provide a simulation framework suitable as a commercial platform, Fien (2008), Varbanov et al. (2004a). In order to identify practical and therefore realistic cost savings, the model must take into account system constraints and variable equipment efficiencies, which requires a dedicated utility modelling package complete with rigorous thermodynamics and a flexible system structure.

PETRONAS, the Malaysian state owned oil and gas company, recognised the economical and operational advantages of building and maintaining utility models of their refineries and processing plants, and commissioned the Industrial Information and Control Centre (I^2C^2) to develop the steam utility package described within this paper. While several other commercial utility modelling packages exist such as ProSteam from KBC, Ariane from ProSim, or Aspen Utilities Operations, PETRONAS decided to use iCON as the simulation platform. iCON is a steady state and dynamic process simulator finely tuned to PETRONAS needs and has been in use since 2003. iCON is based on Virtual Materials Group's VMGSim. The package therefore was to be an 'add on' and thus the core design was the development of utility unit operations and methodology of solving a utility system, rather than steady state solvers, graphical interfaces and so forth which already existed.

This paper details the development of iCON Utility Optimizer (iUO); a utility modelling add on package designed for iCON, and describes some of the issues and difficulties that arose during development.

2. ICON UTILITY OPTIMIZER

As expected from most simulation packages, iUO was designed to perform automatic mass and energy balances, provide a live graphical PFD interface with drag and drop unit operations, as well as a rigorous thermodynamic model for combustion calculations and state properties. What was unique about the iUO package was that by building on an existing chemical process simulator, which was already heavily used by process engineers, installed knowledge of modelling within iCON was retained by these engineers and the learning curve could be substantially shortened. Another advantage of this arrangement was also the ability to model both process and utility systems within the one package, enabling more accurate economic calculations if required.

The PFD of the resulting package is exemplified in Figure 1, which shows a three header closed-loop steam system. The system is built using the custom 'Steam' unit operations developed for the package, and connected using the familiar drag unit operation and drop into flowsheet procedure familiar (and expected by) most process engineers today when using process simulators.

2.1 Mass and Energy Balance

The base iCON distribution is a rigorous process simulation package such that mass and energy balances within a unit operation are automatically closed. If the user supplies specifications which are inconsistent with calculated results, these will be flagged by the software as errors so the user can quickly identify the problem. To further aid in this all flows are represented by colour coded streams, indicating vapour (cyan), liquid (blue), two-phase (pink) and zero-flow (grey). This is a substantial advantage over a spreadsheet based modelling approach which demands considerable user discipline in order to be assured that mass balances are closed and errors are not introduced inadvertently.

2.2 Steam Unit Operations

Steam specific unit operations are built by assembling collections of fundamental process simulation unit operations such as heaters, mixers, balances, and expanders. Using basic process insight, simple models can be built, such as the Simple Boiler in Figure 2, which is two heaters, one splitter, one energy balance and two fraction blocks. By building unit operations in this way we ensure that the entire unit operation's mass and energy balance is complete, as all internal operations must meet the rules of a normal process model.

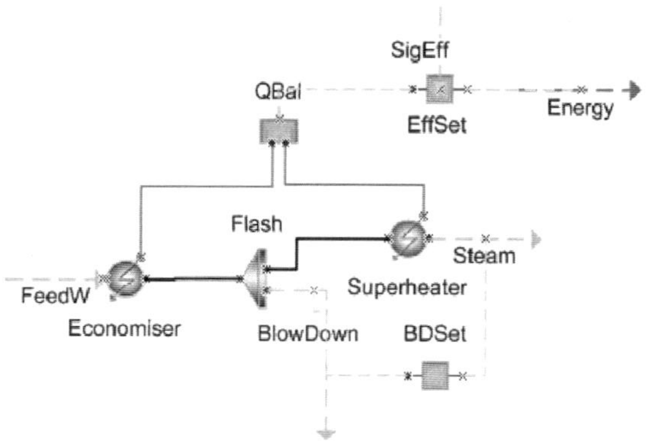

Fig. 2. An exploded view of the Simple Boiler.

A second advantage of this method is the automatic degree of freedom assessment by the iCON solver engine. For example a steam turbine can be specified by flow or by duty, but the same unit operation can fulfill both roles, because the solver will automatically calculate the remaining unknown based on the user's specification.

2.3 Combustion Unit Operations

Combustion unit operations inside the boilers, furnaces and turbines used VMG's Advanced Peng-Robinson (APR) thermodynamic model and a reaction furnace to automatically model the combustion process of the fuel and air mixture specified by the user. These two components automatically complete the combustion stoichiometry and reaction kinetics, enabling accurate estimation of combustion emissions and available heat duty of a given fuel air mix. These were used in the Advanced Boiler, Fired Furnace, Gas Turbine, and other unit operations.

Using the advanced reaction kinetics, the Advanced Gas Turbine predicts NO_x production based on the current operating conditions, and the user can trial varying levels of steam injection to view the $DeNO_x$ effect. Estimation of SO_3 generation for sulfurous fuels is also modelled, such that the acid dew point for a furnace flue gas could be estimated.

Unit operations were built using the same methodology as steam only unit operations, but used split thermodynamic models to model the fuel and steam sections separately, enabling more accurate calculations of state properties.

2.4 Parallel Unit Operations

A common unique feature of utility models is the large number of parallel unit operations, such as steam turbines which operate at the same conditions, but are separated

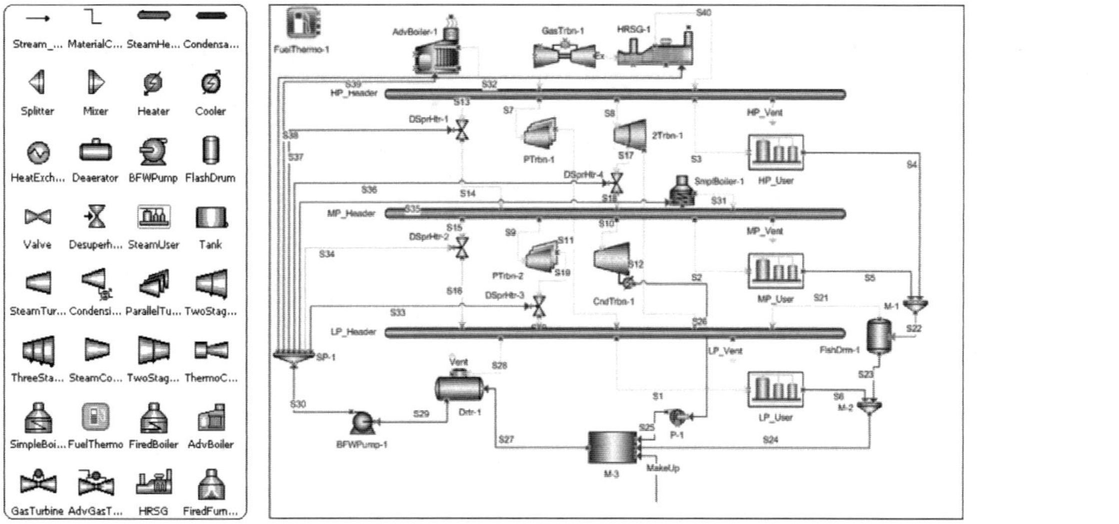

Fig. 1. An example of a steam system comprising of multi-stage turbines, gas turbines, heat recovery steam generators and the advanced boiler. Also shown is the 'Steam' palette for dropping unit operations onto the PFD.

due to sheer size, redundancy, process requirements, or installation requirements. Although a steam turbine may be diagrammatically represented on the PFD as one turbine, there may be in fact ten, twenty or more steam turbines of varying sizes and efficiencies, connected to the same steam temperature and pressure, and all connected into the same header below.

This provided the iUO package with the ability to lump all of these parallel unit operations as one steam turbine, appropriately sized and specified such as to accurately reflect the outlet enthalpy and power of all turbines. The advantage gained here is computational efficiency, because within iCON the steam turbine unit operation required an iterative internal solver to converge the unit. If solved individually this would dominate the computation time when compared to the remainder of the model's operations.

Our solution to the construction of these large groups of steam turbines was to provide an Excel based calculation tool which could connect live data back to iCON, using the iCON ActiveX connection. A user could enter the required number of steam turbines, together with their operating specifications into a preformatted table, collect live temperature and pressure values from iCON, and then calculate an equivalent parallel turbine within Excel. The results could then be automatically exported back to iCON such that in effect, the iCON model only ever solved one turbine for that group, resulting in substantial computational efficiency gains. Figure 3 illustrates a completed steam turbine table.

A second benefit to this arrangement was the ability to control the re-solving of the parallel turbine. A number of case studies run on the models may not affect the incoming enthalpy to this parallel group, such that there was no need to re-solve all the steam turbines at every iteration. When the header temperature or pressure did change, or one of the individual turbines was switched on/off, then the user could resolve the parallel turbine as required.

HP - MP Process Drivers					Equivalent Turbine:		
					Power		3600.00 kW
					Efficiency		70%
					Mass Loss Fraction		0.000
0 Slow Roll = 0.3 Hot = 0.1	Turbine Base Power	Isentropic Efficiency	Steam Flow To Match	Steam Flow In	Steam Flow Out	Turbine Shaftwork	Power Import
1.0	1200.00 kW	70%		16.87 ton/h	16.87 ton/h	1200.00 kW	
1.0	1200.00 kW	70%		16.87 ton/h	16.87 ton/h	1200.00 kW	
1.0	1200.00 kW	70%		16.87 ton/h	16.87 ton/h	1200.00 kW	
			Totals:	50.61 ton/h	50.61 ton/h	3600.00 kW	0.00 kW

Fig. 3. The iUO Excel add in showing the parallel turbine calculation table.

This functionality was built into iUO for steam boilers, steam turbines, steam users (process), furnaces and gas turbines.

3. INDUSTRIAL APPLICATION

iCON Utility Optimizer was benchmarked against two major utility systems: a Methyl Tertiary Butyl Ether (MTBE) refinery and a Liquified Natural Gas (LNG) plant.

Twenty case studies designed to reduce operating costs were run across each model in order to validate the model results. Cases included: reducing the deaerator pressure; optimizing the selection of running steam turbines against electric motors, and installation of Steam Turbo Generators (STGs) to reduce power import requirements. For each case a number of key variables were checked across the model, which included: total running cost; total steam production; header temperatures and let down flows.

As these were hypothetical case studies the values were not compared to actual operating measurements, but were compared with values which were currently being used for a Strategic Energy Review (SER). Typical savings of 1–2% were identified and the results matched easily within predetermined tolerances for the iUO package.

4. UNIQUE CHARACTERISTICS OF UTILITY SYSTEM MODELS

From the outset it would seem that building on a process simulator would be relatively straight forward; the user interface already exists, the thermodynamic models exist or were considered simple, fundamental unit operations already exist, among other benefits. However it quickly became evident that steam utility systems exhibit some unique characteristics that distinguish them from a typical chemical process model.

4.1 Multi-Point Modelling

Modelling for optimization, as was intended with the utility models, requires a new paradigm for both process and utility system models. The model environment must now be able to solve for another operating point, no matter what inputs are changed. For a utilities model, input specifications can be a wide variety of things such as steam temperatures, steam pressures, operating equipment, as well as boiler fuel selection. If any one of these inputs are changed, the resulting solution should be of acceptable accuracy, especially if compared with the actual operating plant. We call this strategy *multi-point* process modelling.

In a multi-point model, all unknowns must be solved by the equations of the model, so that internal specifications are either directly calculated by unit operations or manually entered equations. It is typical that not all internal specifications can be analytically solved, so that a numerical iteration strategy is commonly required. While this adds extra computation time to the model, the model can now represent the plant across a wider range of operating conditions and therefore is now suitable for optimization.

4.2 Single Component Systems

As most process models in the hydrocarbon industry contain multiple components, it was noted that iCON would not automatically infer composition data across a mixer unit operation until mass or molar flows were also entered. This is a default arrangement because calculated compositions (mass or mole fractions) are dependent on flows through the mixer.

However in the case of single component systems such as H_2O in steam systems, once all input stream compositions have been entered, (as 100%) it is logical to assume that the mixer will now calculate that the only option for the output will at that same composition. As shown in Fig. 4, the mixer and input streams have failed to solve even though the unit is fully specified. As the mixer forms an integral part of a utility model (within headers, desuperheaters and deaerators), this problems requires each one these unit operations to have the output composition specified automatically by the package.

4.3 Water and Steam Thermodynamic Model

The thermodynamic model for steam within iCON is an implementation of Steam 1995, and is a modification of the 1995 steam tables available in NIST's Refprop system developed by Virtual Materials Group. This property

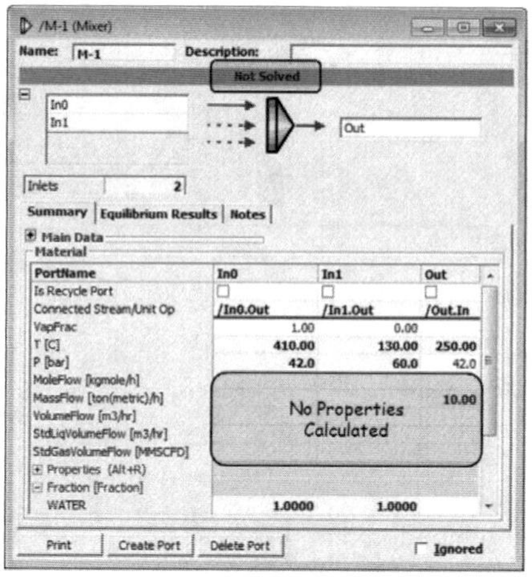

Fig. 4. Example indicating compositions are not automatically inferred on output streams.

package is based on a complex expression of the Helmholtz energy and therefore uses temperature and volume as natural variables instead of pressure and temperature and although very accurate it is not ideally suited for process calculations.

Currently under development for the package are the IAPWS 1997 correlations IAPWS (2007), which include region based polynomial regressions with temperature and pressure pairings. Initial testing has shown these to be up to 14 times faster than the original Steam 1995 implementation.

4.4 Zero Flows

A unique characteristic of steam utility systems is the common situation where the flow of a stream drops to zero. Examples of zero-flow situations include by-pass valves connected to large steam turbines, the substitution of an electric motor instead of a steam turbine, the closing of vents or let-downs when the steam demand changes, or when equipment is shut-down for service.

The difficulty with a zero flows is the subsequent calculation of the thermodynamic properties of the stream, and what can be reliably inferred from the stream. As the flow is zero, the energy of the stream is also zero, which now infers that an energy balance is meaningless. Take for example a mixer with two input ports, and one output port, which is similar in form to a utility system desuperheater.

$$H_{\text{out}} = \frac{H_1 M_1 + H_2 M_2}{M_{\text{out}}}$$

Given arbitrary input pressures and temperatures, and specified outlet flow of zero, $M_{\text{out}} = 0$ (such as in the case where the steam header below requires no steam), the outlet temperature can now be entered as any value within the thermodynamic model limits, and the solution is mathematically valid. If the unit operation were to solve for the outlet enthalpy, the result $H_{\text{out}} = \infty$, causes a thermodynamic exception.

The result of this equation within iCON is shown in Fig. 5 where the mixer appears solved, however both input streams have failed to solve.

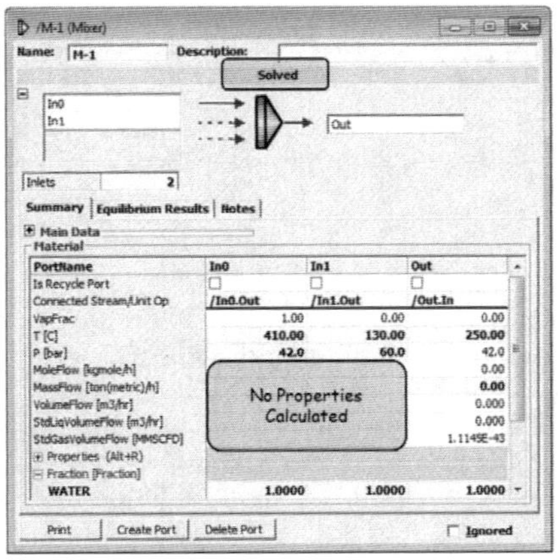

Fig. 5. Example showing zero flow solving issues.

In addition to mixers suffering from zero flow problems, other typical unit operations such as compressors and expanders were also noted to fail to solve with a zero flow. This meant that when a steam turbine was turned off,

$$\text{Power} = M_{\text{out}} \left(H_{\text{in}} - H_{\text{out}}\right), \text{ with } M_{\text{out}} = 0$$

the model would no longer solve as ΔH is underfined. Because the output stream from the expander would not have solved, no information can be propagated subsequently through the flowsheet.

In order to overcome this hurdle our solution was to bias all flows to a small number if a zero flow was encountered entering or exiting a unit operation identified with problems solving with discrete flows. This allowed the thermodynamic model to still calculate the dependent properties of the stream, while being below the threshold identified as a zero flow stream by iCON (identified graphically by a grey stream). This value was also chosen so as to remain within an appropriate dynamic range to avoid numerical issues with the simulator engine. A preferable solution to this would be to remove the equations altogether, and to use a user defined constant or previous solution. This would be possible in an equation based simulator such as Modelica, but not in a modular simulator such as iCON.

5. IMPLEMENTATION IN A CHEMICAL PROCESS SIMULATOR

A critical requirement for the steam utility package is the ability to not only arrange the pre-built unit operations and thereby construct a utility model, but also to implement strategies to ensure efficient and robust solving of these models.

5.1 Closed Loop Steam Systems

The term closed-loop when applied to a steam system context typically refers to the fact that the condensate return is also considered, meaning the complete loop from steam production, through process use, and through condensate collection is modelled. In reality, there are a

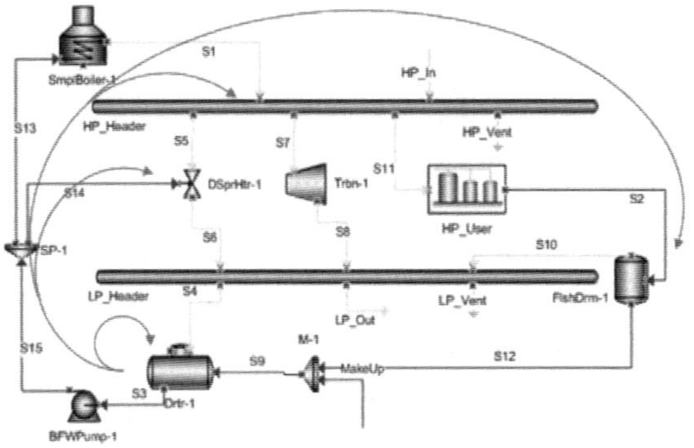

Fig. 6. Two Header steam system with calculation loops shown.

number of loops present within a utility model, as shown in Fig. 6.

In order for the numerical solver to have some chance of establishing a solution, each of these loops must contain an initial estimate of temperature, mass flow, or enthalpy depending on the structure of the loop. All models built with iUO are assumed to be closed loop, thus an iterative recycle/root solver was required for every model.

5.2 Recycle Solver

As detailed in Section 4.1 utility models must be multipoint models, and thus require a number of internal variables to be established iteratively. As with most modular process simulation packages, iCON contains an iterative recycle solver for converging closed loops, in our case using a Broyden based algorithm.

When applying a recycle estimate to a utility model, it was discovered that the direction of the calculation flow (and thus the direction of information flow) must not be constrained to a single direction, as is the default in a process simulator Cota (2003). Taking the steam header temperature as an example, an estimate of temperature provides two solving paths: the first is the outlet temperatures to all equipment connected downstream of the header; and secondly the temperature completes the mass and energy balance of the upstream equipment, dictating the required mass flow of steam (deficit) from the header/boiler above.

Based on the above, the single direction information propagation of a recycle estimate was not sufficient for converging a utility model. Our solution was to implement what is known as a 'controller' in iCON. Simply put this is an iterative non-linear root solver, which adds the estimate to the 'sheet convergence manager'. Controllers can be given a solve level, thus multiple controllers can be solved simultaneously (i.e. a multi-variable root solver), sequentially with different solve levels, or with some combination of the two.

By adding a controller to each variable we had identified as an estimate, together with heuristics for the solve level for each unit operation and a robust non linear solver, we skipped the need to add any recycles to our utility models, regardless of how many loops were created.

5.3 Solver Variable Identification

Key to solving these models so that the user did not inadvertently over specify specifications which would normally vary within typical optimization studies was the identification of degrees of freedom within a utility model. These were variables that initial estimates were required in order to begin solving, and would then be iterated until they converged, in much the same way a recycle works, but using controllers as detailed.

The heuristic we came up with to identify the estimation points follows from typical process simulation modelling. If the variable met the following rules, then it would be estimated:

(1) Expected to change between different optimization runs.
(2) After completing all specifications would still remain unsolved.
(3) If multiple variables remain, choose the one which is expected to vary least.

An example is the steam header unit operation, which would automatically balance (steam in = steam out). The obvious unknown is the amount of steam required from the above header/boiler. However this would not be known until all steam demands exiting the header were known, which required the header temperature to be known. In turn, the header temperature would not be known until the header feed (steam required) was known, and thus we can identify the need for an estimate with initial condition.

The problem is we can estimate either the steam mass flow required, or the header temperature. This was the need for rule 3, as we know from experience the temperature should not vary by more than a few degrees Celsius, while the mass flow may vary widely with process and operating conditions. With numerical conditioning and computational efficiency in mind we estimate the header temperature and thus the header will now iterate to converge the mass balance across it.

Other unit operations required taking a system-wide approach to looking for estimates. One of these was the deaerator, which closes two large loops: one between the lowest steam header and the Boiler Feed Water (BFW) return back to the steam boilers, and the second between the condensate return and the BFW return to the steam boilers. Each loop requires an estimate, and thus two extra estimates are added for each deaerator. This can be seen in Figure 6.

5.4 Solver Level Identification

A feature of the controller solver used for converging model estimates was the ability to parallelize (multi-variable) or constrain to sequential solving for multiple variables. iCON allows a solve level between 0 and 20, with variables on level 20 being solved first.

In order to determine which level an estimate should be placed on, the following heuristic rules were used:

- Lv 10: Variables which contain model wide loops, e.g. condensate return
- Lv 8: Variables which are specific to a unit operation, but influence other unit operations, e.g. header temperature
- Lv 1-5: Variables which are specific to a unit operation, and have no influence on a other unit operations, e.g. boiler air flow

Appropriate gaps between levels were used so that future rules or exceptions could be interleaved.

6. CONCLUSIONS

This paper has introduced the iCON Utility Optimizer utility modelling package and described the advanced functionality built for PETRONAS. Furthermore, it has detailed the unique characteristics of utility systems, and what problems can occur when building them in a package originally designed only for chemical process simulation. Solutions for these problems have been presented, together with heuristics for modelling utility systems within a process simulation environment.

ACKNOWLEDGMENTS:

Financial support to this project from the Industrial Information and Control Centre, Faculty of Engineering, AUT University and The University of Auckland, New Zealand is gratefully acknowledged.

REFERENCES

Raul Cota. *Development of an Open Source Process Simulator.* PhD thesis, University of Calgary, 2003.

Alan Eastwood. ProSteam – A Structured Approach to Steam System Improvement. In *24th IETC*, Houston, 2002.

Alan Eastwood and Chris Bealing. Optimizing the day-to-day operation of utility systems. *Energy Engineering*, 101(1):7–25, 2004.

Diego Fernandez-Polanco and A. Tainshm Richard. OptiSteam reduces operating costs at Spanish Refinery. *Driving Competitive Advantage*, 1Q04, 2004.

Diego Fernandez-Polanco, Alan Eastwood, and Nicola Knight. Real-time utility system optimization. In *Twenty-Sixth Industrial Energy Technology Conference*, Houston, Texas, 2004.

Gert Jan Fien. Make the most of it. *The Chemical Engineer*, 2008.

IAPWS. Revised Release on the IAWPS Industrial Formulation 1997 for the Thermodynamic Properties of Water and Steam, 2007.

W.F. Kenney. *Energy Conservation in Process Industries.* London, 1984.

Hiroyuki Kimura and X. X. Zhu. R-curve concept and its application for industrial energy management. *Industrial & Engineering Chemistry Research*, 39(7):2315–2335, 2000.

B. Linnhoff and E. Hindmarsh. The pinch design method for heat exchanger networks. *Chemical Engineering Science*, 38(5):745–763, 1983.

S. P. Mavromatis and A. C. Kokossis. Conceptual optimisation of utility networks for operational variations–I: Targets and level optimisation. *Chemical Engineering Science*, 53(8):1585–1608, 1998.

P. Varbanov, S. Perry, Y. Makwana, X. X. Zhu, and R. Smith. Top-level analysis of site utility systems. *Chemical Engineering Research and Design*, 82(6):784–795, 2004a.

P. S. Varbanov, S. Doyle, and R. Smith. Modelling and optimization of utility systems. *Chemical Engineering Research and Design*, 82(5):561–578, 2004b.

Simple Procedure for Determining the Scope and Means of Transferring Heat Between Zones of Integrity

E. K. Tamakloe & G. T. Polley
University of Guanajuato, Mexico

As demonstrated by Polley & Heggs (1999) the Pinch Design Method (Linnhoff & Hindmarsh, 1983), despite its elegance, has a number of significant drawbacks. Of these the most serious are that it ignores the process flow-sheet and that it is not easy to apply to problems involving a large number of streams.

These draw-backs can be largely overcome by decomposing the overall design problem into a number of smaller problems that have "stream populations" based upon the process flow sheet (Amidpour & Polley, 1997). This approach mirrors that of Ahmad & Hui (1991) who handled problems in which safety considerations required the separation of sets of streams. These workers introduced the term "zones of integrity" to define the separate sets.

This procedure generally results in simple structures and safer plants. However, there are occasions where transfer of heat between zones can lead to significant energy savings. In this paper we demonstrate how such opportunities can be identified and evaluated.

Identifying the Opportunity

Recall the basic make up of the Grand Composite Curve of a process (an example is shown in Figure 1). This curve is plotted on co-ordinates of "interval temperature" (temperature adjusted by allowable temperature approach) against heat load. It touches the vertical axis (heat load of zero) at the pinch point of the system.

Individual parts of the curve demonstrate whether the process has a surplus of heat or a deficit of heat over a given span of interval temperature (Figure 1). A surplus of heat at a given span can be used to satisfy a deficit present at a lower interval temperature.

Immediately below the Pinch point there is a region in which there is a surplus of heat. Whilst immediately above the Pinch point there is a region in which there is a deficit of heat.

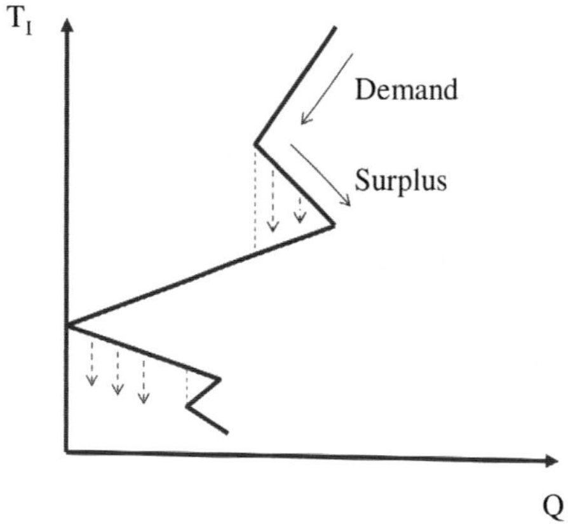

Figure 1. Grand Composite Curve

The possibility of saving energy by transferring heat from one zone of integrity to another can be identified by comparing the Pinch points of the individual zones. If they coincide then energy cannot be saved by linking the two zones.

Identifying the Scope

Now we transfer heat by sending streams from one zone to the other. This could be a hot stream from the zone (zone A) having the higher pinch point into the zone having the lower pinch point (zone B) or a cold stream in the opposite direction.

If we heat a cold stream from zone B to an interval temperature higher the pinch point of zone A we add to the heat demand in zone A. Conversely, if we cool a hot stream from zone A to an interval temperature below the pinch point of zone B we increase the cold utility demand in zone B. So, we see that the temperature span over which these transfers can take place is limited by the two pinch points.

If we transfer too much heat from one zone to another we generate a situation in which we have to transfer heat in the reverse direction at a lower temperature. This situation needs to be avoided. We do so using "utility profiles".

A surplus in one region is used to satisfy a deficit in another. Therefore, the temperatures at which cold utility can be used within a process is given by the regions of the curve in which a surplus does not cover a deficit at a lower temperature. Similarly, the temperatures at which hot utility can be used within process is given by the regions of the curve in which deficit cannot be satisfied by a surplus at a higher temperature. This leads to the "utility profile" (Figure 2).

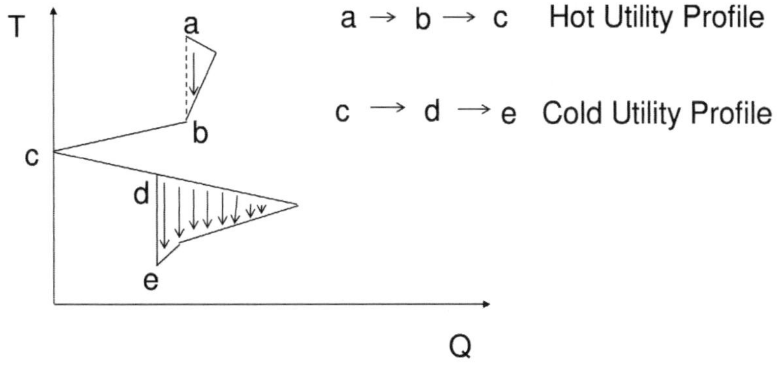

Figure 2. Utility Demand Profiles

We identify the scope for transferring heat between zones by examining the relevant utility profile between the two pinch temperatures. For instance, as shown in Figure 3, the heat that can be absorbed by the zone having the lower pinch point is determined by examining the cold utility profile between the two pinch points.

In Figure 4 the amount of heat that can be provided through a transfer from the zone having the higher pinch is determined by examining the hot utility profile.

The heat that can actually be transferred between the zones is the smaller of the two values.

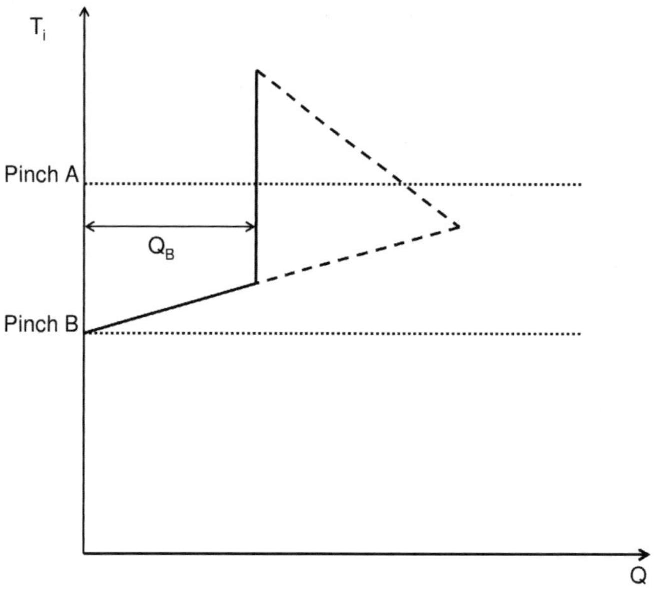

Figure 3. Permissible Transfer of Heat to Zone having Lower Pinch Point

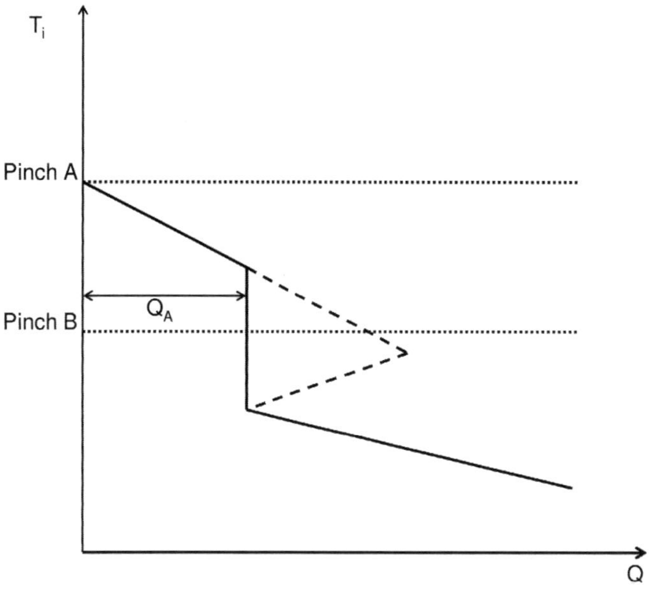

Figure 4. Permissible Transfer of Heat from Zone having Higher Pinch Point

Identifying the Means by Which Energy is Saved

Now the streams present immediately above or below a pinch point are easily identified and these are the streams being used for the transfer of heat between zones.

The amount of heat that can be transferred using a specific stream is simply the heat capacity flow rate of the stream multiplied by the difference in between the two pinch temperatures.

These loads can be brought together in a simple heat load table (which we will term the "transfer load table") where we compare the possible supply from specific hot stream and demand from specific cold stream. Using this table we can identify how much heat can be transferred using a single heat recovery match. It is the lower value of the two stream loads.

So, we easily identify our options for saving energy through inter-zonal transfers in terms of the better matches and the maximum load on these matches.

Example Problem

The pinch point for streams present in Zone A of a process plant is at a temperature interval of 150C and for Zone B it is at a temperature interval of 100C.

Consequently, the hot streams present below the pinch point in Zone A can be used to provide heat for cold streams above the pinch point in Zone B.

The coordinates of the Grand Composite curves are given in Table 1.

Interval Temperature	Q Zone A	Q Zone B
200		3000
150	0	1600
120	1500	2000
100		0
90	1300	
70	1500	
40	1900	
30	2000	

Table 1. Co-ordinates of the Grand Composite Curves

There are 3 hot streams immediately below the pinch in Zone A and the have the following heat capacity rates given in Table 2.

Stream	CP
H1	15
H2	10
H3	25

Table 2. Hot streams immediately below pinch of Zone A

Also there are 2 cold streams immediately above the pinch in Zone B with the following heat capacity rates given in Table 3.

Stream	CP
C1	24
C2	16

Table 3. Cold streams immediately above pinch of Zone B

The Cold Utility Profile for Zone A is shown in Figure 5.

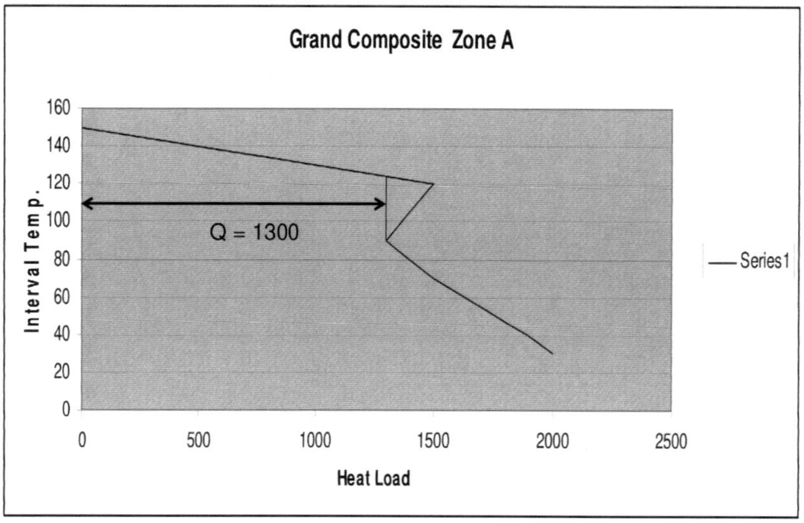

Figure 5. Cold Utility Profile For Zone Having Higher Pinch

We observe that the maximum amount of heat that can be transferred from this zone is 1300.

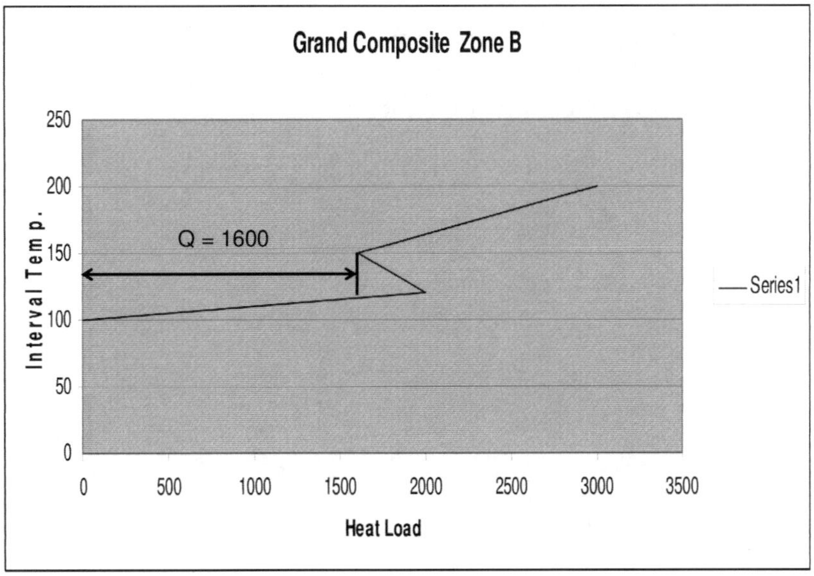

Figure 6. Hot Utility Profile For Zone Having Lower Pinch

The hot utility profile for Zone B is shown in Figure 6. We observe that the maximum amount of heat that can be absorbed by this zone is 1600.

The maximum inter-zonal transfer is the lower of these two values (1300).

We now examine the possible heat recovery matches. This is done in Table 2.

				Hot	Streams	
			No.	1	2	3
			CP	15	10	25
	Cold	Streams	dT	50	50	50
No.	CP	dT	Q	750	500	1250
1	24	50	1200	***750***	***500***	***1200***
2	16	50	800	***750***	***500***	***800***

Table 2. Possible Individual Heat Recovery Matches

The heat capacity flow rates and the temperature span over which heat is transferred (this can be less than the difference between the two pinch points if the supply or target temperature falls between these temperatures) and the permissible heat load is listed for each category of stream. The permissible heat loads are then compared and a matrix giving the lower of the two values developed (bottom right hand corner). The values of the matrix show the possible heat exchange for the individual matches.

We see that we have the options listed in Table 3.

Matches	Saving
C1-H3	1200
C1-H2 plus C2-H3	1300
C1-H1 plus C2-H2	1250

Table 3. Possible Heat Recovery Systems

Transport of Streams

The final question that needs to be addressed is what stream should be piped to the other zone. The answer is dictated by issues such as safety and capital cost.

The over-riding factor is safety. Here the rules are that a flammable stream should not be transported into a non-flameproof zone and that the distances over which hazardous streams are transported should be minimised.

If safety considerations do not dictate choice capital, cost becomes the deciding factor. Capital cost is controlled by size of pipe and material of construction.

Generally we can observe that liquids require significantly smaller pipe diameters than gas or vapour. The transportation of two phase mixtures can be problematic and the distances over which they are moved should be minimised.

References

Ahmad, S. and Hui, D. C. W. "Heat recovery between areas of integrity", Comp. Chem. Eng. 1991, 15(12), 809–832

Amidpour M. & Polley G.T. "Application of Problem Decomposition in Process Integration", TransIChemE. 1997, 75A, 53-63

Linnhoff, B. and Hindmarsh, E. "The pinch design method of heat exchanger networks", Chem. Eng. Sci. 1983, 38(5), 745–763

Polley G.T. & Heggs P.J. "Don't let the "Pinch" pinch you", Chem. Eng. Prog. Dec. 1999, 27-36

Reducing the Load on Cooling Towers Through Revamping of the Individual Coolers

L. Canizalez Davalos, G.T.Polley, M. Riesco Avila & M. Picon Nunez
University of Guanajuato, Mexico

When engineers seek to reduce the thermal load on cooling towers they will often opt for increased heat recovery (via improved heat integration) but will ignore the effect that such action has upon the performance of the individual coolers forming the cooling water network.

The area of the coolers is fixed. The result of reduced overall load is a reduction in the cooling water feed temperature. Whilst improved heat recovery leads to reduced inlet temperatures to the coolers the overall effect will be a fall in the temperature of the process stream leaving the cooler.

The reduction in the load on the cooling tower does not equate with the additional heat recovery being achieved on the plant.

An example of this is given below. In our case study the original duty on the cooling tower was 12.64 MW. Following improved process integration that saved 6.7 MW, the load on the cooling tower was reduced by 5.1 MW (24% less than expected).

In order to achieve the full reduction it is necessary to reduce the over performance of the coolers. This can be achieved in a number of ways: reducing the flow of cooling water through the exchanger (by using either a bypass or by throttling a valve), operating a bypass on the hot side of the exchanger, or by plugging some of the tubes in the exchanger. These options can be used in combination.

This paper presents a systematic procedure for identifying all of the effective options.

Analysis of Ways in which Cooler Performance can be Adjusted

The objective of cooler modification is the restoration of hot stream outlet temperature to a target value.

The temperature at which a stream leaves a heat exchanger is determined by the Effectiveness of the exchanger. This, in turn, is controlled by the ratio of the heat capacity flow-rates of the streams being handled (R) and the number of heat transfer units provided by the exchanger:

$$N_{tu} = \frac{UA}{CP_{\min}} \tag{1}$$

The performance of a cooler can be adjusted in three ways:

1. by adjusting the flow of water through the cooler
2. by operating a bypass on the hot side of the exchanger
3. by reducing the area of the cooler by plugging a fraction of the tubes

Flow reductions result in changes to heat capacity flow-rate. These can be expressed in terms of fractional reduction. So, for the hot stream:

$$CP_h = M(1-\gamma)c_{ph} \tag{2}$$

for the cold stream:

$$CP_c = m(1-\alpha)c_{pc} \tag{3}$$

with the CP ratio being related to the magnitudes of these values:

$$R = \frac{CP_{\min}}{CP_{\max}} \tag{4}$$

The temperature at which the cooling water leaves the cooler relates to actual exchanger outlet temperature rather than hot stream target temperature and is given by:

$$t = t_{in} + \frac{CP_h}{CP_c}(T_{in} - T) \tag{5}$$

The mixed temperature for the hot stream is given by:

$$T_{mix} = (1-\gamma)T + \gamma T_{in} \tag{6}$$

Re-arranging, the required outlet temperature is given by:

$$T = \frac{T_{mix} - \gamma T_{in}}{(1-\gamma)} \tag{7}$$

This mixed temperature equates with the target temperature for the hot stream.

The required temperature changes around the cooler are:

$$\Delta T_h = T_{in} - T \qquad \Delta T_c = t - t_{in} \tag{8}$$

The Effectiveness required to obtain these changes is:

$$\varepsilon = \frac{\Delta T_{max}}{T_{in} - t_{in}} \tag{9}$$

For a multi-pass shell-and-tube heat exchanger, the Number of Transfer Units required to provide a given Effectiveness is given by:

$$N_{tu} = \frac{1}{\left(1+R^2\right)^{1/2}} \ln\left[\frac{2-\varepsilon(1+R-(1+R^2)^{1/2})}{2-\varepsilon(1+R+\left(1+R^2\right)^{1/2})}\right] \tag{10}$$

As noted above, this number is a function of surface area and overall heat transfer coefficient.

The surface area depends on the fraction of tubes that are plugged, and is given by:

$$A = \pi d_o l N_T (1-\beta) \tag{11}$$

Individual film heat transfer coefficients are related to flow fraction and, in the case of the tube-side coefficient, the fraction of tubes plugged. So, for the tube-side we have:

$$h_i = h_{ref,i}\left[(1-\beta)(1-\alpha)\right]^{0.8} \tag{12}$$

and for the shell-side:

$$h_i = h_{ref,i}\left[1-\gamma\right]^{0.6} \tag{13}$$

The reference coefficients are those that the exchanger provided under the original flow conditions.

The overall heat transfer coefficient is then given by:

$$\frac{1}{U}=\frac{1}{h_o}+\frac{d_o}{h_i d_i}+r \qquad (14)$$

The iteration is conveniently undertaken using a Newton-Rapheson procedure for β value at which UA equals required value.

Limit on hot stream bypass fraction.

This outlet temperature for the hot stream falls as the bypass fraction increases. Its value cannot fall below that of the incoming water stream. So:

$$\gamma_{\max}=\frac{T_{mix}-t_{in}}{T_{in}-t_{in}} \qquad (15)$$

This limit can be determined at the start of the calculation.

Limit on reduction in cooling water flow.

Reduction in water velocity within an exchanger will give rise to a reduction in Reynolds Number. This should not be allowed to fall below 10 000. Otherwise fully developed turbulent flow will not exist.

Two factors control the flow velocity: the water flow rate and the fraction of exchanger tubes that are plugged. So, application of this limit has to be incorporated as a check in the calculation of the fraction of tubes plugged.

Procedure

The following procedure identifies all of the options that restore the hot stream outlet temperature of a cooler following a reduction in hot stream inlet temperature.

1. The limit on the hot stream bypass fraction is determined.
2. A search between operation without bypass and this limit is then undertaken. This involves steps 3 to 7 described below.
3. The fraction by which the water flow is reduced is varied between zero and a limiting condition (determined in step 7 below).
4. Required Effectiveness is calculated (from equations listed above)
5. Required Number of Transfer Units is determined.
6. An iterative procedure is applied to determine fraction of plugged tubes that yields the required number of transfer units.

7. If the Reynolds number on the tube-side of the cooler falls below the limiting value of 10 000 the result is unacceptable. The results for further reduced water flows will also be unacceptable.

The results of the analysis are presented in the form of a matrix (Table 1). The columns refer to the fraction of by which the fraction of water flowing through the cooler is reduced. The rows refer to the fraction of hot process stream bypassing the cooler. The elements of the matrix indicate the fraction of tubes that need to be plugged in order to achieve the required performance.

				flow fraction			
Bypass Fraction	**0.000**	**0.011**	**0.021**	**0.032**	**0.043**	**0.054**	**0.064**
0.000	0.5469	0.5462	0.5454	0.5447	0.5439	0.5430	0.5422
0.035	0.5236	0.5228	0.5220	0.5211	0.5202	0.5193	0.5184
0.069	0.4961	0.4952	0.4943	0.4934	0.4924	0.4914	0.4904
0.104	0.4631	0.4621	0.4611	0.4600	0.4589	0.4578	0.4567
0.138	0.4225	0.4213	0.4201	0.4189	0.4176	0.4163	0.4150
0.173	0.3704	0.3691	0.3676	0.3662	0.3647	0.3631	0.3615
0.207	0.3001	0.2984	0.2966	0.2948	0.2929	0.2909	0.2889
0.242	0.1963	0.1939	0.1914	0.1889	0.1862	0.1835	0.1807
0.276	**0.0126**	0.0085	0.0042				

Table 1. Modification Matrix for a specified cooler.

This procedure has been embedded into a simulation of the thermal behaviour of the cooling water network as a whole. (Thermal simulation of networks has been described by Picon Nunez et al, 2010).

If it is assumed that each cooler is to be restored to a specified performance then the thermal conditions for entry to each cooler is set and matrices for each cooler can be generated in a single simulation.

Case Study

A cooling water system has the structure shown in Figure 1. The cooling tower Effectiveness is 0.65 and the wet bulb temperature is 26 C.

The current performance of the individual coolers, in terms of hot stream inlet and outlet temperatures, is shown in Table 2.

Fully details of the dimensions of the coolers, the piping system and the pump characteristics are given in the Appendix.

Current Performance		
Unit	T in	T out
H1	110	41.5
H2	90	44.7
H3	85	46.2
H4	85	41.6
H5	100	44.8
H6	120	40.8
Tower Load = 12.64 MW Water Feed Temp. = 30.6 C T_{water} return = 39.3 C		

Table 2. Current Performance

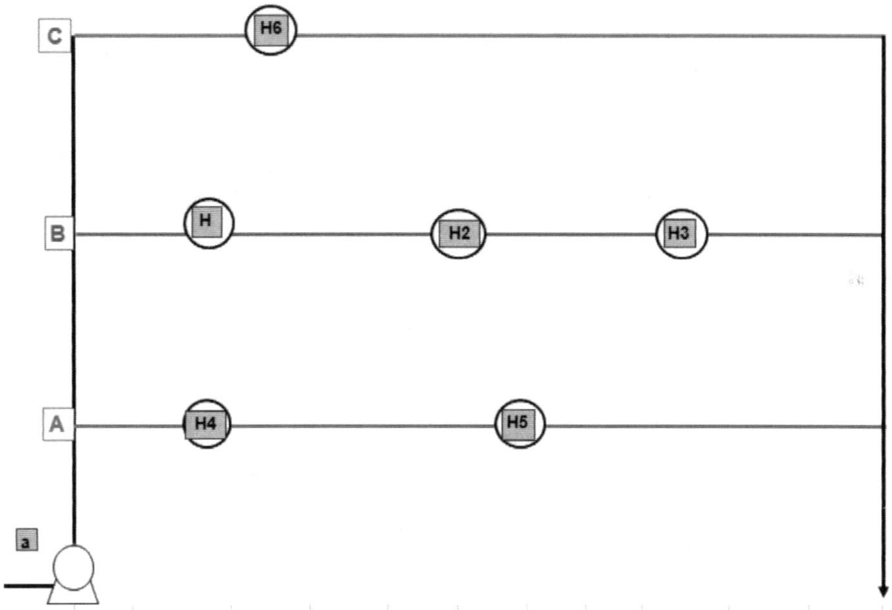

Figure 1. Cooling Water Network

Following modifications to the plants heat recovery system the performance of the plant changes to that given in Table 3. The hot stream entry temperatures are those following improved heat recovery. The exit temperatures are found by simulation.

We now consider the modifications necessary to restore each cooler to its original hot stream exit condition. Each individual branch is now considered in turn.

Following Increased Heat Recovery on the Plant			
Unit	T in (new)	***T out***	Tout (previously)
H1	70	***34.4***	41.5
H2	65	***36.7***	44.7
H3	65	***37.9***	46.2
H4	65	***36.1***	41.6
H5	75	***38.2***	44.8
H6	75	***34.0***	40.8
Load on Tower = 7.54 MW $T_{water\ return}$ = 33.9 C Water feed temp. = 28.8 C			

Table 3. Revised Performance

Branch A contains coolers H4 and H5. The modification matrices for the individual coolers are given in Tables 4 and 5. Recall that the values of water reduction and hot stream bypass are related to limiting values.

	flow fraction							
Bypass Fraction	**0.0000**	**0.0107**	**0.0215**	**0.0322**	**0.0430**	**0.0537**	**0.0645**	**0.0752**
0.0000	0.3959	0.3956	0.3952	0.3948	0.3944	0.3941	0.3937	0.3933
0.0345	0.3664	0.3660	0.3656	0.3652	0.3648	0.3644	0.3639	0.3635
0.0690	0.3318	0.3314	0.3309	0.3305	0.3300	0.3295	0.3291	0.3286
0.1035	0.2905	0.2900	0.2895	0.2890	0.2885	0.2880	0.2875	0.2869
0.1380	0.2400	0.2394	0.2389	0.2383	0.2377	0.2371	0.2365	0.2359
0.1725	0.1761	0.1755	0.1748	0.1742	0.1735	0.1728	0.1721	0.1714
0.2071	**0.0915**	0.0908	0.0900	0.0891	0.0883	0.0875	0.0866	0.0858

Table 4. Modification Matrix: H4

Our first observation is that the cooler response is insensitive to cooling water flow-rate. This is because the water side heat transfer coefficient is very much higher than that encountered on the process side. Furthermore, the process side has the lower heat capacity flow-rate. Consequently, the Number of Transfer Units (and hence, Effectiveness) of the cooler is insensitive to water side conditions.

The better means of reducing cooler performance are plugging of tubes and operation of a bypass on the hot stream.

	flow fraction							
Bypass Fraction	**0.0000**	**0.0107**	**0.0215**	**0.0322**	**0.0430**	**0.0537**	**0.0645**	**0.0752**
0.0000	0.3917	0.3911	0.3904	0.3898	0.3892	0.3885	0.3878	0.3871
0.0345	0.3624	0.3618	0.3611	0.3604	0.3597	0.3590	0.3583	0.3575
0.0690	0.3282	0.3275	0.3268	0.3260	0.3252	0.3244	0.3236	0.3228
0.1035	0.2874	0.2866	0.2858	0.2849	0.2840	0.2831	0.2822	0.2813
0.1380	0.2374	0.2365	0.2355	0.2346	0.2336	0.2325	0.2315	0.2304
0.1725	0.1741	0.1730	0.1719	0.1707	0.1695	0.1683	0.1670	0.1658
0.2071	**0.0897**	0.0883	0.0869	0.0854	0.0840	0.0825	0.0809	0.0793

Table 5. Modification Matrix: H5

For both coolers identified option is to 10 % of its tubes and bypass 21% of the hot flow.

Branch B contains coolers H1, H2 and H3. The modification matrices for the individual coolers are given in Tables 6 to 8.

Bypass	flow fraction							
Fraction	**0.000**	**0.011**	**0.021**	**0.032**	**0.043**	**0.054**	**0.064**	**0.075**
0.000	0.5469	0.5462	0.5454	0.5447	0.5439	0.5430	0.5422	0.5413
0.035	0.5236	0.5228	0.5220	0.5211	0.5202	0.5193	0.5184	0.5174
0.069	0.4961	0.4952	0.4943	0.4934	0.4924	0.4914	0.4904	0.4893
0.104	0.4631	0.4621	0.4611	0.4600	0.4589	0.4578	0.4567	0.4554
0.138	0.4225	0.4213	0.4201	0.4189	0.4176	0.4163	0.4150	0.4136
0.173	0.3704	0.3691	0.3676	0.3662	0.3647	0.3631	0.3615	0.3598
0.207	0.3001	0.2984	0.2966	0.2948	0.2929	0.2909	0.2889	0.2868
0.242	0.1963	0.1939	0.1914	0.1889	0.1862	0.1835	0.1807	0.1778
0.276	**0.0126**	0.0085	0.0042					

Table 6. Modification Matrix: H1

Bypass Fraction	**0.0000**	**0.0107**	**0.0215**	**0.0322**	**0.0430**	**0.0537**	**0.0645**	**0.0752**
0.0000	0.6654	0.6653	0.6652	0.6651	0.6650	0.6649	0.6648	0.6647
0.0345	0.6486	0.6485	0.6484	0.6483	0.6482	0.6481	0.6480	0.6478
0.0690	0.6290	0.6289	0.6288	0.6286	0.6285	0.6284	0.6282	0.6281
0.1035	0.6056	0.6054	0.6053	0.6051	0.6050	0.6049	0.6047	0.6046
0.1380	0.5769	0.5768	0.5766	0.5765	0.5763	0.5761	0.5760	0.5758
0.1725	0.5408	0.5406	0.5404	0.5403	0.5401	0.5399	0.5397	0.5395
0.2071	0.4932	0.4930	0.4928	0.4926	0.4924	0.4921	0.4919	0.4917
0.2416	0.4262	0.4259	0.4256	0.4254	0.4251	0.4248	0.4245	0.4242
0.2761	0.3198	0.3194	0.3190	0.3186	0.3182	0.3178	0.3173	0.3169
0.3106	**0.0963**	0.0955	0.0947	0.0938	0.0930	0.0921	0.0913	0.0904

Table 7. Modification Matrix: H2

The identified modifications are to bypass 30% of the hot stream around cooler H1; plug 10% of the tubes and bypass 31% of the flow around cooler H2, and plug 33% of the tube and bypass 31% of the hot stream around cooler H3.

					flow fraction			
Bypass Fraction	**0.0000**	**0.0107**	**0.0215**	**0.0322**	**0.0430**	**0.0537**	**0.0645**	**0.0752**
0.0000	0.7655	0.7653	0.7652	0.7650	0.7648	0.7646	0.7644	0.7642
0.0345	0.7538	0.7536	0.7534	0.7532	0.7530	0.7528	0.7526	0.7524
0.0690	0.7400	0.7398	0.7396	0.7394	0.7392	0.7390	0.7387	0.7385
0.1035	0.7235	0.7233	0.7230	0.7228	0.7226	0.7223	0.7221	0.7218
0.1380	0.7032	0.7029	0.7027	0.7024	0.7021	0.7018	0.7015	0.7012
0.1725	0.6773	0.6770	0.6767	0.6764	0.6761	0.6757	0.6754	0.6751
0.2071	0.6428	0.6425	0.6421	0.6417	0.6413	0.6409	0.6405	0.6401
0.2416	0.5934	0.5929	0.5924	0.5919	0.5914	0.5909	0.5904	0.5899
0.2761	0.5126	0.5119	0.5112	0.5105	0.5097	0.5090	0.5082	0.5074
0.3106	**0.3297**	0.3281	0.3265	0.3249	0.3232	0.3214	0.3196	0.3178

Table 8. Modification Matrix: H3

Finally, the modification matrix for cooler H6 is given in Table 9.

					flow fraction			
Bypass Fraction	**0.0000**	**0.0107**	**0.0215**	**0.0322**	**0.0430**	**0.0537**	**0.0645**	**0.0752**
0.0000	0.4827	0.4825	0.4822	0.4819	0.4817	0.4814	0.4811	0.4809
0.0345	0.4627	0.4624	0.4621	0.4618	0.4616	0.4613	0.4610	0.4607
0.0690	0.4391	0.4388	0.4385	0.4382	0.4379	0.4376	0.4373	0.4370
0.1035	0.4110	0.4107	0.4103	0.4100	0.4096	0.4093	0.4089	0.4086
0.1380	0.3765	0.3761	0.3757	0.3754	0.3750	0.3746	0.3742	0.3738
0.1725	0.3328	0.3324	0.3319	0.3315	0.3310	0.3306	0.3301	0.3296
0.2071	0.2748	0.2743	0.2738	0.2732	0.2727	0.2721	0.2715	0.2710
0.2416	0.1920	0.1913	0.1906	0.1899	0.1891	0.1884	0.1876	0.1869
0.2761	**0.0560**	0.0550	0.0539	0.0529	0.0518	0.0506	0.0495	0.0483

Table 9. Modification Matrix: H6

The modification required for this unit is to plug 6% of the tubes and bypass 28% of the hot stream.

Following these modifications the system performance given in Table 10 was achieved.

The load on the cooling tower was reduced to 5.82 MW. This gave a saving of 6.62 MW (against the load reduction due to increased heat recovery within the plant of 6.67 MW).

The reduction in load resulted in a fall in cooling water feed temperature (from 28.8 C to 28.2 C). This drop was not (for ease of computation) allowed for during the derivation of the modification matrices.

Following Modification of Coolers			
Unit	T in (new)	***T out***	Tout (previously)
H1	70	***43.4***	41.5
H2	65	***43.9***	44.7
H3	65	***45.2***	46.2
H4	65	***41.4***	41.6
H5	75	***44.9***	44.8
H6	75	***43.6***	40.8
Load on Tower = 5.82 MW $T_{water\ return}$ = 32.1 C Water feed temp. = 28.2 C			

Table 10. Revised Performance

Conclusions

Improvement to the heat recovery in process plants is sometimes undertaken with the objective of reducing the load on the factory's cooling water system. The response of the individual coolers to these changes is rarely considered. In this work we have demonstrated that improvements in heat recovery do not fully transmit to reductions in cooling tower load.

Significant reductions in cooling tower load can often be made by raising the temperatures at which the process streams leave the individual coolers. These options should not be overlooked.

References

Picon Nunez M., Fuentes-Silva A. L., & Canizales-Davalos L. "Shortcut performance method for assessing the flexibility of cooling systems", PRES10, Prague, August, 2010

Case Study Details

Cooling Tower:	Effectiveness	0.65
	Wet Bulb Temperature	26 C
Pump Characteristic:	$P = 150 + 2.6\,V - 274\,V^2$	kPa

Piping:

Branch	Diam.	Length	Elevn.	k valve
Feed	0.35	200	0	0.5
A	0.25	120	5	0.5
B	0.2	170	5	0.5
C	0.2	180	10	0.5
Return	0.35	200	0	15

Heat Exchangers:

Unit		H6-C	H1-B	H2-B	H3-B	H4-A	H5A
k valve		0.5	0.5	0.5	0.5	0.5	0.5
No. Tubes		800	820	450	800	400	850
No. Passes		2	2	2	2	2	2
Bypass-water fraction		0	0	0	0	0	0
plug		0	0	0	0	0	0
Ref. Flow	kg/s	85	85	85	85	85	85
Ref. htc water	W/m^2K	6570	6700	10830	6570	11900	6500
Bypass-hot		0	0	0	0	0	0
Ref. Flow	kg/s	14	20	12	23	14	18
Ref. htc hot	W/m^2K	1035	495	510	680	650	375
heat cap. hot		2600	2250	2100	1800	2800	2200
length	m	6.1	6.1	6.1	6.1	6.1	6.1
tube i.d.	m	0.0148	0.0148	0.0148	0.0148	0.0148	0.0148
tube o.d.	m	0.01905	0.01905	0.01905	0.01905	0.01905	0.01905
Rd	m^2K/W	0.00224	0.0005	0.0005	0.0005	0.0005	0.001

Advances in Heavy Oil Upgrading

The UOP slurry-hydrocracking process converts the residue from conventional crude oils and bitumen to higher value liquid products. This talk discusses recent advances in the slurry hydrocracking technology. The development of advanced analytical characterization techniques have been necessary to further progress the technology. More specifically, advances in FT-ICR-MS, polarized light microscopy, XRD and other analytical technologies have provided additional insights into the slurry hydrocracking process. The process of mesophase growth has been monitored using polarized light microscopy coupled with image analysis and advanced statistical processing techniques. A highly automated and flexible pilot plant unit has been designed and constructed to perform studies to support and develop the technology. The pilot plant's integrated fractionation and product recycle sections have provided valuable data.

Novel Additives to Improve Resid Quality

Parag Shah
Refinery Technical Lead
Dorf Ketal Chemicals (I) Pvt. Ltd

P.N. Ramaswamy
Technical Director
Dorf Ketal Chemicals LLC

Matthew Knight
President & CEO
Dorf Ketal Chemicals LLC

Prepared for Presentation at the 2011 Spring National Meeting
Chicago, IL, March, 15, 2011

Novel Additives to Improve Resid Quality

Parag Shah

Refinery Technical Lead

Dorf Ketal Chemicals (I) Pvt. Ltd

P.N. Ramaswamy

Technical Director

Dorf Ketal Chemicals LLC

Matthew Knight

President & CEO

Dorf Ketal Chemicals LLC

Abstract

With an increasing focus on resid upgrade and bottom of the barrel conversion over the years due to economic advantages, refiners are increasingly processing heavy crudes that yield higher residue content. These crudes are also higher in salt, metal & sulfur content. Often in these cases, there are issues with resid quality in terms of metal content such as Iron (Fe), Calcium (Ca), Sodium (Na) etc. The presence of sulfur/H_2S also poses problems that interfere in chemical treatment for removal of some of the mentioned metals above. Several refineries hence blend these heavier crude oils with lighter, sweeter feed-stocks to help meet resid quality.

Crude oils containing calcium naphthenates are a good example for the issue at hand. These crudes originate from various areas. Some of them are from West Africa (Kuito, Doba), the North Sea (Heidrun, Captain), China (Shengli, Bohai Bay) and Indonesia (Serang). The problems posed by high calcium crudes are emulsion stability within the desalter, scaling problems caused by high levels of calcium in the desalter effluent water and corrosion due to a low pH. The high metal content generally makes the crudes conductive which results in high amperage draws of the desalters. Doba crude is high in acids as well as metals which poses further problems during processing. Hence the best way in which to process this crude is with crudes of high API and low sulfur content.

There are several issues that co-exist and it becomes difficult to develop a distinctly clear strategy in resolving these issues. There are metals removal programs that work well under a controlled environment. These products that are also acidic in nature pose corrosion issues and often this limit treat rates. This paper will discuss in detail various issues related with resid quality when processing such type of crude oils in a refinery. There are interesting case studies available where these additives have provided good resid upgrading opportunities due to better performance.

Introduction

It has been recognized globally that crude oil demand is increasing, and as the price per barrel of light and sweet crude increases, the margins for fuel product sale are decreasing. Hence the processing of heavy, difficult crudes is becoming a necessity rather than a desire and the challenge is ever-present. The primary locations from where these crudes originate are Venezuela, Canada, California, China and Africa.

As is evident from Figure 1, the amount of crude oil being produced by different regions across the world is forecasted to increase over the coming years. With Latin America already producing a large quantity of heavy crude oil, it is one of the only regions that is keeping heavy crude oil production constant over the coming years. Production in the Middle East, North America and Africa is predicted to increase significantly highlighting the need for changes in refinery operation and/ or infrastructure.

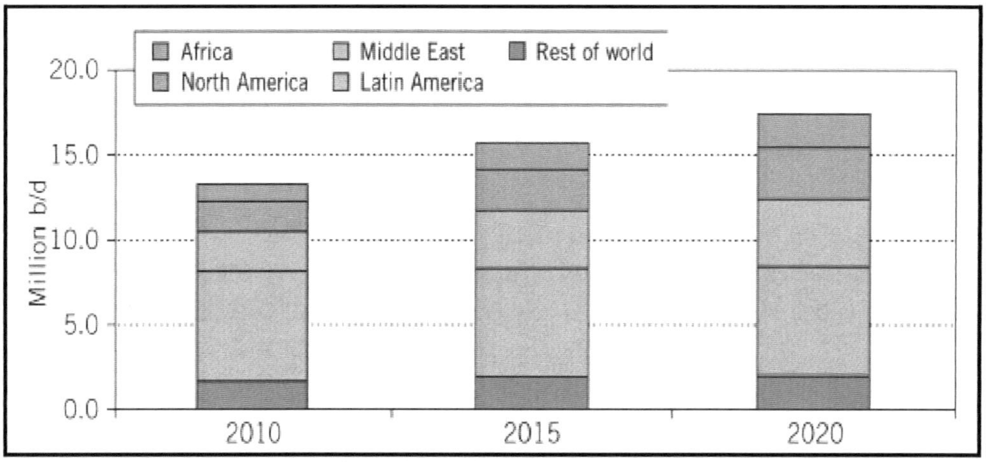

Figure 1: Heavy Crude Oil Production (Oil & Gas Journal, 2010)

The demand for crude oil is increasing as per the consumption data in Figure 2. However, if we compare the amount at which consumption is increasing and production is increasing, demand is surpassing supply.

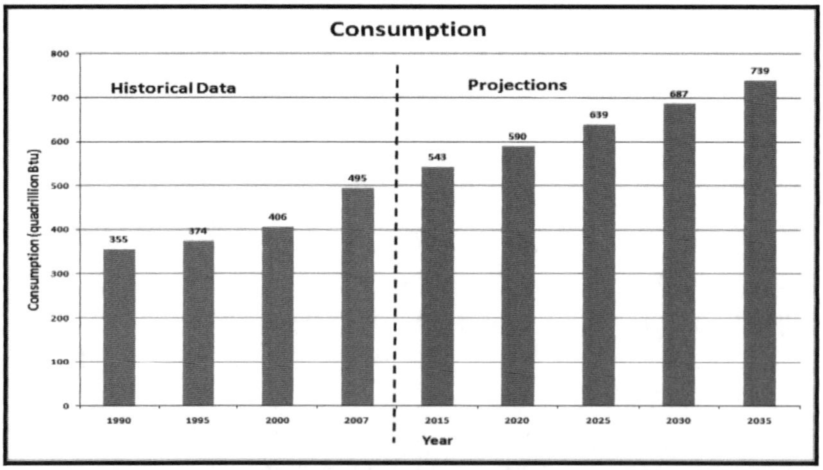

Figure 2: World Marketed Energy Consumption, 1990-2035 (EIA, 2009)

Figure 3 below shows the current energy reserves globally. The Middle East is still the number one in terms of reserves, with the rest of the world combined less than this amount.

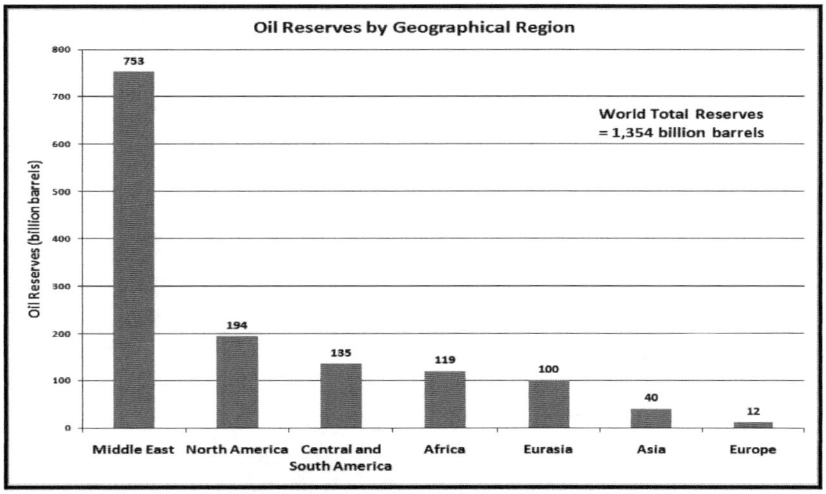

Figure 3: Oil Reserves by Geographical Region (Oil & Gas Journal, 2009)

Increases in heavy crude processing will in turn increase the production of residue which will put a greater emphasis on resid upgrading processes e.g. hydrocracking, catalytic cracking, delayed coking etc. Although over long periods the average API number of all crudes produced has not changed much, the response of OPEC (Organization of the Petroleum Exporting Countries), especially Saudi Arabia to increase production has been possible only through increasing production of heavier, sour crudes (Bacon & Tordo, Crude Oil Price Differentials and Differences in Oil Qualities: A Statistical Analysis, 2005).

Heavy crudes have a number of properties that are common as the main contributors to their characteristics namely asphaltene content, metal content, TAN value etc. The table below comprises the properties of a number of crudes that are heavy.

Table 1: Properties of some Heavy Crudes

Crude	**Country**	**API°**	**BS&W**	**Sulfur**	**Asph.**	**TAN**	**Ni**	**V**	**Fe**
			vol%	%	%	mgKOH/g	ppm	Ppm	ppm
Eocene	Kuwait	18.7	0.25	3.97	6.4	0.20	21	56	60.96
Grane	Norway	18.7	0.33	0.826	1.16	2.2	5.4	16	2.4
Alba	UK	19.3	0.05	1.26	0.7	1.42	8	37	2
Kuito	Angola	19.7	0.30	0.05	1.87	2.41	43.8	38.7	48.5
Marlim	Brazil	19.9	0.05	0.79	4.1	1.21	18.9	30.2	6
Maya	Mexico	20.2	0.50	3.556	10.0	0.06	28	57.9	314.8
Lokele	Cameroon	20.2	0.1	0.45	0.4	2.7	29.3	3.5	16
Doba	Chad	21.1	0.50	0.12	0.06	5.18	9.38	0.48	0.49
Ras Gharib	Egypt	21.5	0.45	3.62	7.5	0.33	33	171.2	218.1
PDVSA 22	Venezuela	21.7	0.70	2.15	5.0	1.46	21	56.6	297.6

Increased Coking

It has been observed over the years that the crude oil imports of heavy crude are ever increasing into the United States. Figures shows that for November 2010, the top five sources of US crude oil imports were Canada, Mexico, Saudi Arabia, Venezuela and Nigeria with Canada being the top importer (US EIA, 2010).

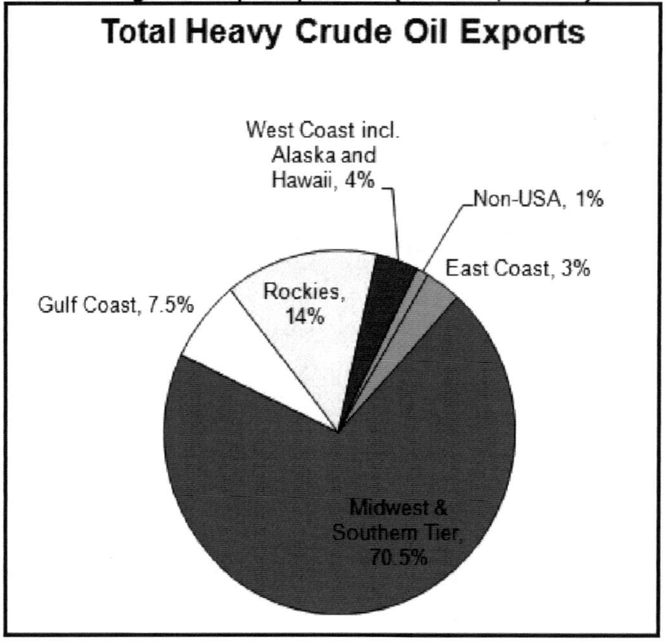

Figure 4: Total Heavy Crude Oil Exports from Canada (Total Crude Exports by Destination, Annual; National Energy Board, Canada, 2009)

A large proportion of the crudes from Canada are from Tar Sands that have properties that make the crudes difficult to process. One of these is a high Conradson Carbon Residue (CCR). This results in an increased coke make, but on top of this, added coke deposition within the furnaces. This leads to high fouling rates and limitations on throughput. Since thermal cracking is one of the first steps of resid upgrading, it is vital to control this fouling at an early stage; prolonging the process reduces profitability.

For this, Dorf Ketal have developed additives that have been proven to reduce fouling rates. They were tested by an international oil major at their Research and Development site using Hot Liquid Process Simulator equipment.

The results are shown in Figure 5. From the tests it is evident that Dorf Ketal Additive No. 3 helped reduce fouling rates by over 65% following a period of four hours of accelerated fouling conditions. Considering the traffic of heavy Canadian Oil, this is a step-forward in helping refiners within the United States and globally.

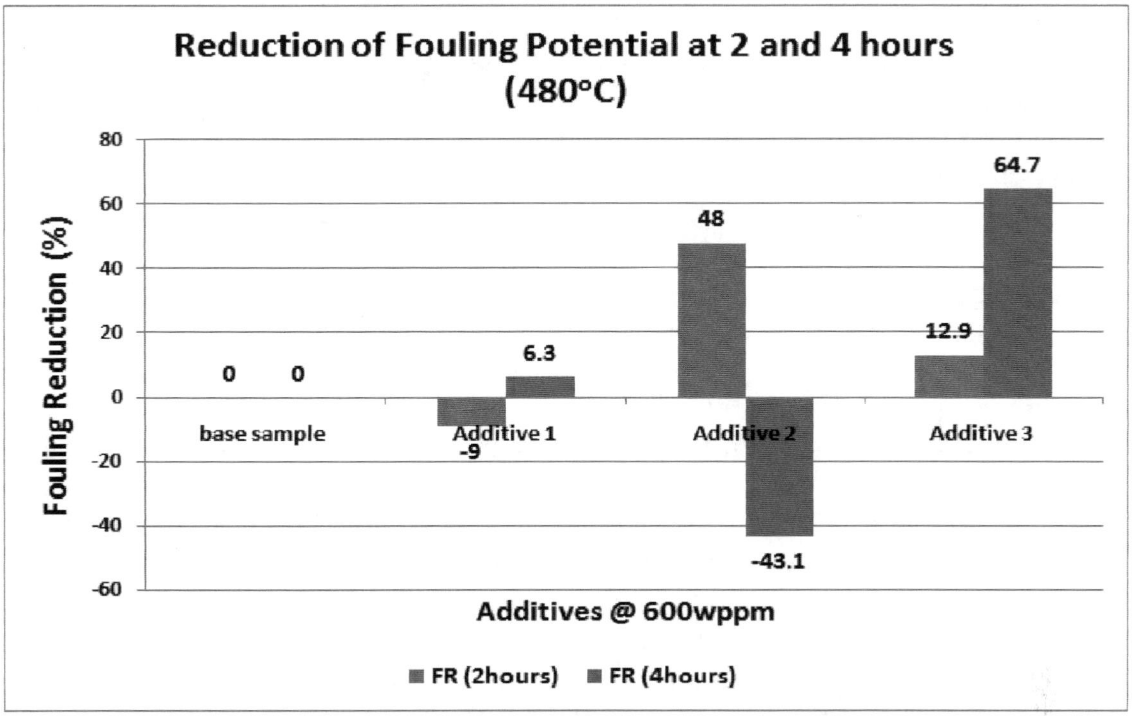

Figure 5: Fouling Rate Reductions with Dorf Ketal Additives

High Metals

There have been a number of strategies executed to determine how best to process a heavy crude with high metal content. The process is made step by step for each unit. The first begins at the blending stage. When processing crudes of a high metal content, one might think that blending with crudes of a much less metallic content would be beneficial so that the total conductivity of the crude reduces. However, studies have shown that when high metal crudes are blended in this way, the conductivity remains and even increases.

Desalter operation is affected by high metal crudes as the conductivity of the blend increases. This has been seen at a number of sites processing crudes of a metallic nature. Crudes containing calcium naphthenates have an added impact on desalter operation which shall be discussed in detail. Crudes with high metal content are shown below.

Table 2: Crudes with high metal content

Characteristics	API Gravity	BS&W	TAN	Ni	V
UOM	°API	%vol	mgKOH/g	ppm	Ppm
Leona 22	21.70	0.32	0.91	62.60	267.50
Leona 24	23.72	0.10	0.85	45.00	185.50
Meri	17.86	0.62	1.51	69.70	287.50
Nowrooz	20.28	0.10	1.09	36.20	137.20
Sorroosh	19.40	0.80	0.44	39.50	119.00

Metal content is usually determined through the summation of nickel and vanadium. Even a small percentage of these crudes within the crude blend increase the total metal content significantly. The active sites of these metal elements result in an increase in conductivity because of which the amperage of the desalter increases. This distresses the desalter causing poor dehydration and increased oil in brine values.

A number of parameters affect desalter performance e.g. crude type, temperature, pressure, mix-valve pressure drop, interface level, wash water quantity, wash water quality and demulsifier. Increase in amperage takes place because of:

- ✓ Increase in metal value of blend
- ✓ Increase in wash water rate
- ✓ Increase in temperature by 2-3°C

For minimizing oil and grease in brine, desalter operation with respect to electric field should be optimum. Voltage makes water droplet coalescence possible (Pruneda, Escobedo, & Vazquez, 2005) hence the voltage within the desalter should be high. However during processing of crudes with high metal content the electric field is diminished due to the conductivity of the metals. This hinders the agglomeration of water molecules since conductivity is dispersed. This is especially critical when the density difference is minimal between heavy crudes and water.

An increase in Desalter Temperature results in increasing conductivity in the desalters; during the processing of high metal crudes there has been cases of an increase in the desalter inlet temperature of around 2-3°C. Increase in desalter temperature and presence of higher amount of metal content in the desalter results in the increase in the amperage value of Desalters.

The iron in the crude oil concentrates in the residual fraction and can affect downstream operations processing the residue. For those refineries making anode grade coke, the contaminant iron can result in product downgrade (Clarke, 2008). For refineries processing atmospheric residue in their fluid catalytic cracking unit, iron contamination can increase the rate of catalyst deactivation.

To mitigate the negative impacts stated above when processing crudes of a high metal content, a metals removal additive has been engineered by Dorf Ketal. The chemical is injected upstream of the desalter mix-valve and has dual-functionality. The first aspect is a metal removing agent; this is a blend of organic acids and because of this property, a corrosion inhibitor provides the second aspect of the additive to protect the desalter lines. The metals will react with the organic acids to form a water soluble salt which is removed along with the brine water. The effect is a significant reduction in metal load on downstream units and catalysts.

At some sites, desalters are installed upstream of resid cracking units to further remove metals from the feed. This is particularly necessary as mentioned for catalyst protection, and hence another location where the metal removing agent can be used.

Calcium Naphthenates in Crude

Calcium naphthenates cause an array of disturbances to normal operations. Some crudes that contain calcium naphthenates are Doba, Kuito, Heidrun, Ciao Fao Dien and Shengli. Calcium naphthenates tend to accumulated at the interface within the desalter. This results in water carryover and increased oil in brine which causes a negative impact on the waste water treatment plant.

There are also concerns within the crude unit with regards to corrosion since most high Calcium naphthenate crudes also hold an increased TAN. The organic acid loading in the overhead system can also increase neutralizer demand and there is an increased risk of high temperature naphthenic acid attack.

There is a chance of catalyst deactivation if resids are fed to the FCC/RFCC or hydrocracking units. If the resids are fed to the coker, then chances are that anode grade coke will be off specification due to the excess metals. Furthermore, new specifications arising for fuel oil have limitations on calcium content as per ISO-8217 regulations.

Table 3: ISO-8217 guidelines for Residual Fuels

Parameter	Unit	Limit	RMA [a]	RMB	RMD	RME	RMG				RMK		
			10	30	80	180	180	380	500	700	380	500	700
Viscosity at 50°C	mm²/s	Max	10.00	30.00	80.00	180.0	180.0	380.0	500.0	700.0	380.0	500.0	700.0
Density at 15°C	kg/m³	Max	920.0	960.0	975.0	991.0	991.0				1010.0		
Micro Carbon Residue	% m/m	Max	2.50	10.00	14.00	15.00	18.00				20.00		
Aluminium + Silicon	mg/kg	Max	25	40		50	60						
Sodium	mg/kg	Max	50	100		50	100						
Ash	% m/m	Max	0.040	0.070			0.100				0.150		
Vanadium	mg/kg	Max	50	150			350				450		
CCAI	-	Max	850	860			870						
Water	% V/V	Max	0.30	0.50									
Pour point (upper) [b], Summer	°C	Max	6		30								
Pour point (upper) [b], Winter	°C	Max	0		30								
Flash point	°C	Min	60.0										
Sulphur [c]	% m/m	Max	Statutory requirements										
Total Sediment, aged	% m/m	Max	0.10										
Acid Number [e]	mgKOH/g	Max	2.5										
Used lubricating oils (ULO): Calcium and Zinc; or Calcium and Phosphorus	mg/kg	-	The fuel shall be free from ULO, and shall be considered to contain ULO when either one of the following conditions is met: Calcium > 30 and zinc >15; or Calcium > 30 and phosphorus > 15.										
Hydrogen sulphide [d]	mg/kg	Max	2.00										
a	This residual marine fuel grade is formerly DMC distillate under ISO 8217:2005.												
b	Purchasers shall ensure that this pour point is suitable for the equipment on board, especially in cold climates.												
c	The purchaser shall define the maximum sulphur content according to the relevant statutory requirements.												
d	Effective only from 1 July 2012.												
e	Strong acids are not acceptable, even at levels not detectable by the standard test methods for SAN. As acid numbers below the values stated in the table do not guarantee that the fuels are free from problems associated with the presence of acidic compounds, it is the responsibility of the supplier and the purchaser to agree upon an acceptable acid number.												

Doba crude has 400ppm calcium on average and with regular desalting hardware and additives, only approximately 15% of this can be removed. This will result in 340ppm calcium being carried downstream which will affect the coke quality, resid fuel oil quality, FCCU catalysts and aggravate fouling. Hence, to capitalize fully on the potential gains from processing Doba, certain amendments need to be made. The following case study discusses Doba processing at a refinery in Asia.

Table 4: Doba Crude Properties

Parameters	**Value**
Gravity, °API	21.1
Specific Gravity	0.927
Sulfur (wt%)	0.12
Nitrogen (ppm)	1890
Pour Point (°C)	-1.5
Acid Number (mgKOH/g)	5.18
Nickel (ppm)	9.38
Vanadium (ppm)	0.48

The main problem to the refiner with Doba crude was the presence of heavy metals such as Calcium and Iron at high concentrations. These metals caused various downstream fouling of the effluent treatment plant and catalyst poisoning of the Resid Fluid Catalytic Cracker and so it became necessary to remove high Calcium coming with the feed during the processing of Doba crude. A treatment program was installed to aid in calcium removal for better downstream operation.

The key to a good treatment program is to treat the desalter with calcium removal agent and a demulsifier to rapidly resolve the emulsion pad / rag layer. A corrosion inhibitor is used to protect low pH areas that develop during the application of calcium removal agents.

- Calcium in residue (atmospheric column bottoms) had a 15% reduction thus Dorf Ketal was able to maintain <20 ppm of Calcium in residue crude
- Decalcification efficiency increased to >85% with the help of Calcium removal agent (CRA) injection
- Brine pH was consistently maintained between safe operating levels of 5.5 to 6.5 as per the graph given below.

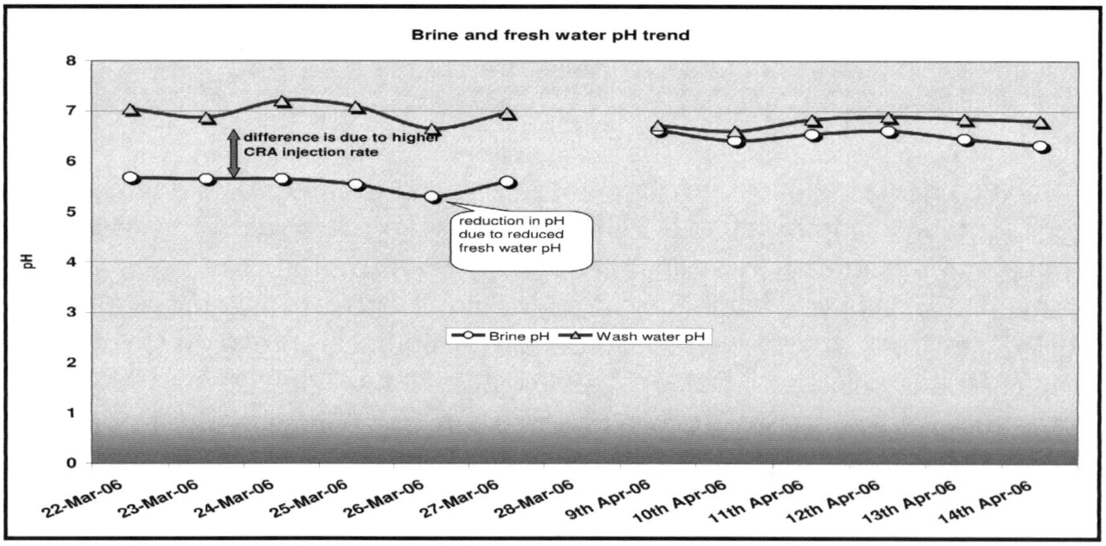

Figure 6: Brine and Wash water pH trend during DK CRA treatment

Prior to the treatment program, a number of upsets were observed in the desalter resulting in water carryover with desalted crude, reflected by sudden rise in amperage and drop in voltage. Also, a large quantity of oil under-carry with desalter effluent water (1.5 – 4%) was seen. Such desalter upsets adversely affect system reliability as corrosion rates increase in the crude column overhead section due to salt carryover and also the load on the downstream effluent treatment plant increases due to large quantity of oil under-carry with desalter effluent water which requires higher treatment rates at the effluent treatment plant.

Based on the above observations, Dorf Ketal recommended the use of demulsifier during 2nd stage of CRA treatment. Dorf Ketal demulsifier was a tailor made product keeping in mind the above two requirements to process high metal/ TAN crudes and avoid sludge or oil carryover. Various stages of Portable Electric Desalter (PED) studies had also been conducted to determine the effect of our desalting aid on current crudes being treated at the refinery. Water-oil emulsion resolving capability and sludge separation rates in terms of reduced COD of effluent water improved.

Figure 7: Brine Quality

Effluent water quality with 7 % Doba before Dorf Ketal CRA and Demulsifier addition and effluent water quality with Dorf Ketal CRA and desalting aid technology. After injection, there was clear water indicating no sludge generation during Doba processing.

Injection of CRA without corrosion inhibitor can cause corrosion in the desalter internals and in the wash water piping. Dorf Ketal CRA program contains corrosion inhibitor to handle the corrosion issues due to acidity generated in the wash water and brine. A good monitoring system is also a key to success in controlling the corrosion in this loop. Dorf Ketal suggested advance corrosion probes which can give real time measurement of corrosion at any point of time during Doba treatment. This helps to proactively control corrosion rates in brine heat exchangers and wash water injection.

The Corrosion Inhibitor performance was being monitored using special corrosion probes placed in CRA injection line and desalter effluent water line. Corrosion probe readings were taken before and after the Doba treatment with Dorf Ketal CRA. The corrosion rate has been maintained at 1.88mpy in both the CRA injection line and brine water line i.e. within the target corrosion rate of <3mpy.

The refinery is now able to handle crudes with high calcium naphthenate content whilst keeping their hardware in good condition. Furthermore, crudes like Doba are $5-$10 cheaper per barrel than other crudes so the margins would increase.

High Calcium Naphthenates and TAN

Another property of heavy crudes is often a high total acid number (TAN). The naphthenic acids within the crude can cause severe corrosion at high temperatures. This phenomenon can also occur when crudes high in calcium naphthenates are processed, as they dissociate to naphthenic acid. High Tan crudes are projected to take an increasing share of the rising volume of global oil production and are sold at discounted prices compared to the marker crudes (Bacon & Tordo, Crude Oil Price Differentials and Differences in Oil Qualities: A Statistical Analysis, 2005).

One such crude is Ciao Fao Dien (CFD) which was processed with the help of a calcium removal agent at a refinery in Asia. The crude properties are given below.

Table 5: CFD Crude Properties

Parameter	Value
Density (g/cm3)	0.9199
API (°)	22.3
Total Sulfur (%wt)	0.18
BS&W (%vol)	0.10
TAN (mgKOH/g)	2.3
Calcium (ppm)	150
Viscosity @ 40°C (cSt)	65.42
Filterable Solids (ptb)	16.0
Conductivity @ 1000ppm in Isopar L (ps/m)	5000
Salt (ptb)	4.0

Organic acid contained within Inhibited Calcium Removal Agent (ICRA) will sequester Calcium Naphthenate. Calcium Naphthenate will be hydrolyzed into calcium and Naphthenic acid. Calcium then will react with acid group of ICRA and will form water-soluble salt, which can be easily removed from desalter. Reaction kinetics will favor the reaction of calcium and the acid group of ICRA compared to Naphthenic acid. Naphthenic acid is soluble in oil phase and hence will go along with the crude to downstream unit, which is protected by high temperature corrosion inhibitor (HTCI). Low pH is expected upstream of acid injection; this was taken care of by water soluble inhibitor present in ICRA.

The main focus area was to get maximum calcium removal without affecting desalter performance with minimal dosage. Try-line status was checked twice in one shift (8 hours) to check for impact of CFD in desalter performance. This has helped tackle a few small upsets during processing which was taken care by available field team. The maximum % of CFD crude processed during the period was up to 5%.

Please find attached table showing performance of Dorf Ketal ICRA during CFD processing.

Table 6: During CFD processing

Date	CFD in Blend	Ca in Feed	Ca Removal Efficiency		Salt Removal Efficiency		Dehydration Efficiency		Oil in Brine		COD
	%	ppm	%		%		%		Ppm		Ppm
Desalter Set			1	2	1	2	1	2	1	2	Brine-ETP
KPI			≥70%		≥95%		≥70%		≤100ppm		<1200ppm
05-Sep	3.1	5.8	93.10		96.64	97.01	95.04	95.05	40.0	37.0	929.0
06-Sep	3.1	8.5	80.00	90.59	97.07	96.22	94.10	95.10	59.0	50.0	920.0
07-Sep	5.0	8.4	76.19	78.57	96.20	95.23	95.04	95.05	50.0	41.0	912.0

Table 7: Brine Water sample analysis

Desalted Crude	Na (ppm)	Ca (ppm)	pH	% CFD in Blend
Sample 1	435.25	103.04	7.12	2.60
Sample 2	361.40	97.26	7.87	3.00
Sample 3	88.80	16.04	6.76	2.60
Sample 4	120.15	27.09	6.81	3.00
Sample 5	340.86	78.57	6.95	2.60
Sample 6	149.91	30.59	7.10	2.60
Sample 7	66.35	12.13	7.17	3.00

Table 8: Crude Analysis

Sample	Ca (ppm)	Ca Removal Efficiency (%)	% CFD in Blend
Raw Crude	22.00	-	5.00
Desalted Crude (1)	3.00	86.36	5.00
Desalted Crude (2)	1.00	95.45	5.00

It was seen that with ICRA, the refiner achieved more than 75% Ca removal and no adverse effect was observed in desalter downstream system or desalter performance.

High TAN

As mentioned earlier, a further property of heavy crudes is usually a high total acid number (TAN). TAN includes a number of naturally found components e.g. mineral acids, phenols, hydrogen sulfide, mercaptans and carbon dioxide but within TAN it is specifically the naphthenic acids that are corrosive at high temperatures, hence there has been disputes regarding the reliability of TAN for crude discounts. It is known that crudes are priced based on their properties. 'World Bank' correlates discount as a function of API gravity, sulfur content and TAN. Currently the categories are such that <0.5 TAN holds no discount and above this, each point of excess TAN lowers the price of crude by $0.051 per dollar of Brent. The excess TAN is defined as TAN of crude oil minus TAN of Brent (0.07mgKOH/g). Given that the increase in demand has been mainly driven by transportation fuel, the higher proportion of heavier, sour crudes on the market has increased the cost different between the light and the heavy crudes and also between sweet and sour crudes (Bacon & Tordo, Crude Oil Price Differentials and Differences in Oil Qualities: A Statistical Analysis, 2005).

The high temperature corrosion inhibitor allows processing of this discounted crude by protecting the metallurgy of the downstream units. At high temperatures, the naphthenic acids attack the metal surface. Naphthenic acid corrosion occurs from 200°C - 420°C since below this temperature the naphthenic acids are too heavy to react and above this temperature they break down to lighter acids. Specifically in jet fuel and kerosene, naphthenic acids can be problematic.

Current solutions are:

- ✓ Bring the average TAN to <1
- ✓ Continuous injection of high temperature corrosion inhibitor
- ✓ Upgrade material of construction to a higher chrome and/or molybdenum in areas susceptible to high temperature corrosion.

The patent-pending high temperature corrosion inhibitor of Dorf Ketal, has been used successfully at sites where the TAN number is high, and together with this, the sulfur content of the blend is low. Many additives rely on the refiner keeping the sulfur content at a higher level to contribute to metallurgy surface protection, but when the sulfur content is lower than usual, the chemical will not perform to its consistent standard. The case study below discusses the high temperature corrosion inhibitor application in further detail.

A refinery was suffering from severe metal loss in its lines made of the following metallurgies:

- Atmospheric residue line till entrance of vacuum furnace is AISI 317L
- Furnace Tubes: P5 (5% Cr)
- Furnace outlet till vacuum tower: AISI 316

Following high corrosion rates with an earlier treatment program, the refiner decided to use Dorf Ketal's HTCI. The flow was kept continuous and a constant outlet temperature was maintained (414 to 418°C). Figure 8 shows how the mass loss of the P5 metallurgy furnace tubes reduced significantly with Dorf Ketal's treatment program.

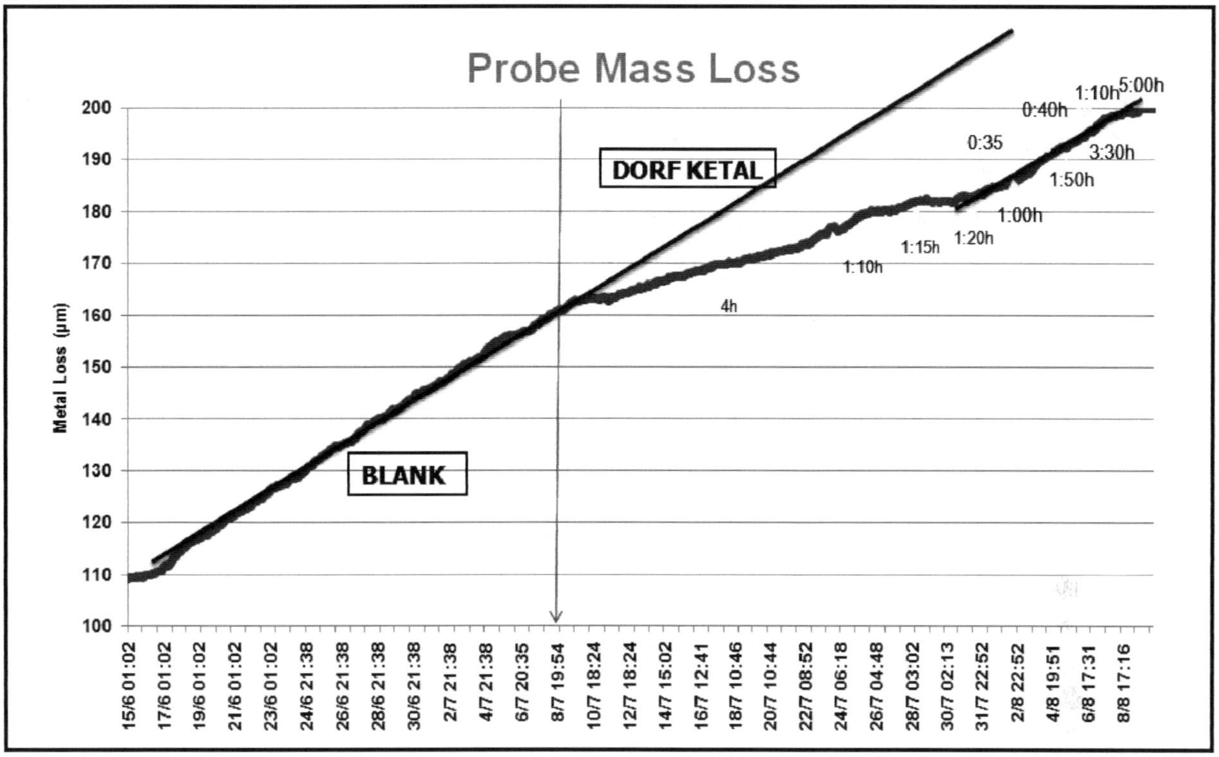

Figure 8: Improved performance with High Temperature Corrosion Inhibitor

High Phenols in Desalter Wash Water

When processing heavy crudes, it is likely that the amount of phenolic water being produced and then used internally will increase. Phenolic water contains cracked components as well as ammonia and H_2S which is usually stripped before use in other units within the site e.g. as desalter wash water. These properties contribute to a high pH which is known to make stable oil and water emulsions. This stable emulsion results in:

- Increased salt slippage along with desalted crude
- Increased water carry over due to poor emulsion resolution
- Increased Oil and grease under carry along with brine water

It is recommended that the desalter wash water be maintained from pH 5.5-6 since the higher the wash water pH, the higher the solubility of ammonia in crude during desalting operations. The pH and NH_3 content of the wash water can be minimized by further stripping. However, without this operation, wash water that is carried along with the desalted crude is bound to carry NH_3, which gets finally liberated in the crude column overhead. This NH_3 content results in the NH_4Cl deposition in the crude column overhead system due to reduced margin between the crude column top temperature and NH_4Cl desublimation temperature.

Hence, desalter acidification is an effective means of resolving difficult emulsions in desalters due to increased pH levels in the desalter wash water as seen in Figure 9. Acidification will help convert oil coated filterable solids into water soluble compounds and can be helpful in reducing solid loading at ETP. Removal of iron sulfide to specified target levels is feasible as long as the stoichiometric demand of organic acid is met.

However with 6 pH of acidified wash water, there will not be sufficient free acid available to break FeS bond. Thus organic acid will not react with the carbides, carbines, coke particles, silica or FeS present in the crude. Acidification of the desalter's inlet wash water to an appropriate pH, improves emulsion resolution, thus reducing water and solids carryover along with desalted crude.

The pH of the desalter wash water can be controlled within required limit with the help of injection of Organic Acid in the wash water going to desalter.

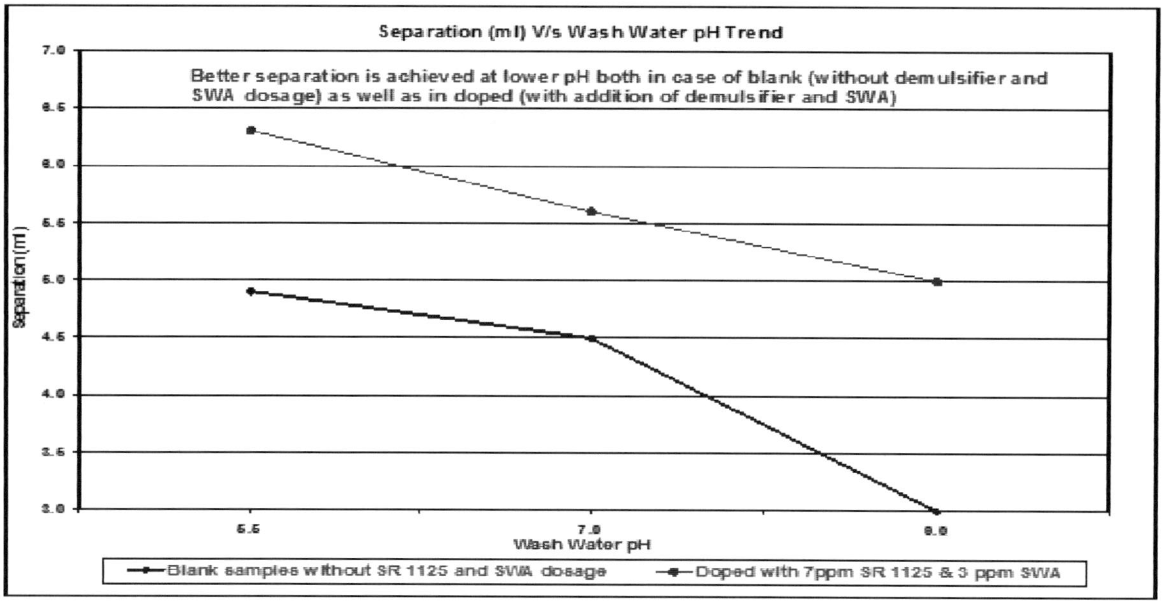

Figure 9: Effect of pH on Separation

Table 9: Results of Acid Injection into Desalter Wash Water

	Before Acid Injection	After Acid Injection	Difference	% Change
Brine pH	8.31	7.2	1.11	13.36
BS&W O/L	0.26	0.15	0.11	42.31
Salt O/L	0.9	0.25	0.65	72.22
Oil in Brine	34.376	28	6.376	18.55
Effluent pH	8.41	7.14	1.27	15.10
Phenols	79	55	24	30.38

Heavy Crudes with High Nitrogen and Salt Content

Often heavy crudes contain a high amount of nitrogen (<1800ppm) and when coupled with high salt content as feed to a FCCU/ RFCCU, this is destined to cause problems. At a refinery in the Middle East, this issue was seen and caused restrictions in processing. The incoming nitrogen and sodium chlorides remained with the vacuum residue feed to the cracking unit and upon cracking, formed ammonium chloride salts. Ammonium chloride salts deposit within the main fractionator on the trays and cause plugging and corrosion issues leading to an increase in pressure drop.

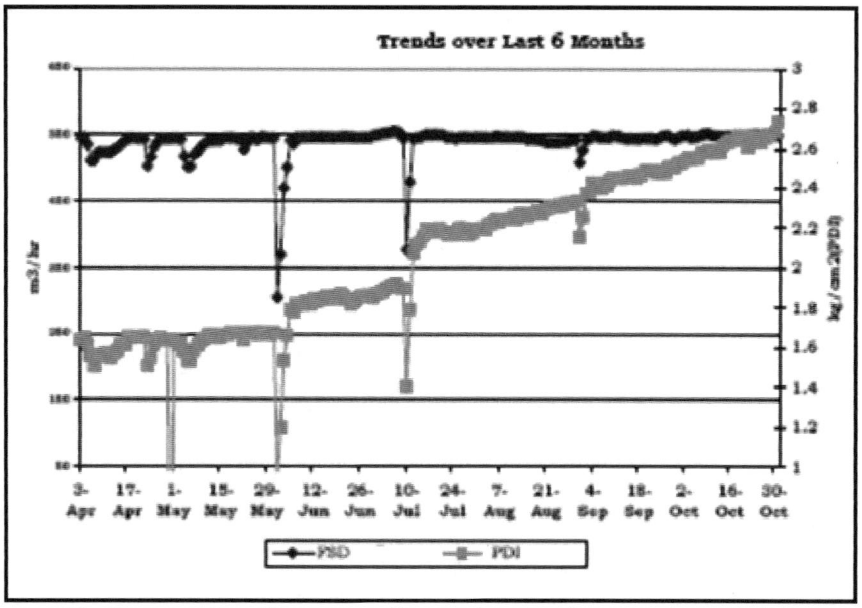

Figure 10: Increasing Pressure Drop due to Ammonium Salt Deposits

A conventional method to remove the salts is through water washing of the column or lowering the top temperature of the tower to condense water on the upper trays. The salts are soluble in water and are removed one of the side draws. While this method is effective in removing the water-soluble salts, water washing incurs additional operating costs and potential difficulties. These costs include reduced throughput and re-processing costs of off-specification products produced during water washes both of which result in huge revenue loss.
As a solution, Dorf Ketal has engineered an additive for dispersing solid salt particles. The additive inhibits the build-up of these deposits by keeping them dispersed when used on a continuous basis. Keeping the salts in dispersed phase will prevent the salt particles from agglomerating.

Conclusion

It is evident that heavy crude processing is not trouble-free for the majority of refineries constructed over the years. Newer refineries can incorporate changes to allow processing of certain types of crude (e.g. upgrade metallurgy for high TAN) but otherwise it is an uphill battle to cater to the ever-growing heavy crude market that is prevalent and will further develop in quantity. Hence the need for additives that allows refiners to process heavy crudes with relative ease and help increase profit margins.

References

1. Bacon, R., & Tordo, S. (2005). *Crude Oil Price Differentials and Differences in Oil Qualities: A Statistical Analysis.* Washington, D.C.: Energy Sector Management Assistance Program.

2. Clarke, B. (2008). Impact of Feed Properties and Operating Parameters on Delayed Coker Petcoke Quality. *2nd Asia BBTC.* Kuala Lumpur: Foster Wheeler.

3. EIA. (2009, November). *World Marketed Energy Consumption, 1990-2035: History: EIA, International Energy Statistics database.* Retrieved January 2011, from www.eia.gov/emeu/international.

4. Oil & Gas Journal. (2009). "Worldwide Look at Reserves and Production". *Oil & Gas Journal , 105* (48), 22-23.

5. Oil & Gas Journal. (2010). Crude Demand to Increase Feed Quality. *Oil & Gas Journal , 108* (46).

6. Pruneda, E. F., Escobedo, E. R., & Vazquez, F. J. (2005). Optimum Temperature in the Electrostatic Desalting of Maya Crude Oil. *J. Mex. Chem. Soc , 49* (1), 14-19.

7. *Total Crude Exports by Destination, Annual; National Energy Board, Canada.* (2009). Retrieved January 30, 2011, from National Energy Board: http://www.neb.gc.ca/clf-nsi/rnrgynfmtn/sttstc/crdlndptrlmprdct/ttlcrdlxprtdstntn-eng.html

8. US EIA. (2010, November). *Independent Statistics and Analysis.* Retrieved January 30, 2011, from US Energy Information Administration: http://www.eia.doe.gov/pub/oil_gas/petroleum/data_publications/company_level_imports/current/import.html

A NEW CONTINUOUS COKING PROCESS

Douglas Sullivan, P.E.
Lidcon, Inc., Spring, Texas

And

James Speight, Ph.D., DSc
CD &W, Inc. Laramie, Wyoming

Abstract

A new process for continuous coking and decoking of heavy oils and vacuum residues is described. The process can produce higher liquid yields and lower coke production. The reactor configuration enhances the liberation of volatiles and increases the speed of carbonization of pre-coke materials.

Introduction

Processing vacuum residua and other heavy feedstocks to recover volatiles and to produce petroleum coke have been important aspects of oil refining for over 70 years. Today modern refineries employ one of two types of coking processes, either fluidized bed coking or delayed coking (1, 2). This is likely to continue into the foreseeable future and of the 130+ coking units in service worldwide, approximately 90% are delayed cokers.

Both delayed coking and fluid coking types of coking processes have received many upgrades over the years, but they remain mature technologies (3). Although optimization of the coking process has been advocated and put into practice (4, 5, 6), no new coking process has emerged that is decidedly superior to these existing coking technologies. With the introduction of the new reactor design as described here, the advent of a new era of coking technology could be near.

The new coking process (Figure 1) described herein, accepts heavy feedstocks (including vacuum residua) from a conventional residuum heater in a coking plant and continuously discharges volatiles and dry petroleum coke particles. The process promotes a rapid recovery of volatiles from the feedstocks thereby increasing the yield of condensable volatiles with a concurrent reduction in coke formation. Moreover, the new process also promotes rapid carbonization of any produced coke thereby producing a coke of uniform composition and uniform size of the coke particles.

Process Flow

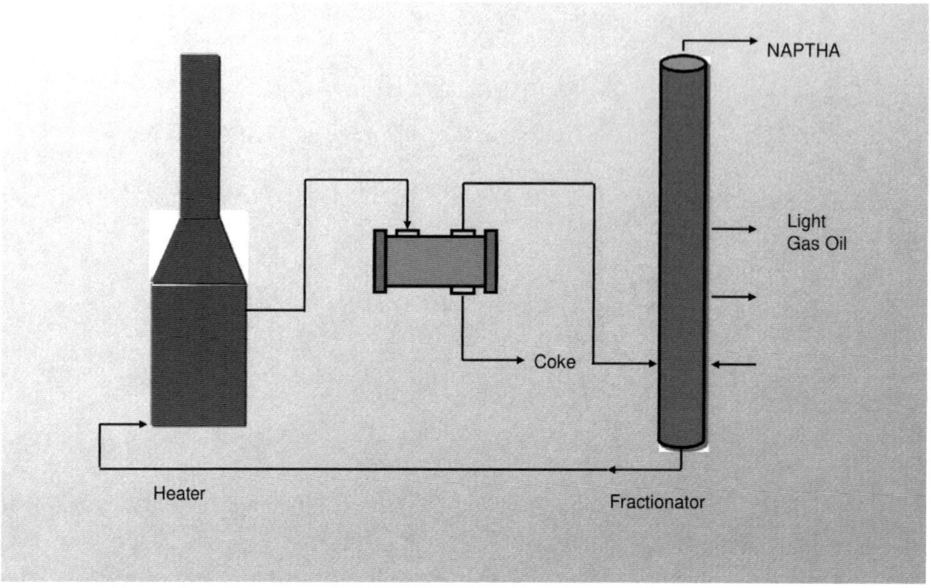

Figure 1: Simplified Flow Scheme of New Coking Process

It has been demonstrated (8) that heated thin films of Athabasca vacuum residue (AVR) produce a higher rate of recovery of volatiles and a lower production of coke than would be possible when processing large agglomerated masses as happens in delayed coking. The new process takes advantage of this phenomenon by using a kneading and mixing action to continuously expose new residuum surface to the vapor space producing a faster and more complete removal of volatiles from the residuum and the produced petroleum coke. It was also demonstrated on AVR that as much as 5% additional liquid can be recovered and the amount of coke production is reduced and that not only are more valuable liquids recoverable, the liquids are likely to be richer in middle distillates. Laboratory coking tests (3) on Arabian heavy crude that had short vapor residence coupled with a residuum residence time long enough to produce solid coke, increased liquid yields an average of 7.9 wt. %.

Process Description

The reactor/devolatilizer receives feed of heavy oil or residuum from a fired heater at a temperature of 850–1000 °F, similar to the heater discharge temperature from the heater in a delayed coke plant. The feed flows directly into one end of the horizontal reactor/ devolatilizer. A kneading/mixing action inside the reactor/devolatilizer causes new surfaces of the residuum mass to be continuously exposed to the gas phase several times per second. This creation of new liquid-gas interface enhances the rapid mass transfer of volatiles into the gas phase similar to what happens with thin films.

The liberated volatiles are then rapidly cooled in a quench vessel to retard possible degradation by secondary cracking. With the rapid reduction of volatiles content in the residuum mass, the carbonization reaction rates are accelerated, enabling continuous and rapid production of solid petroleum coke particles. The rapid release of volatiles also reduces the opportunity for the volatiles to react with the hot residuum and become locked in the coke. In order to reduce secondary cracking of the volatiles after they are released into the vapor phase, the vapor flows immediately (0.1-0.2 seconds) into the quench section after leaving the reactor. In the quench section the temperature of the vapor is reduced 125-175 ^{o}F by countercurrent flow against gas oil (or residuum feedstock) liquid droplets in a low pressure drop vapor-liquid contactor. The gas oil quench liquid also scrubs entrained solids out of the vapor stream and reduces downstream fouling.

Concurrently with the carbonization reactions and the formation of coke, some cracking of side chains off the larger molecules likely occurs (1, 2) producing lower boiling fractions that make a rapid escape to the quench vessel. Volatiles may also be generated after the solid coke is formed but, in the delayed coking process, many of these late forming volatiles remain trapped in the coke. The new process continuously breaks up the coke particles and promotes the release of these late forming cracked volatiles, allowing them to escape into the vapor phase.

Reactor Size

Because the reactor/devolatilizer (Figure 2) occupies less space than delayed coking drums, it can be easily retrofitted into existing delayed coke plants. The reactor can be mounted at grade or on an elevated platform.

Assuming a 5 minute residence time, the reactor in Figure 2 can process 2,000 Tons/Day (T/D) of solid coke. If the feed stream to this coker has a Conradson carbon residue (CCR) of 20-25, the residuum feed rate could be approximately 40,000 barrels/day (B/D). If the feedstock is a tar sands residuum with a CCR of 50, the feed rate would be approximately 20,000 B/D

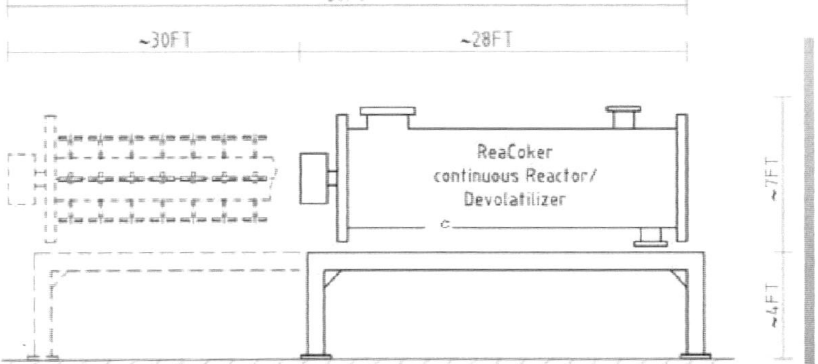

Figure 2: Dimensions of a Continuous Coking Reactor

Processing Considerations

Processing of hot residua can present potential operational hurdles. Some of these potential difficulties could be high viscosity feedstocks and thermal products, bearing and seal life, foaming, and a long period to complete coke formation. A robust design to successfully deal with these issues is currently under development.

High Viscosity

As the residuum and solids travel the length of the reactor, the volatile content decreases while the coking reactions proceed. The viscosity of the residuum mass increases as it travels and becomes a highly viscous plastic phase, eventually becoming solid coke. Athabasca residuum coking test (9) reached a maximum viscosity near 100,000 cp.

The proposed reactor/devolatilizer is of the same design genre as reactors and devolatilizers used in the chemical processing industries. In some of these services, viscous chemical wastes and elastomers are being processed at viscosities up to 15,000,000 cp. Given this experience, it is unlikely that viscosity will present a problem in the reactor.

Bearing and Seal Life

To deal with high temperature, the bearings and seals will be mounted outboard of the internal processing volume. A liquid cooling circuit for the bearings and a continuous purge flow for the seals protects against wear and damage. Since the rotor speed is only 25-30 rpm, long life is expected for the bearings and seals.

Foaming

Some heavy feedstocks may be more likely to exhibit foaming than others. It is known that falling pressure in the coker is most responsible for the generation and propagation of foam. The new continuous coker operates at a constant pressure therefore it is less likely to experience foaming. If foaming were to occur, the action of the rotating members will break up foam and, if necessary, antifoam agents could easily be added in the quench section.

Reaction Time

There is frequently a misconception about the time required to achieve complete liquid-to-solid conversation in cokers. Laboratory tests heating thin films of Athabasca vacuum residuum showed that the reaction time to convert all the liquid phase to solid coke was quite small and at 936 ^{o}F complete conversation of resid to solid coke was achieved in less than 30 seconds (8, 9, 10).

The new reactor/devolatilizer may take slightly longer to achieve complete solid coke production. Therefore, the current design is based upon a five minute retention time of solids, but the reactor can function at higher throughput rates if the coke formation is completed earlier. The reactor can continue to perform at below design feed rates all the way down to zero feed rate. The turndown ratio is essentially infinite.

New Process Benefits

The new process delivers a number of benefits compared to delayed coking. The substantial benefits that accrue to the user of the process are yield improvement, lower capital cost, energy savings, better environmental stewardship, and safety.

Yield Improvement

The most important benefit of the new coking process is the improvement of liquid yield. For certain residuum it may be possible to achieve a liquid yield improvement near 10%. But, it only takes a 4% improvement in liquid yield to provide a 100% annual return on a retrofitting investment. Figure 3 shows how the quench chamber is positioned on top of the reactor/devolatilizer. The short gas path minimizes the time to quench the hot gas and preserve the quality of the resulting liquid yield improvement.

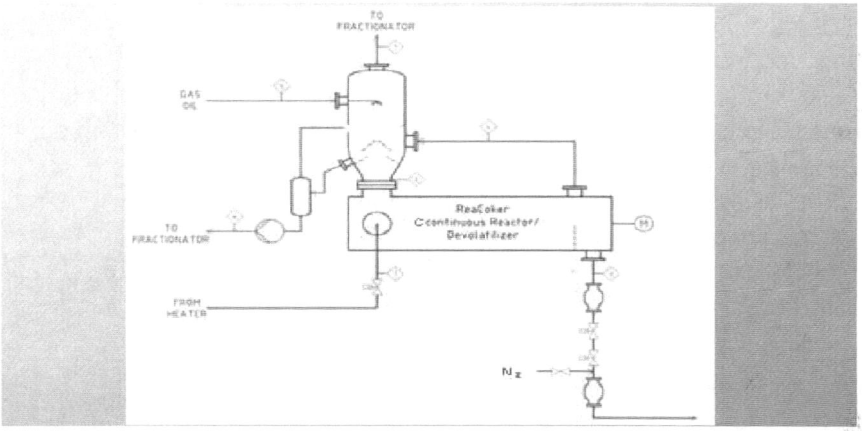

Figure 3: Reactor/Devolatilizer and Quench Chamber

Capital Cost

Because of the elimination of several parts of a delayed coking plant, the estimated capital cost of a coke plant using the new process will be 75-85 % of the cost of a new delayed coke plant. The estimated cost to convert an existing delayed coke plant to the new process is $600-900 per barrels per day (B/D) of coker feed rate capacity for residuum feedstocks of 25-30 CCR.

Energy Savings

In addition to the recovery of additional and more valuable volatiles than in delayed coking operations, energy savings by the new process is also considerable. The consumption of energy and utilities is less than the delayed coking process because steam, cutting water, and quenching water are not required. And since there is no quenching of the coke in the new process, energy from the hot coke can also be recovered. The overall annual saving in energy for a refinery using the new process instead of the delayed coking process can be $100-200/B/D of coker feed, depending on the size of the coker and the CCR of the residuum feedstock.

Better Environmental Stewardship

The new process is continuous therefore it is never opened to the atmosphere. There is no cutting procedure as in the delayed coking process where high-pressure water is used to cut the coke out of the open drums. No vapor and no coke particles are released into the atmosphere by the new process.

Safety

Fewer workers are required to operate the new process coke plant. And since the process is continuous and not open to the atmosphere, workers are not exposed to coke plant atmospheric contaminants. Whereas, delayed coking plants have both solid carbon particles and dangerous vapors in the coke plant atmospheric environment, the potential of coker fires and explosions is virtually eliminated in the new process.

Summary

The new continuous coking process gives oil refiners a new technology for treating residuum that can increase recovery of valuable hydrocarbons and make the coker plant more efficient and easier to operate.

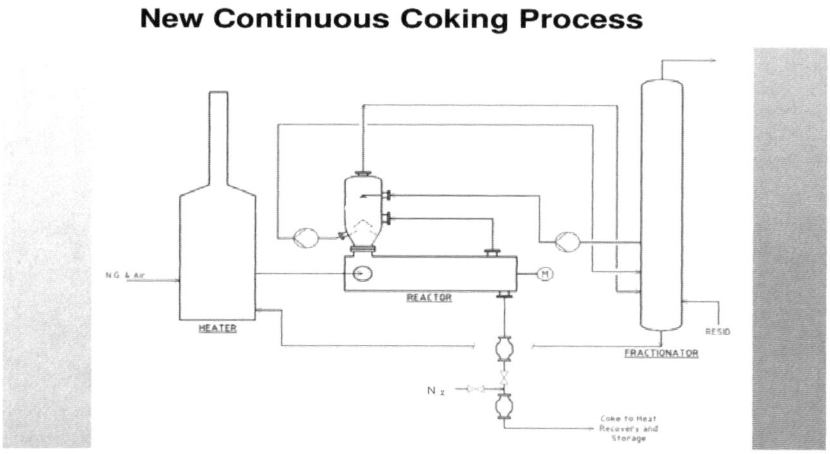

Figure 4: Flow Scheme of New Coking Plant

References

1. Speight, James G. (2007), The Chemistry and Technology of Petroleum, 4th Edition, CRC Press, Taylor & Francis Group, Boca Raton, Florida.
2. Speight, James G. (2011), The Refinery of the Future, Gulf Professional Publishing, Elsevier, Oxford, United Kingdom.

3. Wiehe, Irwin A. (2008), *Process Chemistry of Petroleum Macromolecules*, CRC Press, Boca Raton, FL.
4. Elliott, J. D. (2003), "Optimize Coker Operations," *Hydrocarbon Processing*, Sept. 85-90.
5. Elliott, J. D. (2004), "Fine-Tune Your Delayed Coker: Obstacles and Objectives," *Hydrocarbon Processing*, Sept. 83-90.
6. Elliott, J. D. (2005), "Delayed Coking: Recycle & Zero Recycle Concepts and Considerations," *World Refining*, Jan./Feb.16-21.
7. Aminu, M. O., Janet A, Elliot, William C. McCaffrey, C., and Murray R Gray (2004), "Fluid Properties at Coking Process Conditions," *Ind. Eng. Chem. Res.*, 43, 2929-2935.
8. Gray, Murray G., Tuyet Le, William C. McCaffrey, Franco Berruti, Suresh Soundararajan, Edward Chan, Iftikhar Huq, and Clint Thorne (2001), "Coupling of Mass Transfer and Reaction in Coking of Thin Films of an Athabasca Vacuum Residue," *Ind. Eng. Chem. Res.*, 40, 3317-3324.
9. Gray, Murray R., Zisheng Zhang, William C. McCaffrey, Iftikhar Huq, Lisa Boddez, Zhenghe Xu, , and Janet A. Elliott (2003), "Measurement of Adhesive Forces during Coking of Athabasca Vacuum Residue," *Ind. Eng. Chem. Res.*, 42, 3549-3554.
10. Gray, Murray R, William C.McCaffrey, Iftikhar Huq, and Tuyet Le (2004), "Kinetics of Cracking and Devolatilization during Coking of Athabasca Residues," *Ind. Eng. Chem. Res.*, 43, 5438-5445.
11. Hanson, D. W., N.P. Lieberman, and E.T .Lieberman (1999),"De-entrainment and Washing of Flash-Zone Vapors in Heavy Oil Fractionators," *Hydrocarbon Processing,* July, 55-60.

(64d) Real-Time Optimization Strategies for the Delayed Coking Unit

Peyton C. Richmond and Chandra K. Soni, Chemical Engineering Department, Lamar University, Beaumont, TX

The Delayed Coker Unit (DCU) stands as one of the most profitable units in the refinery due to high profit margins on heavy crude oil refining. However, because of the semi-continuous nature of this upgrading process, the DCU separation process is perturbed frequently. Thus, this unit has a very complex relationship between the controlled variables and the manipulated variables, and it can be very difficult to control and optimize.

In this paper we study the control and optimization of a typical DCU obtained from the literature. The unit is modeled using a dynamic simulation package and it is optimized using an equation oriented Real-time Optimization (RTO) package. The DCU dynamic simulation is configured to run under the control of an industrial Distributed Control System (DCS) in order to provide a realistic platform to study the associated optimization problem.

There are a number of recycle streams including a small part of Coke Drum vapors that get recycled into the fresh feed. This internal (hidden) recycle depends on the effect of the other recycle streams and greatly affects the plant economics. In a high capacity DCU plant, small changes in the recycle rates can cause a huge financial penalty, so it is important to both maintain optimum recycle rates and keep the products within specifications.

See more of this Session: Heavy Oil and Bottom of the Barrel Processing
See more of this Group/Topical: Topical 7: 14th Topical on Refinery Processing

Chemistry and Engine Performance of Diesel Fuels Derived from Oil Sands and Oil Shale Crude

Bruce Bunting, Oak Ridge National Laboratory, Knoxville, TN

ABSTRACT

Over the past 5 years, ORNL has performed engine tests with approximately 100 diesel range fuels in a homogeneous charge compression ignition (HCCI) configuration, including fuels derived from oil sands and oil shale crudes. Overall, the fuels varied in cetane, boiling point distribution, and chemical composition, and some contained nitrogen, oxygen, or olefins. Some of the fuels met the main specifications for ASTM #2 diesel fuel, but others were blended to go beyond the specifications in order to gain a wider and clearer understanding of fuel chemistry and property effects. All fuels were run in an HCCI engine at 1800 rpm, in the load range of 2.5 to 4.5 bar IMEP, using intake air heating for combustion phasing control, and at a compression ratio of 10.6. The engine response to fuel variables has been analyzed statistically. Generally, the engine responded well to fuels with moderate cetane, lower specific gravity, lower T90, and lower aromatics.

FUELS EVALUATED

Forty-seven fuels were selected from the 100 fuels tested for further analysis for this report, divided into 4 groups. This data has been published previously in separate studies (ref. 1-7). These references provide more information about the individual fuels, how they were obtained, and also serve as acknowledgement of the other researchers who were involved in the procurement and evaluation of the fuels. This paper is a new analysis of data combined from these papers with a particular focus on fuels derived from heavy oil in the diesel range. The 1st group of selected fuels includes seven #2 ULSD diesel fuels meeting ASTM specifications and providing a baseline for the study. The 2nd group includes 16 #2 ULSD fuels which were blended to go beyond ASTM specifications for boiling points, cetane, and aromatics to provide further study of these variables. The 3rd group is composed of 17 fuels derived from 100% oil sands crude, and includes both finished fuels and refinery intermediates in order to study a broad range of chemistry and properties. The 4th group includes 7 narrow cut fractions covering diesel boiling range, derived from oil shale crude, and which were partially hydrotreated to retain olefin, oxygen, and nitrogen content. These four groups of fuels were obtained, analyzed, and evaluated over the time period of 2008 to 2011 and the chemistry and property analysis for the fuels was performed by a number of laboratories. Table 1 summarizes the range of properties and chemistry for these fuels, and the reader is referred to the original references (1-7) for more detail. Starred values in this table were assumed, not measured, based on normal refinery practice of excluding olefins and nitrogen- and oxygen-containing species from commercial diesel fuel.

Table 1, Range of properties and chemistry for the 4 fuel groups included in this report.

FUEL GROUP	cetane	T10, C	T50, C	T90, C	total saturates, vol%	mono aromatics, vol%	poly aromatics, vol%	total aromatics, vol%	oxygen, wt%	iodine number	nitrogen, ppm	specific gravity
conventional #2	44.8 / 73.0	191 / 212	246 /261	301 / 324	61.7 / 99.1	0.9 / 23.3	0.0 / 15.0	0.9 / 36.4	0**	0**	0**	0.770 / 0.851
special #2	28.5 / 55.0	171 / 262	193 / 288	249 / 346	46.8 / 82.1	15.5 / 44.0	1.0 / 33.5	17.8 / 53.2	0**	0**	0**	0.804 / 0.888
oil sands derived	32.3 / 55.0	172 / 272	193 / 318	226 / 363	60.0 / 84.6	10.6 / 29.8	0.3 / 9.0	15.3 / 38.7	0**	0**	0**	0.819 / 0.888
oil shale derived	38.3 / 43.9	114 / 298	154 / 319	184 / 336	66.3 / 74.9	19.5 / 30.9	0.0 / 8.4	22.9 / 30.9	0.70 / 0.86	9.8 / 19.7	2529 / 9266	0.788 / 0.873
ASTM D975-S15	40 min			282 / 338				35 max				

Correlations between fuel variables are important because they help define how one characteristic is related to another, how fuel properties and chemistry can be manipulated in blending operations, and to identify highly correlated variables, which should not be carried into statistical analysis of results. Table 2 shows correlation coefficients between the fuel variables in Table 1.

Table 2, Correlation of properties and chemistry in fuels.

	cetane	T10, C	T50, C	T90, C	total saturates, vol%	mono aromatics, vol%	poly aromatics, vol%	total aromatics, vol%	oxygen, wt%	iodine number	nitrogen, ppm	specific gravity
cetane	1.00											
T10, C	0.27	1.00										
T50, C	0.44	0.91	1.00									
T90, C	0.29	0.63	0.79	1.00								
total saturates, wt%	0.51	-0.02	0.02	-0.01	1.00							
mono aromatics, wt%	-0.73	-0.36	-0.39	-0.30	-0.66	1.00						
poly aromatics, wt%	0.07	0.38	0.39	0.36	-0.64	-0.13	1.00					
total aromatics, wt%	-0.53	0.00	-0.03	0.02	-0.99	0.69	0.62	1.00				
oxygen, wt%	-0.12	0.19	0.01	-0.30	0.03	0.00	-0.11	-0.08	1.00			
iodine number	-0.10	0.35	0.15	-0.16	0.02	-0.03	-0.06	-0.07	0.96	1.00		
nitrogen, ppm	-0.09	0.42	0.22	-0.09	0.03	-0.05	-0.04	-0.07	0.92	0.99	1.00	
specific gravity	-0.15	0.73	0.74	0.57	-0.57	0.17	0.59	0.57	0.01	0.13	0.18	1.00

In this table, correlation coefficients at or above 0.7 are highlighted. It is apparent, across all these fuels, that cetane is mainly reverse correlated to mono aromatics, that T10 and T50 and T50 and T90 are correlated, that saturates and aromatics are reverse correlated, and that nitrogen, oxygen, and iodine number are all strongly correlated. These three properties were only present in the oil shale derived fuels, due to the partial hydrotreating, and because of the strong correlation between them, cannot be evaluated for independent effects with this data. Variables selected for subsequent analysis include cetane, T10 and T90, mono aromatics, poly aromatics, iodine number, and specific gravity. Iodine number is a measure of unsaturated bonds in HC molecules and is so named because of the method used. It also is a surrogate variable for oxygen and nitrogen in this data set as mentioned previously. Cetane was modeled using the other variables, using AVL Cameo®, an engine mapping and modeling software. A second order model with interactions and automatic selection and exclusion of terms gave an R^2 of 0.51 and a standard deviation of 4.9. The modeling process selected only the term mono aromatics. This indicates, that across all these fuels, mono aromatics is the major factor affecting cetane, and that T10, T90, poly aromatics, iodine number, and specific gravity do not show a statistically significant effect on cetane. Two fuels were eliminated from this analysis. One was a Fischer Tropsch diesel fuel with 73 cetane, eliminated because it exerted an excessive leverage on the regression process. The other was a fuel with 28.7 cetane and only 18.5% mono aromatics, which was identified as an outlier in the statistical process. This fuel was blended using non-traditional chemistries including light isoparaffins. The relatively poor quality of the overall model may be due to effects of other fuel variables not included in this analysis and also due to the number of different labs measuring the cetane of these fuels over the time frame of these experiments.

DESCRIPTION OF ENGINE

The engine used for this study is a simple, homogeneous charge compression ignition (HCCI) engine converted from a 517 cc single cylinder commercial diesel engine. The engine has been used as a basis for many studies (ref. 1-7) and will not be explained here beyond the following description. In converting the engine to HCCI, the diesel fuel injector was removed and a heated intake port atomizing fuel injector was added. This injector is operated at a temperature up to 375°C and produces a fine fog of fuel which mixes with the incoming air before entering the cylinder. Combustion phasing is controlled by heating the intake air between room temperature and 320°C. The engine operates un-throttled. Compression ratio has also been lowered from 20 to 10.6 to maintain an optimum range of combustion phasing with diesel type fuels. Fuel rate is controlled with a small metering pump. Valve timing is stock from the original diesel engine. In operation, air-fuel ratio (AFR) is controlled both by intake air temperature and fuel rate. Combustion phasing is controlled by intake air temperature and AFR and engine power output is controlled by fuel rate. The engine was rebuilt several times during these experiments and this may also contribute to some variation in the data. The simple method of combustion control employed in this engine places a great emphasis on fuel chemistry and kinetics and also makes modeling of the engine very simple. The same kinetic processes occur in more complex engines, but are distributed in time and space due to the fuel injection process.

EXPERIMENTAL PLAN

Each fuel was run at fixed fuel rates, with intake temperature varied to control combustion phasing, and data acquired included temperatures and pressures, gaseous and smoke emissions, and combustion analysis using an AVL Indicom® system. This results in a timing sweep for the fuels, which can be studied to evaluate optimum performance or tradeoffs between various emission parameters. For some fuels, several fuel rates were run, and for others only one. Overall, there are 975 data points covering the 47 fuels; engine variables range from 3 to 14.7 grams/minute fuel rate, 0.7 to 4.2 bar IMEP, and 112 to 400 °C intake air temperature. Over this range, NOx varies from 0 to 448 ppm and smoke number from 0 to 5.2.

DATA ANALYSIS

Data is analyzed by using statistics, with fuel properties and chemistry and engine control targets as input variables. Because of the simplicity of the engine it can be modeled with a minimum of variables. In this case, we chose combustion phasing (crank angle for 50% burn, known as MFB50) and torque output (indicated mean effective pressure, IMEP, bar) as input variables. In our nomenclature, 360° crank angle is TDC of the combustion cycle. Using these engine related variables also allows the model to be exercised at constant combustion phasing and power levels. For fuel variables, we chose cetane, mono aromatics, poly aromatics, specific gravity, T90, and iodine number (also correlated to oxygen and nitrogen).

Data is analyzed using the statistics analysis package in AVL Cameo®, which is intended for experimental design, mapping, optimization, and control map generation for combustion engines. Models are constructed as second order polynomials with cross terms, using automatic selection of model terms at a 95% significance level and a forward/backwards selection method. For NO_x and smoke, a square root transform is also applied to the experimental data to prevent the prediction of negative values. This problem arises with HCCI combustion because many operating points exhibit near zero NO_x and smoke. AVL Cameo® tracks the experimental design space (convex hull) to ensure that the models are not extrapolated beyond the range of experimental data. We have found that fuel characteristics very rarely fall into an orthogonal experimental design, and it is very important to track the experimental design space to avoid extrapolation of the models beyond the range of the experimental data. We should also note that a number of models with about equal quality of statistical fit can be constructed from the data in order to assist in the analysis desired. The analysis presented here is intended to provide a broad view of the analysis techniques and overall fuel effects on HCCI engine performance. The independent variables we study in this report include emissions (NO_x, smoke, CO, and HC), fuel economy (ISFC) and control information (intake temperature).

Figure 1 (shown after the conclusions) compares the measured response to modeled response for each of the data points. The models provide reasonable R^2 values, ranging from 0.69 to 0.81.

The overall responses of the variables of interest to the model independent variables are shown in Figure 2 (shown after the conclusions). In this figure, intake temperature, smoke, HC, CO, NOx, and ISFC are listed on the y-axes, and MFB50, IMEP, specific gravity, iodine number, % polyaromatics, % monoaromatics, T90, and cetane are listed on the x-axes. The horizontal green bars on each chart indicate the range of the experimental data and thus define the region where the model can be considered valid. The green dashed lines on the charts indicate a 95% confidence interval for the prediction. Intake temperature provides an indication of how easy a fuel is to ignite, but is not of direct concern as long as it falls within the range of engine control capability. Smoke is minimized by low engine load (IMEP) and by fuels with low specific gravity, low mono aromatics, and low cetane. NOx decreases at retarded combustion phasing and low load and for fuels with high specific gravity, low iodine number (probably because iodine number is highly correlated to fuel nitrogen in this data set), and low aromatics, low T90, and high cetane. Fuel consumption (ISFC) shows optimum values based on combustion phasing, load, high specific gravity, low poly aromatics, low T90, and approximately 40 cetane. Some of these trends are intuitive and others are harder to explain and may arise from the non-orthogonal experimental design space and other characteristics of the individual fuels.

CONCLUSIONS

In the analysis of the fuels for this report, and for the variables used to characterize the fuels, there was nothing to distinguish fuels derived from oil sands or oil shale when compared to more conventional fuels. Cetane was found to correlate only to % monoaromatics. Engine performance and emissions showed tradeoffs with specific gravity, aromatics, and cetane. Fuel consumption was found to improve with fuels of low specific gravity, low poly aromatics, and low T90 with a cetane of about 40. These models could be applied to optimization of engine and fuels for specific sets of emissions constraints, but results would apply only to this specific style of engine at the speed-load point selected. The statistical models used for analyzing the data provide a broad overview of trends but would need further refinement or additional data if used for optimizing the engine or a fuel formulation. Even with the large number of fuels and engine data points used in this study and the fuel variables assessed, the models provide only adequate to good statistical fit. Fuel formulations can rarely be designed in an orthogonal manner, except for a few variables, making it very important to track experimental design space when using models. Additionally, a fuel can influence engine performance both by a direct chemical effect and also because of a match or mismatch to optimum engine operating condition.

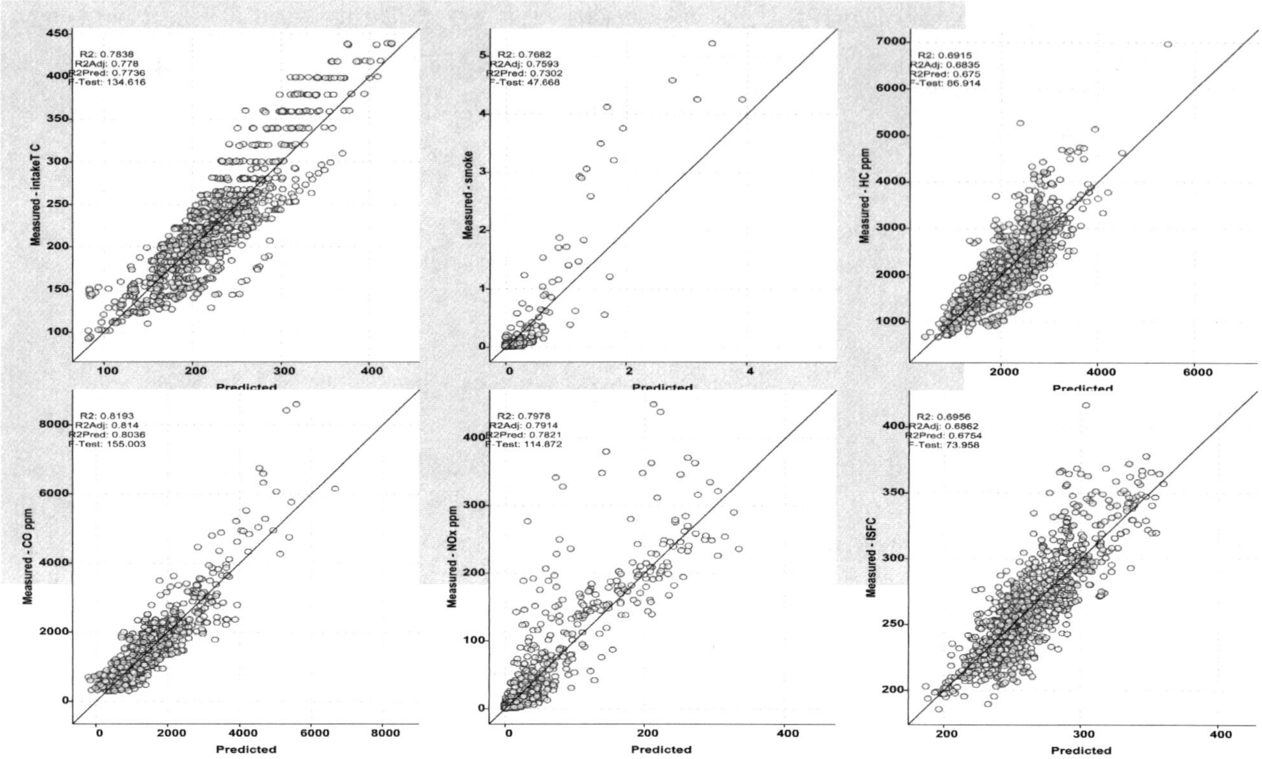

Figure 1, Measured vs. predicted response for 6 main variables of interest: intake temperature, smoke, HC, CO, NOx, and ISFC. demonstrating model ability to reproduce experimental data.

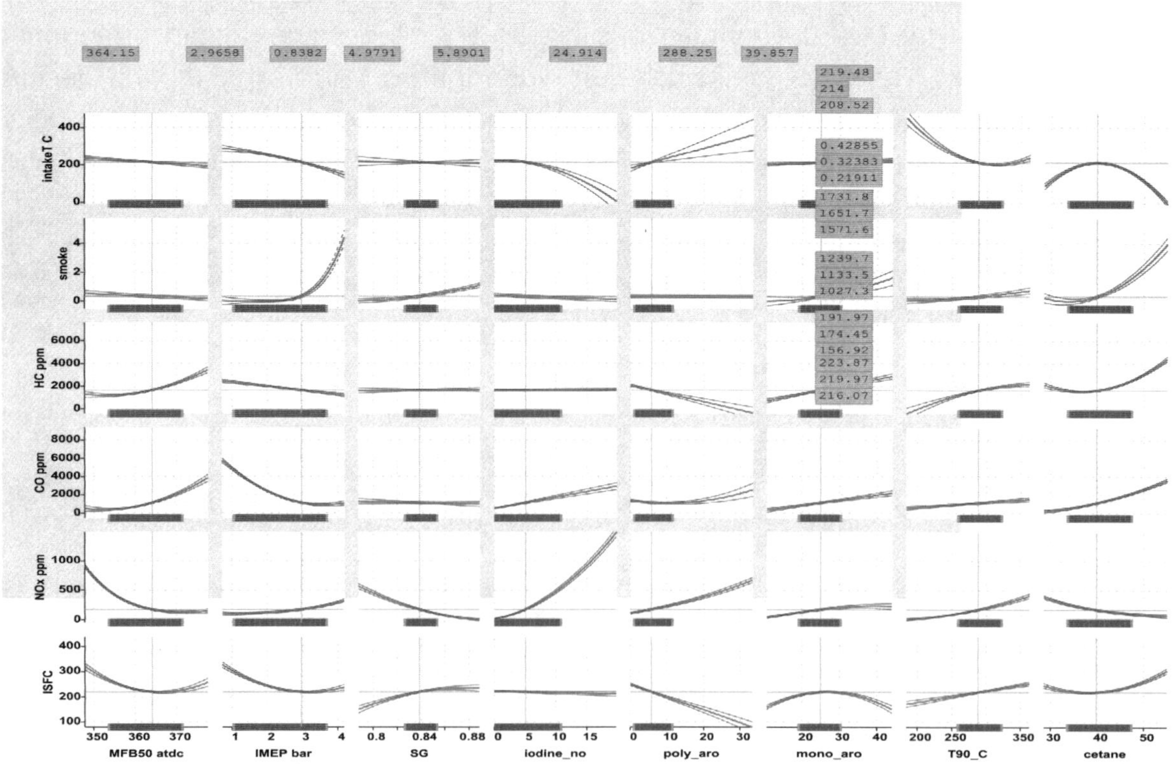

Figure 2, Response of engine variables to model independent variables, when each fuel variable is set to an overall average value.

REFERENCES

1. Bruce G. Bunting, Romain Lardet, and Robert W. Crawford, Statistical Overview of 5 Years of HCCI Fuel and Engine Data from ORNL, 2010 DOE DEER Conference, 09/2010.
2. Bruce G. Bunting, Scott J. Eaton, and Robert W. Crawford, Rincon Ranch Consulting, Performance Evaluation and Optimization of Diesel Fuel Properties and Chemistry in an HCCI Engine, SAE paper 2009-01-2645.
3. Scott J. Eaton, Bruce G. Bunting, Samuel A. Lewis, and Craig Fairbridge, Effect of Narrow Cut Oil Shale Derived Distillates on HCCI Engine Performance, SAE paper 2009-01-2646.
4. Bruce G. Bunting, Scott J. Eaton, Robert W. Crawford, Yi Xu, Les Wolf, Shankar Kumar, Don Stanton, and Howard Fang, Performance of Biodiesel Blends of Different FAME Distributions in HCCI Combustion, SAE paper 2009-01-1342.
5. Bruce G. Bunting, Scott J. Eaton, John Storey, Craig Fairbridge, Ken Mitchell, Robert W. Crawford, Tom Gallant, Jim Franz, and Mikhail Alnaijar, The Chemistry, Properties, and HCCI Combustion Behavior of Refinery Streams Derived from Canadian Oil Sands Crude, SAE paper 2008-01-2406.
6. Bruce G. Bunting, Robert W. Crawford, Les R. Wolf, and Yi Xu, The Relationships of Diesel Fuel Properties, Chemistry, and HCCI Engine Performance as Determined by Principal Components Analysis, SAE paper 2007-01-4059.
7. James P. Szybist and Bruce G. Bunting, Cetane Number and Engine Speed Effects on Diesel HCCI Performance and Emissions, SAE paper 2005-01-3723.

(111a) Simulation In Early Stage Process Development

Joe Schroer, Clearwaterbay Technology, Inc., Pomona, CA

Simulation tools are frequently used in process development. When properly applied, they can provide powerful results, such as the evaluation of many flowsheet alternatives in a short amount of time, and numerical calculation and representation of the complex chemical system. When used for process development, there are challenges. One concern is the limited availability of the properties of new chemical systems and mixture properties for previously not evaluated systems. Key questions that help to focus the efforts are "What is the simulation objective?", and "What is the process development objective?"

See more of this Session: Best Practices Utilizing Simulation Tools, Pre and Post-Project
See more of this Group/Topical: Topical 7: 14th Topical on Refinery Processing

(111b) Understanding Design and Operational Issues Using Dynamic Simulations

Dipanjan Bhattacharya[1], Manish Jain[1] and Ramachandra P. Tekumalla[2], (1)Bechtel Oil and Gas, Houston, TX, (2)Advanced Simulation & Analysis, Bechtel Corporation, Houston, TX

Simulation has traditionally been used for steady state process design. In reality, many process design controlling scenarios are dictated by transient conditions. Dynamic simulation effectively predicts these transient situations during process design and helps in reducing future operational and start-up risks. This article goes through multiple case studies where dynamic simulation has been effectively used to design equipments as well as predicting methods to minimize operational problems. The article also shows cases where dynamic simulation validates design calculations conducted by traditional methods.

See more of this Session: Best Practices Utilizing Simulation Tools, Pre and Post-Project
See more of this Group/Topical: Topical 7: 14th Topical on Refinery Processing

(111c) Business Justification for the Investment In Dynamic Simulators

Martin Berutti, MYNAH Technologies, Chesterfield, MO

Almost everyone who has used a dynamic simulator for testing automation system configuration, validating process design, or training operators claim that they received great value from the simulation tools. But how do users build a solid economic justification for this investment? This paper will show how to justify the investment in dynamic simulation by increasing the Net Present Value (NPV) and Internal Rate of Return (IRR) of the total automation system capital investment. Case studies are shown to demonstrate the operational effectiveness and project efficiency that influence these calculations, plus examples how the simulation tools were applied.

See more of this Session: Best Practices Utilizing Simulation Tools, Pre and Post-Project
See more of this Group/Topical: Topical 7: 14th Topical on Refinery Processing

Operational Inputs into the Mechanical Integrity Program

F. Russ Davis, CSP
Aon Energy Risk Engineering
6455 South Shore Blvd. Suite 400, League City, TX 77573
russ.davis@aon.com

Prepared for Presentation at
American Institute of Chemical Engineers
2011 Spring Meeting
7th Global Congress on Process Safety
Chicago, Illinois
March 13-16, 2011

UNPUBLISHED

Operational Inputs into the Mechanical Integrity Program

F. Russ Davis, CSP
Aon Energy Risk Engineering
6455 South Shore Blvd. Suite 400, League City, TX 77573
russ.davis@aon.com

Keywords: mechanical integrity (MI), corrosive, temperature, pressure, pH

Abstract

A predictive mechanical integrity (MI) program is based on the following:

Damage mechanisms that affect process equipment

Rate at which degradation progresses

Ability of the inspection department to identify and correct equipment damage before equipment failure

Operational temperature, pressure, flow and composition are process variables that greatly affect process equipment life cycles.

A risk-based inspection program or any proactive inspection program should establish mechanical integrity (MI) plans for all equipment included in the program. An equipment plan should be based on anticipated damage, suitable inspection techniques capable of assuring the reliability of the process equipment in order to reduce the risk of operating the equipment to the lowest practicable level. Equipment plans are created based on the chemicals in the operational environment and the damage that may be anticipated from those chemicals.

Any time that the operational environment changes, the damage that affects the process equipment may also change. This paper will explore the process parameters and their affect on process equipment and the mechanical integrity (MI) program. How changes made to operational parameters such as temperature, flow, pH, pressure and process stream consistency will be reviewed and discussed. This paper will result in recommendations on how to incorporate process changes and their affect into the mechanical integrity (MI) program.

1. Introduction

The primary purpose of a mechanical integrity (MI) program is to prevent critical equipment failures leading to a loss of containment of highly hazardous chemicals. Critical equipment should be defined as equipment whose failure could reasonably and directly lead to the occurrence of a hazardous event. Equipment failures have resulted in very serious fires and explosions with injury, loss of life and significant impact to corporation assets. Recent accident investigations such as the 2005 BP Texas City and the 2009 Silver Eagle Refinery explosions have listed failure of the mechanical integrity (MI) program as root cause or as contributing causes.

Note: Do not add page numbers. Do not refer to page numbers when referencing different portions of the paper

The mechanical integrity (MI) program should be predicated on a risk basis. Equipment should be inspected, tested, and maintained based on the risk associated with operating the equipment to personnel, the environment and to the fiscal well being of the company. Mechanical integrity risk is defined as the product of the consequence of a failure and the likelihood that the failure will occur. Mechanical integrity programs impact the likelihood of failure by recognizing a damage mechanism, evaluating the effects of the damage mechanism on the process equipment and inspecting and testing the equipment on a frequency that will predict the life cycle of the equipment. Equipment is then repaired or replaced prior to hazardous failure. The damage mechanism part of the equation is the operational environment. The operational environment includes, but is not limited to:

- Chemicals in the process stream
- Temperature of operations
- Pressure of operations
- Flow/erosivity of the process stream
- pH of the process stream

Changes to the operational environment can have an impact on the ability of the mechanical integrity department to predict the life-cycle of the equipment. All changes to operational parameters should be conveyed to the mechanical integrity (MI) department. The mechanical integrity (MI) department can then assess the impact to process equipment and adjust equipment inspection and testing plans if needed. Operations departments should have a system in place to recognize process changes and communicate the changes to the mechanical integrity (MI) department.

2. Risk-Based Mechanical Integrity (RBMI) Program Overview

In a risk-based mechanical integrity (RBMI) program each equipment item receives a risk evaluation to determine the criticality of the equipment within the inspection and testing program. RBMI programs are primarily directed at fixed equipment but the concept can be used to determine the relative risk ranking for any process equipment or system. To determine the relative risk of an equipment failure the consequence of the failure is compared with the likelihood of the failure. Likelihood of failure (LOF) is determined by analyzing all damage mechanisms that can adversely affect the equipment item dependent on the material of construction of the item. Damage mechanisms are most closely associated with the operating parameters in which the equipment resides. A simplified example is carbon steel equipment operating in a process stream with water and oxygen making equipment susceptible to corrosion.

Temperature and flow rates can have a significant effect on the rate of corrosion and the anticipated life of the equipment. Higher temperatures can speed corrosion rates in some environments. Higher operating temperatures contribute to their own damage mechanisms such as creep, thermal shock and high temperature hydrogen attack.

A risk-based analysis is performed using data supplied by the operations department. Anticipated damage mechanisms and assumed corrosion rates are developed based on that operational data. Inspection plans are then developed for high risk equipment. If there are significant changes in operational parameters from those used to perform the initial risk assessment the RBMI department may have to re-evaluate the risk analysis and reformulate inspection and testing plans.

3. Operational Parameter Changes can affect the Mechanical Integrity (MI) Program

3.1 Erosion/Erosion-corrosion

Erosion is the removal of surface material as a result of impingement by solids, liquids, or vapors. All typical metals and refractory are susceptible to some degree to erosion.

Erosion-corrosion is the damage that occurs when corrosion contributes to the erosion process by removing protective films or scale and exposing the metal surface to further corrosion. The erosion process then impinges on the metal surface and removes the metal or refractory.

Erosion-corrosion is the more common of the two mechanisms. The erosion element can be increased by the corrosiveness of the operating environment. Increased velocity of the process stream will increase the rate of metal loss due to the mechanism. Increasing the corrosivity of the process stream will also increase the metal loss. Increases in temperature can also be a contributor to a higher than anticipated corrosion rate.

The mechanical integrity (MI) department should understand what equipment is susceptible to erosion-corrosion in the process. Operations personnel should be part of the process analysis so the correct process data can be used to evaluate risk of equipment failure. Operations should provide worst-case process information. Operational conditions such as temperature and flow rates used to determine the likelihood of equipment failure should be based on the worst possible operating conditions. When temperatures, flow rates and pH levels change the mechanical integrity (MI) department should be informed. Systems for making sure information flows from operations to the mechanical integrity (MI) department will be discussed in later paragraphs.

3.2 Corrosion Fatigue

Corrosion fatigue is cracking that develops in a corrosive environment when cyclic loading is present. Cracking can often initiate from pit locations caused by corrosion. The cyclic loading then propagates the cracks. Deaerators and boilers are examples of equipment that is susceptible to this damage mechanism. Inspection plans are typically established based on plant history or industry average. If operations is experiencing a larger than normal cyclic operations (i.e., shutdowns and startups), the mechanical integrity (MI) department should reassess the frequency between inspections. Corrosion fatigue cracking and crack propagation is directly tied to the number of cycles experienced by the equipment.

3.3 Chloride Stress Corrosion Cracking (Cl SCC)

Chloride Stress Corrosion Cracking (Cl SCC) is cracking caused by chlorides in the operating environment, primarily affecting 300 series stainless steel materials and effecting other alloys to a lesser degree. Cl SCC occurs under the combined action of tensile stress, temperature and aqueous chloride. Tensile stresses can be either induced due to operating environment or from residual manufacturing stresses. An increase in temperature increases the likelihood of cracking. Cracking usually occurs at a temperature of 140°F or higher but cracking has been found in environments where temperatures were lower than 140°F.

If operating temperatures are lower than 100°F, Cl SCC may not have been a major factor when the mechanical integrity (MI) department assessed the damage mechanism associated with this process equipment. Equipment inspection plans will be specified that may not include an emphasis of testing for cracking. If operating temperatures are raised equipment assessment should be re-evaluated. The same can be said for chloride concentration. When chloride concentrations are raised cracking can occur more aggressively.

Operational changes can have a severe impact on process equipment where chlorides are present. Changes in feedstock pH or percentage of chlorides and especially in temperature can make equipment susceptible to Cl SCC. A raise in process temperature can lead to chloride cracking or accelerate the Cl SCC damage mechanism. The water used for cleaning or as a part of the process can also have an impact on the amount of chlorides that can affect the equipment. Using water with high chloride content can cause Cl SCC in process equipment over time. Idling of process equipment containing chlorides, without clearing, can allow chloride concentration to build up leading to Cl SCC.

3.4 Caustic Corrosion

Caustic corrosion can be localized or general corrosion depending upon the concentration and strength. The caustic can be a constituent of the process stream or used for cleaning process equipment. Caustic corrosion primarily affects carbon steels, low alloy steels and 300 series stainless steels. Caustic corrosion can cause an accelerated loss of material leading to failures and loss of containment.

Process equipment preventative maintenance and inspection plans are typically established based on operating conditions. The use of caustic for cleaning equipment should be analyzed by the mechanical integrity (MI) department. Inspection plans can be established that recognize that there can be caustic corrosion present.

Problems arise when operations either adds caustic to streams where it was not previously found or starts using a cleaning substance containing caustic. The mechanical integrity (MI) department should be notified so that inspection plans can be modified.

3.5 Caustic Stress Corrosion Cracking

Caustic stress corrosion cracking is caused by embrittlement of susceptible materials in the presence of stress and elevated temperature. Increasing caustic concentrations and temperature increase the likelihood of cracking, but cracking has been found in concentrations as low as 50 ppm. Stresses can be either residual from manufacturing or induced from the operating pressures or installation. Carbon steel, low alloy steels and 300 series stainless steels are especially susceptible to caustic stress corrosion cracking.

Operating temperature is a critical factor used by the mechanical integrity (MI) department when equipment inspection plans are developed. It is understood that higher temperatures accelerate caustic cracking. When operating temperatures are raised or when temperature excursions occur the mechanical integrity (MI) department should re-evaluate frequency of inspection and testing for caustic cracking. When caustic concentrations are changed the mechanical integrity (MI) department should re-evaluate equipment inspection plans. As can be seen in the service graph from API 571[1] below, the susceptibility of carbon steel to caustic cracking is dependent on caustic concentration and temperature.

CAUSTIC SODA SERVICE GRAPH

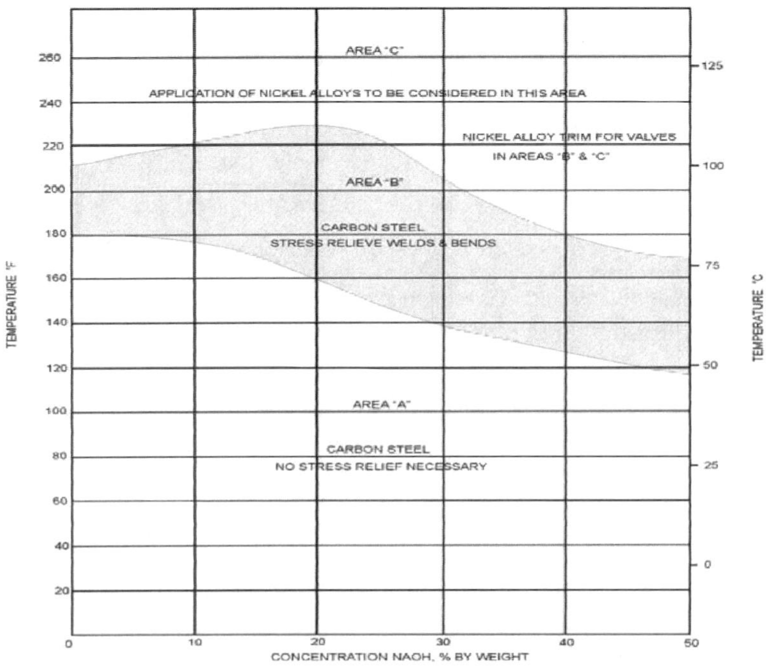

3.6 Microbial Corrosion (MIC)

Microbial corrosion, often referred to as microbiologically influenced corrosion (MIC), is a form of corrosion caused by living organisms. Microbial corrosion is usually found in aqueous environments especially prevalent where water flow is stagnant or very low. Microbial corrosion can attack many materials including carbon steel, stainless steel, aluminum and copper and some nickel based alloys. Microbial corrosion can be found in heat exchangers, storage tanks and piping where the flow is low or there are times of stagnant flow or when the water is not properly treated.

In many plant organizations operations is in control of water treating. The mechanical integrity (MI) department should be aware that microbial corrosion is a possible damage mechanism in systems where water is present and should be made aware of all changes to the process. If equipment is going to be left idle for a period of time the mechanical integrity (MI) department should be part of the discussion on how best to leave equipment idle. Operations should be the first to know when there are problems with water treating systems. If the treatment system is lost or if the treatment chemical is changed the mechanical integrity (MI) department should be notified. The mechanical integrity (MI) department can then re-assess equipment testing and inspection plans where needed.

3.7 Hydrogen Embrittlement

Hydrogen embrittlement is a loss of ductility in high strength steels due to penetration by atomic hydrogen. Hydrogen embrittlement can lead to brittle cracking and failure of process equipment. Hydrogen embrittlement can affect carbon steel, low alloy steels, some stainless steel and also some nickel based alloys. The three (3) factors that must be present for hydrogen embrittlement to occur are:

1. Stress must be present; either residual stress from manufacture or applied stress from the process
2. Hydrogen must be present at the critical concentration within the material
3. Steel or alloy must be susceptible to hydrogen embrittlement

Services where hydrogen embrittlement can be found include wet H_2S or HF acid services. Temperature is a critical element in hydrogen embrittlement. Hydrogen embrittlement is most likely to occur at room temperature operations. Hydrogen embrittlement is less likely to occur at higher temperatures, above 180°F.

Changes in operating temperature can have an impact on the possibility of hydrogen embrittlement occurrence. If operating temperatures are lowered the mechanical integrity (MI) department should be notified and equipment plans re-assessed. Equipment that was operating above the hydrogen embrittlement temperature may now need to be inspected and inspection plans should be changed.

3.8 Wet H_2S Damage

Wet H_2S damage includes the following:

- Hydrogen blistering
- Hydrogen induced cracking
- Stress oriented hydrogen induced cracking (SOHIC)
- Sulfide stress corrosion cracking (SSC)

Wet H_2S affects carbon steel and low alloy steels. Process stream pH, H_2S levels, contaminants, and temperature are all critical factors that affect process equipments susceptibility to wet H_2S damage.

Operational changes should always be relayed to the mechanical integrity (MI) department. When the pH of the feed stock changes the damage mechanism affecting process equipment can quickly change as well. Free water content changes can also make equipment susceptible to wet H_2S damage. Operations should work in conjunction with the mechanical integrity (MI) department when free water is noted in equipment where it is not usually found or when the free water content increases. When free water in liquid phase is present the likelihood of damage occurring is much higher when H_2S is present. The operations or process department monitors the daily parameters of the process. This group should recognize and be aware of the possibility of equipment damage occurring when parameters such as amount of free water in the process change.

3.9 Amine Stress Corrosion Cracking

Amine stress corrosion cracking can be found in Amine Scrubber Systems where H_2S and CO_2 are removed. Carbon steel and low alloy steels are susceptible to amine stress corrosion cracking. Stress, amine concentration and temperature are the critical factors for determining if material is susceptible to amine stress corrosion cracking. Stresses can be residual from manufacturing or induced from operations. Increase in temperature increases the severity and likelihood of amine stress corrosion cracking. Lower amine concentrations are less susceptible to amine stress corrosion cracking.

Amine concentration and operational temperature of the amine process are parameters monitored by the operations/process department. Changes made to the operation parameters can alter whether equipment is susceptible to amine stress corrosion cracking. When process parameters change the mechanical integrity (MI) department should be informed because equipment plans may need to be re-assessed.

3.10 Creep Damage

Equipment operated at high temperature can slowly deform over time due to stress. The deformation can eventually lead to equipment rupture. All metals and alloys are susceptible to creep damage. Equipment inspection and testing plans are based on the operating temperature and the time in service. Operating pressure is a critical factor in creep damage as a stress inducer. High pressures induce more stress into the equipment. API 571 notes that an increase of about 25°F, or a 15% increase in stress, can shorten equipment remaining life by as much as 50%.

If operations run a furnace at higher temperatures or at higher pressures the mechanical integrity (MI) department should be consulted. The equipment inspection and test frequency may need to be changed dramatically.

3.11 Thermal Fatigue Cracking

Thermal fatigue cracking can occur due to a thermal shock. Metals will expand and contract when affected by temperature differences. If hot metal comes in contact with a cold process stream rapid contraction can lead to cracking. Process upsets that send cold material to hot equipment should quickly be reported to the mechanical integrity (MI) department. Equipment testing and inspection plans may need to be revised.

4. Communications – Operational Information to Mechanical Integrity

Section 3 discussed some typical damage mechanisms experienced in refineries and chemical plants and how operational parameters can affect these damage mechanisms. By understanding the effects operational changes or upsets can have on process equipment it becomes clear that the mechanical integrity (MI) department should receive and analyze operational data. There are various ways that the mechanical integrity (MI) department should receive this information. The following sections discuss various options.

4.1 Process Trending and Monitoring

The operations department typically has a system in place for monitoring process trends such as Aspen's IP21. The mechanical integrity (MI) department should be trained in the use of process monitoring systems. When trends are developed to monitor critical equipment parameters the mechanical integrity (MI) department can quickly receive information that may have an effect on process equipment. When temperatures and pressures are changed the mechanical integrity (MI) department should be able to spot these changes. Possible damaging effects can be assessed and inspection and test plans can be modified if necessary.

4.2 Operational Changes

In too many facilities the mechanical integrity department is not a required approver of process changes. The mechanical integrity (MI) department should always be required to review process changes that may affect equipment. Higher or lower temperatures, higher or lower pressures, and higher or lower flow rates can significantly affect equipment life. As we have seen in the discussion on various damage mechanisms, each mechanism has critical factors that affect the possibility and severity of the damage that can be experienced by process equipment. When operational parameters change equipment that may not have been susceptible to a specific damage mechanism can now experience severe damage leading to equipment failure and loss of containment. Changes that the operations department may not necessarily see as adverse can affect the life of process equipment. Changes in feed pH can greatly affect the corrosion rate experienced by process equipment. Changes in sulfur or hydrogen content of raw materials can cause cracking in process equipment. Higher flow rates of abrasive processes leads to high erosion/corrosion rates. Operations should ensure that the mechanical integrity (MI) department is a required reviewer of all process changes.

4.3 Operational Upsets

Operational upsets are typically noted in operational reports. These reports should be forwarded to the mechanical integrity (MI) department and used in the evaluation of equipment test and inspection plans. Temperature excursions, pH excursions, and pressure excursions should always be relayed to the mechanical integrity (MI) department. Water breakthrough or build-up in processes where it is not normally found should be noted as an operational deviance and this information should be transmitted to the mechanical integrity (MI) department. Quality testing information may not be recognized as a factor that affects process equipment but in some cases factors affecting quality such as un-reacted chemical breakthrough in the process may allow corrosive materials into equipment where it is not normally found. The mechanical integrity (MI) department should be made aware of these quality issues in order to analyze possible damage effects.

5. Path Forward

Risk-based mechanical integrity systems evaluate the possible damage mechanisms that may adversely affect process equipment. Inspection and testing plans are developed to detect equipment damage and to repair or replace equipment prior to failure and loss of containment. These equipment plans are based on information provided by the operations group. When operational parameters change process equipment damage mechanism will change as well. The mechanical integrity (MI) department should be informed about any changes that will alter the basis for the equipment risk analysis and the inspection and testing plans developed for the process equipment. A system should be instituted that will convey operational information to the mechanical integrity (MI) department. The system should contain administrative controls such as a management of change (MOC) system as well as the on-line systems such as IP 21 that Operation uses to track process trends.

Plant operational designs should consult the mechanical integrity (MI) department for information when new units are designed or modified. Typically if the operational group needs the information so does the mechanical integrity (MI) department. It is important that adequate temperature, pH, flow and pressure instrumentation be located to provide sufficient data. The flow of data from the process to the mechanical integrity (MI) department should be considered in plant design.

Awareness training for operational personnel should be part of every operator's and engineer's required training matrix. Operations personnel need to understand how process changes could affect equipment. Operations personnel and the mechanical integrity (MI) department need to understand where process information comes from and how it is available. The mechanical integrity (MI) department should be trained in the use of the process trending software.

Safer plants, equipment reliability and less downtime are the primary benefits of a proactive Risk-based Mechanical Integrity (RBMI) program. The risk based mechanical integrity (RBMI) program is evergreen. Process information is the critical parameter in the design and maintenance of the RBMI program.

References

[1] API 571, "Damage Mechanisms Affecting Fixed Equipment in the Refining Industry", 1st ed. December 2003.

[2] API 580, "Risk-Based Inspection", 2nd ed. November, 2009.

[3] API 581, "Risk-Based Inspection Base Resource Documents", 1st ed., October 2000.

From Process Simulation to Operator Training - A Success Story of Business Improvement Enabled by Simulation Technologies

Rajat Basu, Yuon Chiu
Honeywell Specialty Materials, Buffalo, NY
Eddy Yap
Honeywell Process Solutions, Houston, TX

Honeywell

Honeywell SM and ACS: A Winning Linkage

Honeywell

Honeywell Specialty Materials

Honeywell Specialty Materials is a \$4.14 billion world-leader in supplying high performance chemicals. The Geismar site is a leader in fluorine technology and provides acids and non-ozone depleting refrigerants to a broad customer base.

Honeywell Automation and Control

Honeywell Automation and Control Solutions is a \$12.6 billion business applying sensing and control expertise that helps create safer, more comfortable, more secure and more productive environments. Honeywell is the leader in providing value added industrial solutions to the chemical marketplace.

Honeywell - A world leader in fluorine technology that applies leading control technology to create a safer, more comfortable, secure, productive and responsive solution for the chemical marketplace

We are owner-operators as well as solution providers

Lifecycle for Chemical Process Modeling

Honeywell

Helps Process Eng'rs

Steady State Simulation

- Design new plants/unit operations
- Reduce Operating Costs
- Increase Capacity
- Reduce Cycle Times
- Define Conditions for New Products
- "What-If Scenarios"
- Improve Quality
- Train Optimization
- Reduce Extractables
- Capture Expert knowledge

Helps Mfg Eng'rs

Dynamic Simulation

- Optimize Transitions
- Reduce Waste
- Test New Control Systems
- Improve Quality (Reduce Variability)
- Improve Loop Tuning
- Reduce Recovery Time
- Multivariate Control
- Inferential Sensors
- Product by Process

Operations Simulation

- Off-Line Training (Simulator)
- Operator Certification
- Error Trapping-Locate & Diagnose
- Upset/Start-Up/Shutdown Simulator

Increasing Knowledge and Reach

Impact Depends on Knowledge and Reach/Depth

Development of operator training simulator

Honeywell

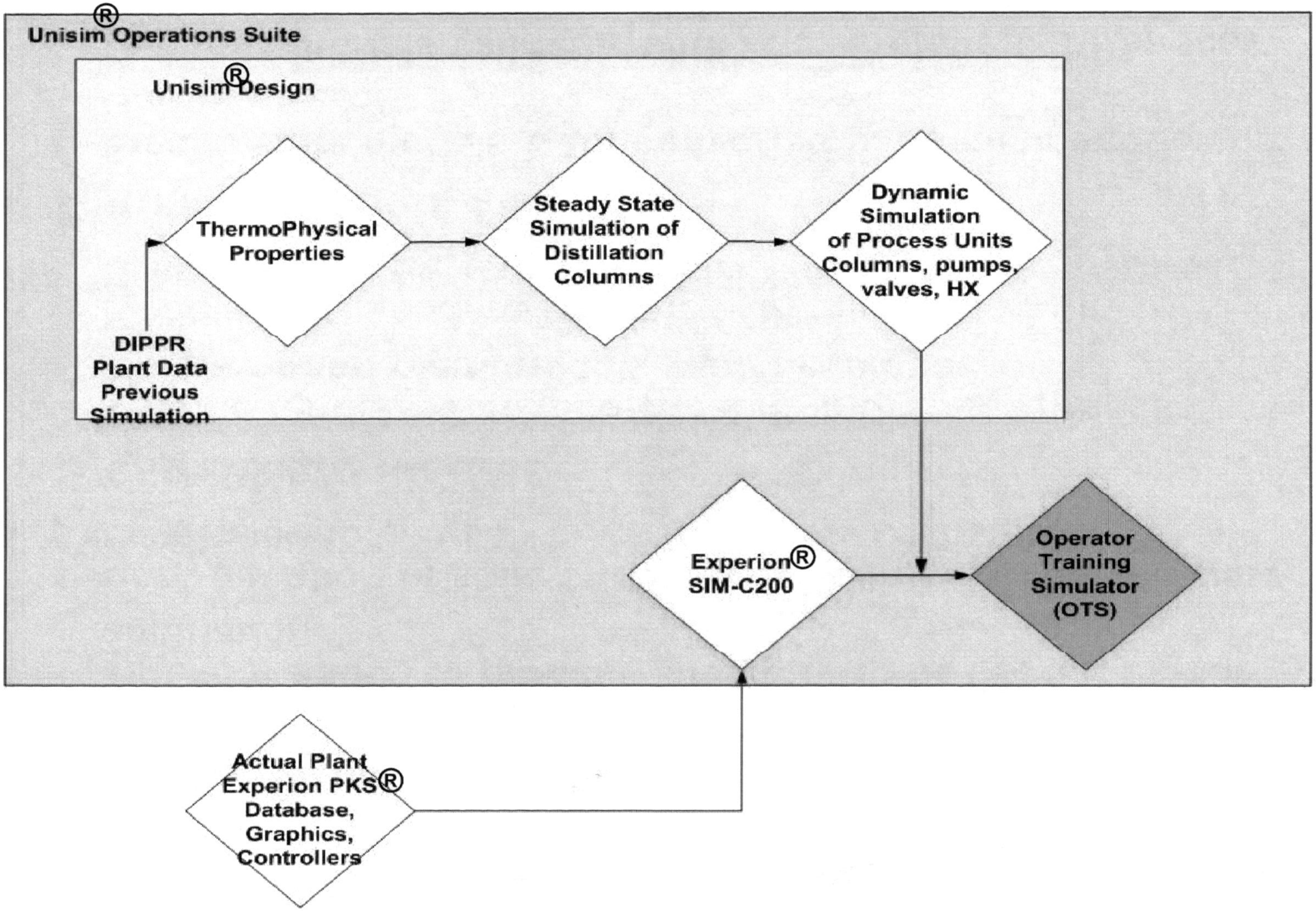

Model Development

Honeywell

- Thermodynamic Models –
  - Thermodynamic properties were measured and correlated and later added to process simulator to be used for plant simulation
  - Both equation of state and activity coefficient models were developed to be used in various parts of the plant
- Steady State Models
  - Thermo models were incorporated in UniSim® process simulator and plant models were developed
  - Models were validated on-line with plant data from UNIFORMANCE® data historian
- Dynamic Models
  - Steady state models were converted to dynamic models in UniSim
  - Dynamic models validated with known plant disturbances

Steady State on-line Model Validation

Honeywell

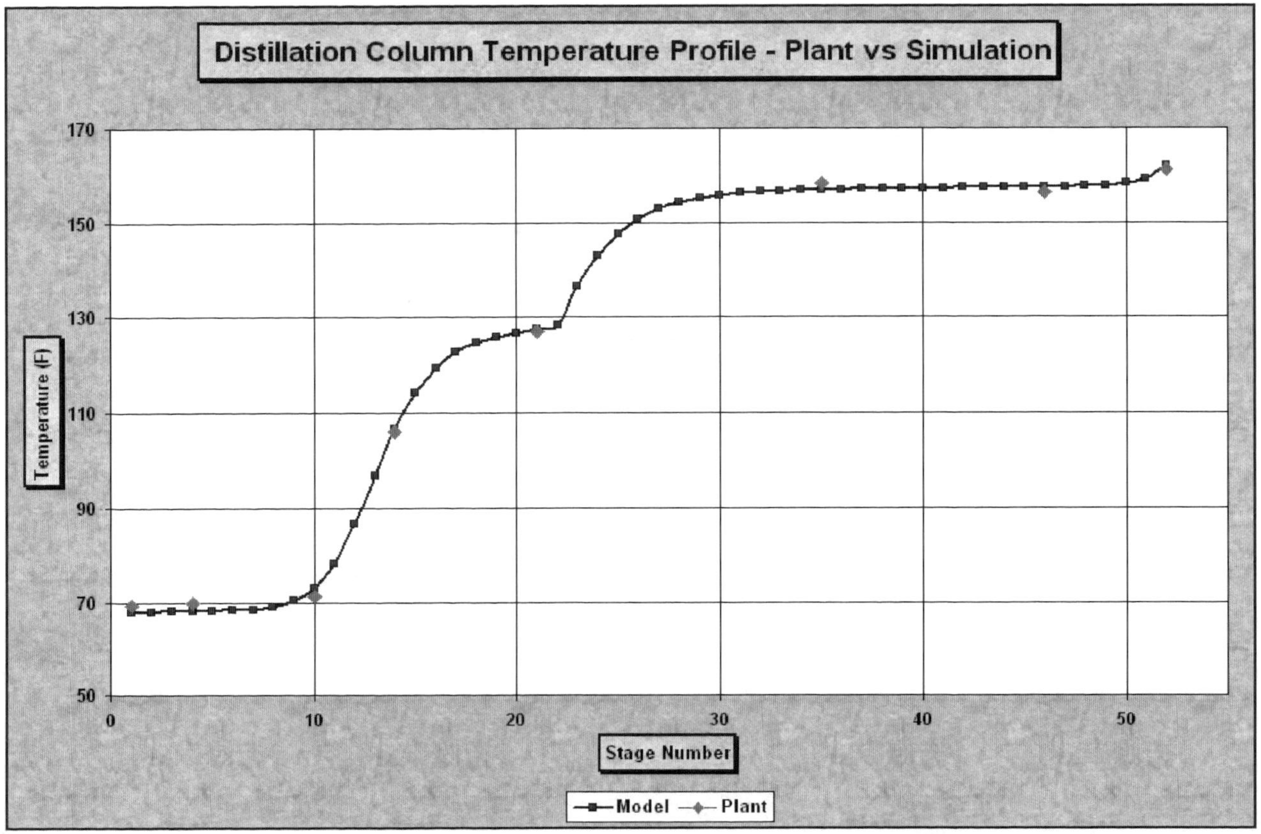

On-line plant data from data historian is validated with model calculation – aids plant operation and troubleshooting

UniSim®-Experion® PKS C200 Architecture

Honeywell

UniSim® – Experion® PKS C200

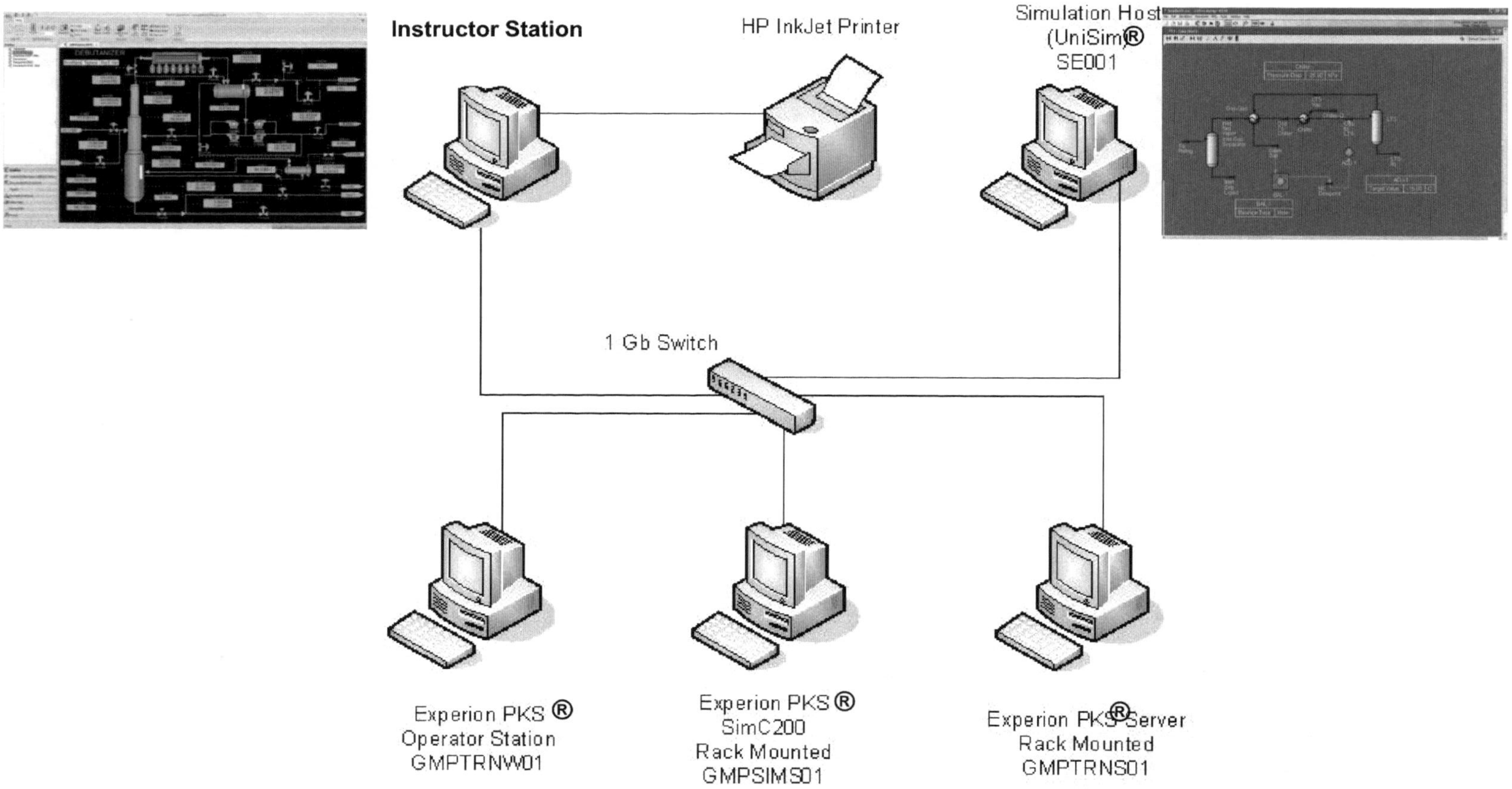

Maintenance of OTS

Honeywell

- Advantages of Virtualization[1]
  - Lower Cost – Multiple Virtual Machines (VM) can run on a single server
  - Greater flexibility – Users can add, remove or change computers and network components at will.
  - Better manageability – Snapshots provide a mechanism to roll back and forth between various backups. This substantially lowers the risk of upgrades and other changes.
  - Easier migration – Data centers can migrate to new hardware by simply copying VMs to a new server.
  - Easier remote support – Easier to duplicate identical system at remote location or easier remote access to the system.

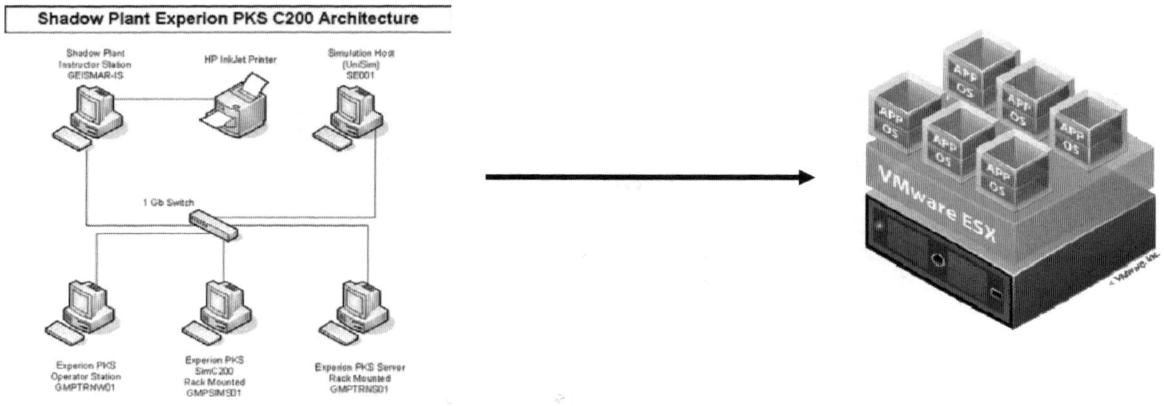

1. Virtualization of Operator Training & Production Systems at Honeywell Specialty Materials by John McIlwain

How Operator training is working

Honeywell

- **Effectiveness and Benefits of OTS has been well documented[2]**
  - Operator Training System (OTS) applied to Fluorocarbon Plant
    - First Principles Models built in UniSim® coupled with Experion® System perfectly mimics actual control environment
    - Built-in and Custom Scenarios train operators for Start-Up, Shut-Down, plus Abnormal Situations.
  - Results:
    - Reduced time to qualify an operator from >18 months to six months.
    - Reduced start-up time by 20 hours on average.
    - Used OTS in virtual mode to commission new EPKS versions.

OTS simulates virtual plants and helps operators

2. The Role of Simulator Technology In Operator Training Programs, Jose Alamo, Nov. 2010 Control Engineering

How a Training Simulator Helps Operations

Honeywell

- Highly Effective Training:
  - Looks & behaves just like the real plant (Like a Flight Simulator !)
  - No risk to failure
  - Endlessly repeatable
  - Visual: compare performance with best in class.
  - Motivates: fun to do
  - Builds confidence
- Very Specific & Practical Training:
  - Address start-up, shut-down, and recurring problems
  - Demonstrate impact of instrument & equipment failures.
- Develop better procedures:
  - Test improvements before they go on-line
  - Check impact of anticipated problems: power or service failures, etc.
- Identify and Solve Process Control Issues:
  - Makes control issues obvious
  - Allows us to test new strategies off-line

Operator Training System in Use

Honeywell

Instructors training operator using UniSim®/Experion® training screens

Increasing Operator Confidence in Procedures & Best Practices

Projected Financial Benefits

Honeywell

- **Estimated Benefits:**
  - Start-Up Speed: Gain 1 day production each product switch; Develop better procedures off-line. Std Work → Improve Std Work
  - Disturbance response improvement - Gain 7-10 hours production each major disturbance
  - Significant $$ benefit payback in about 1-2 years
  - This is in addition to the benefits obtained from steady-state and dynamic simulations used for various plant operations improvement

Conclusion

Honeywell

- This presentation highlights the development, validation and use of plant models from steady state to dynamic and then to operator training module development
- The advantage of steady state and dynamic in one simulation platform (UniSim®) helped ease the conversion from steady-state to dynamic
- In every stage there are advantages to the engineering operations
- Ultimately these models helped develop the Operator Training System for the plant
- Significant $$ benefit attained by improving operations in the plant

Acknowledgements

- Sanjoy Saha – Honeywell ACS
- Laurie Wang – Honeywell ACS
- Fred Gabriel – SM Plant Engineer
- John McIlwain – SM Control Engineer
- Charles Shivers – SM Operator training specialist
- Wendy Cart – SM operator training specialist

(111f) Use of Process Simulators for Control Studies and Operator Training at RasGas

Abdulla Al-Subaey[1], Ferdi Pangan[1], John W. Womack[2] and **Ahmet Pekediz**[3], (1)Ras Gas Company Limited, Doha, Qatar, (2)Markham Dynamics Corporation, Princeton, NJ, (3)Honeywell Process Solutions, London, ON, Canada

Ras Gas Company Limited (Ras Gas) operates LNG production as well as pipeline quality gas production facilities in the state of Qatar. RasGas has made extensive use of Process Simulators (POTS) during pre-project commissioning and later during normal plant operation. The primary use is for Distributed Control System (DCS) training of central control room Operators. The paper explains the training use of Process Simulator made by RasGas, and provides examples of other benefits.

See more of this Session: Best Practices Utilizing Simulation Tools, Pre and Post-Project
See more of this Group/Topical: Topical 7: 14th Topical on Refinery Processing

Evaluating the Heat Exchanger Technologies Available for Use in Refinery Pre-Heat Trains

G.T.Polley, E. Tamikloe & M. Picon Nunez
University of Guanajuato, Mexico

Fouling behaviour is a crucial consideration in the selection and design of heat exchangers to be used in refinery pre-heat trains.

We make the following observations regarding fouling associated with the heating of crude oil:

1. Deposition on an existing deposit occurs more readily than deposition on a clean metal surface.
2. The fouling rate increases with increase in wall temperature
3. The fouling rate decreases with increase in wall shear stress

We then use these observations to evaluate the following range of technologies.

Heat Exchanger Cleaning

Given the first observation we conclude that when cleaning of heat exchangers it is important to remove all of the existing deposits and any corrosion layer on the tubes.

The best option for cleaning on the tube-side is hydro-drilling.

For the cleaning of heat exchanger shells and Compabloc exchangers chemical cleaning enhanced by the use of ultrasound appears to be the best option (McCartney, Petrophase 2010).

Placing Crude on the Shell-side of Exchangers

Provided a high level of wall shear can be maintained there is a significant wall temperature advantage in using the shell-side of the exchanger to heat the crude.

However, care must be taken in the selection of the baffle geometry. Work on identifying baffle geometry that provides the most uniform flow distribution through tube bundles is underway. This work has shown that there can be significant regions of stagnant flow if a poor baffle geometry is used. A segmental baffle is preferred. The cut should be in the range 20-30% and the spacing should be such that the window and cross-flow areas are approximately the same.

The need for a high level of wall shear means that shell diameters of less than 1 metre are favoured. The use of large shell diameter, even with close baffle spacing, is to be avoided.

Helical baffles only have a heat transfer advantage over segmental baffles at very low shell diameters (0.3 m or less) and large baffle angles. The use of large baffle angles at large shell diameters leads to both low velocities and low heat transfer coefficients. The use of low baffle angles does not lead to better heat transfer than can be achieved using segmental baffles.

Tube Inserts

Heat transfer rates within exchanger tubes can be significantly enhanced through the installation of tube inserts. A range of insert types is available: wire mesh (HiTRAN), twisted tapes, helical coils, rotating coils (TUROTAL) and vibrating springs (SPIRELF).

Wire Mesh Inserts

Wire mesh inserts typically operate in the Reynolds Number range 500 to 4000. Under these conditions it is possible to obtain large increases in heat transfer coefficient without incurring high pressure drop.

The action of the loops in contact with the tube wall results in significant movement of fluid away from the tube wall. The shear at these points is high.

They can often be used to provide a highly cost effective revamp for one particular heat recovery match widely used in pre-heat trains – a match between vacuum residues and crude oil made prior to the de-salter. The traditional approach to exchanger design for this match is to place the residues (that have relatively high viscosity) on the shell-side of the unit. By switching the residues to the tube-side and installing wire mesh inserts two things are gained The heat transfer from the residues is significantly enhanced through the action of the inserts and the heat transfer to the crude is enhanced because it flows through the tube bundle rather than through the exchanger tubes.

With regard to using these inserts to mitigate fouling on the crude side of exchangers there is one significant worry. The heat transfer enhancement is due to the action of the wire loops at the wall of the tube. Should fouling not be totally suppressed these loops will become embedded in deposit. The action of the loop will then be stopped and fouling will accelerate.

Turbotal™

Turbotal inserts consist of a rotating wire coil drive through flow across a vane situated at the entry to the tube. They have to be correctly orientated in the flow direction.

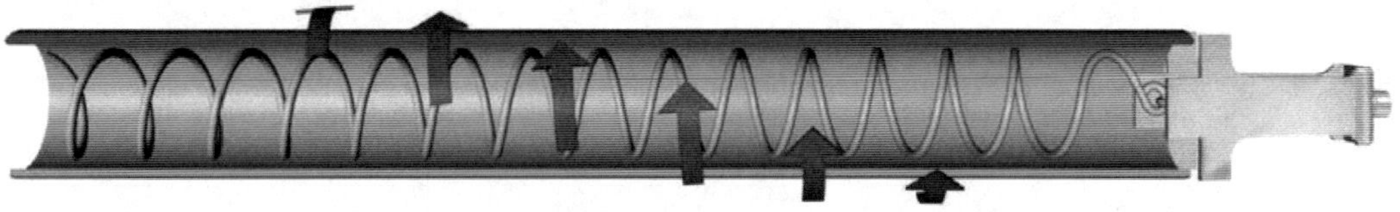

The units operate over a limited velocity range (0.7 to 1.6 m/s) and it is recommended that they be replaced every two years. This recommendation has resulted in significant limitation in their acceptance by refiners. However, the benefits of using these inserts are significant.

These units provide significant heat transfer enhancement.

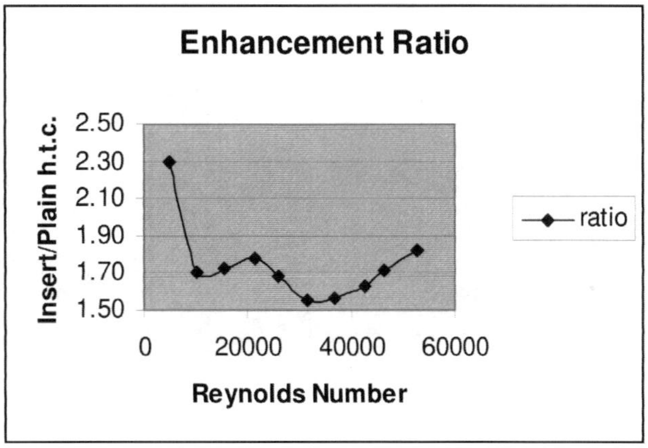

During the early stages of deposit development the mitigation effect is solely associated with the improvements heat transfer rates (which lead to reduced wall temperature) and to increased fluid shear. Once, the deposit has filled the gap between the tube wall and the coil a further mitigating effect comes into operation. The action of the coil appears to remove fresh deposit such that there is no further growth. This results in a predictable limit on the fouling resistance. This limit can be observed in the plant data shown below:

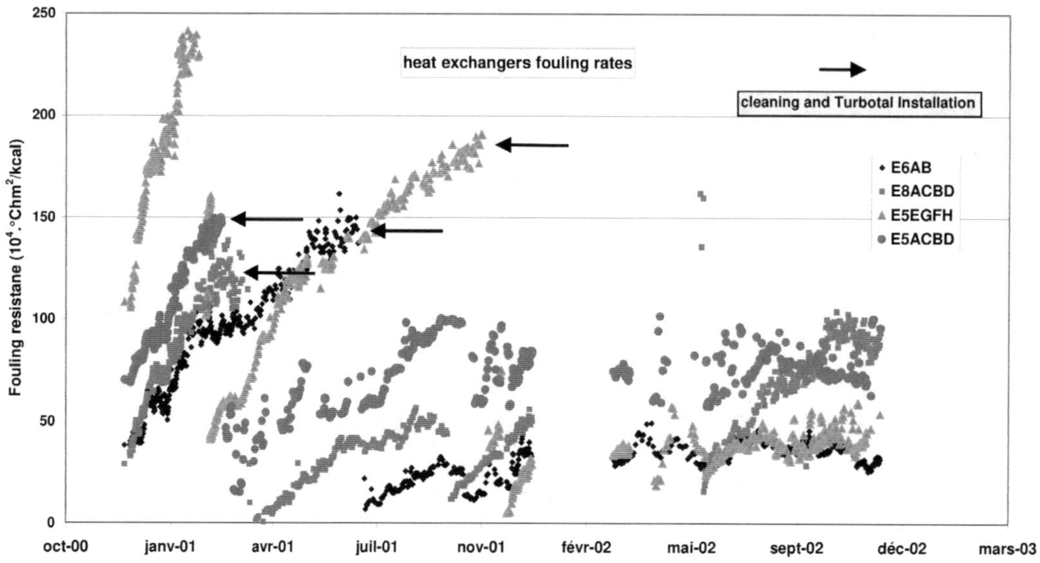

The limiting resistance is around 0.004 m^2K/W. In some refineries this level of fouling can be achieved in less than three months operation. In these circumstances the economic justification of using these inserts is usually large.

Spirelf™

Spirelf inserts consist of springs that can both move backwards and forwards and in an orbital manner within the tube.

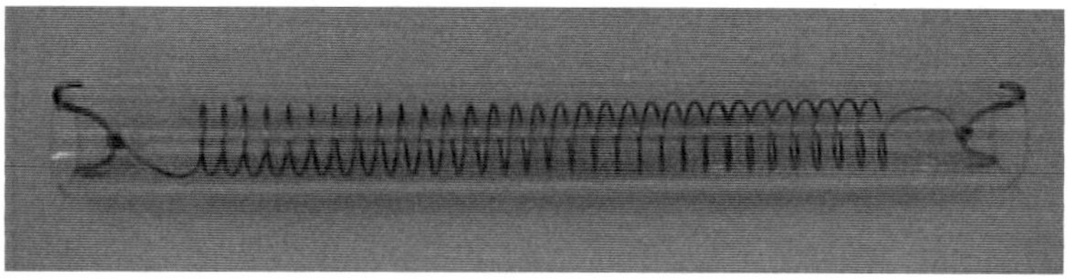

They provide significant heat transfer enhancement. This, coupled with the additional shear generated within the tube by the movement of the inserts results in significant reduction in fouling.

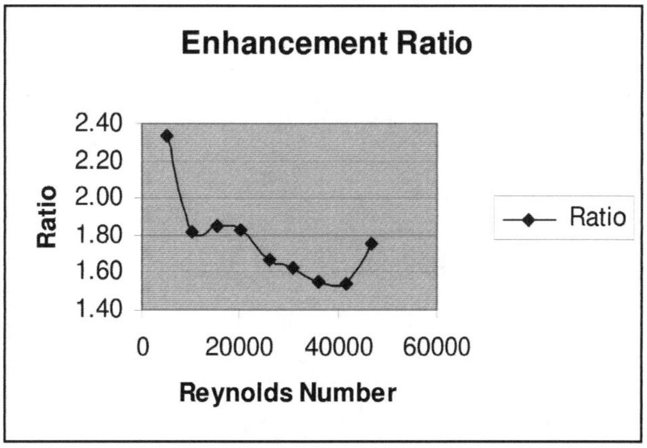

The inserts can operate over a wider velocity range than that specified for Turbotal. They also have a significantly longer operating life.

These inserts have been widely applied in refineries world-wide.

Compabloc^{TM}

Compabloc exchangers are welded plate units constructed from stainless steel or high alloys. The channels formed by the plates provide for counter-current flow throughout the unit. High shear rates (a factor of 5 greater than achievable in a shell-and-tube exchanger) are generated. This high shear leads to significant enhancement in heat transfer coefficients.

Since, both hot and cold side heat transfer coefficients are enhanced their use can result in increased wall temperature and increased chemical reaction fouling rates. Under turndown conditions, one unit was found to have higher fouling rates than encountered in a shell-and-tube exchanger on the same duty. Consequently consideration of fouling has to be made as an integral part of exchanger design and operation.

With regard to design, units can be set up with multiple pass arrangements (and the hot and cold streams do not have to have the same number of passes). The number of passes on the hot side can be smaller than the number of passes on the cold side with the objective of increasing the heat transfer coefficient on the cold side (or reducing that on the hot side) in order to reduce wall temperature.

In terms of operation care needs to be taken to maintain high shear rates under all operating conditions. Failure to do so can result in significant increase in fouling rates:

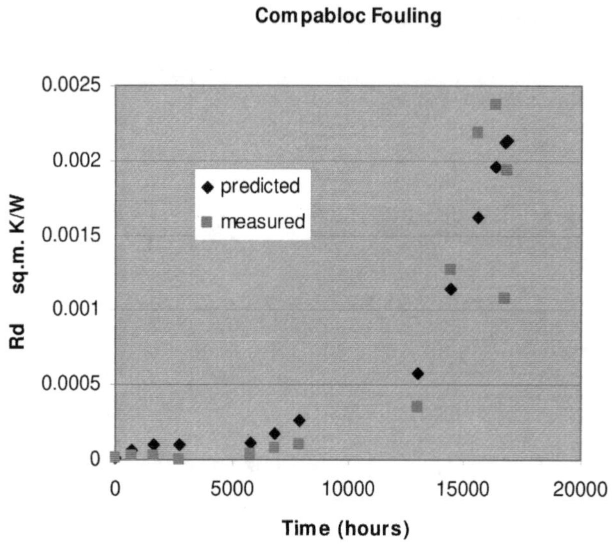

This plant had two Compabloc units. For one operating scenario (covering the first 8000 hours of operation) a single unit was employed. A second operating scenario required the use of two units. Unfortunately, the units were set up in parallel. Bringing the second unit into operation resulted in a halving of the flow through the original unit. So, wall shear stress conditions of typically 70 Pa were reduced to around 17 Pa. The result was a very sharp increase in fouling rate.

If the second unit had been set up in series with the first (with a bypass being used in the first operating period) the reduction in shear would not have occurred and the very low fouling rate encountered during the first operating period would have been maintained.

Conclusions

If we apply simple considerations of the interaction between the thermo-hydraulic behaviour of heat exchangers and the principle factors controlling fouling rates we can achieve better equipment designs and more efficient plant operation.

AIChE Paper Number

Increased fouling potential and Mitigation Plans for Heavy sour crude oils

Parag Shah

Refinery Technical Lead

Dorf Ketal Chemicals (I) Pvt. Ltd

P.N. Ramaswamy

Technical Director

Dorf Ketal Chemicals LLC

Matthew Knight

President & CEO

Dorf Ketal Chemicals LLC

Prepared for Presentation at the 2011 Spring National Meeting

Chicago, IL, March 16, 2011

Increased fouling potential and Mitigation Plans for Heavy sour crude oils

Parag Shah

Refinery Technical Lead

Dorf Ketal Chemicals (I) Pvt. Ltd

P.N. Ramaswamy

Technical Director

Dorf Ketal Chemicals LLC

Matthew Knight

President & CEO

Dorf Ketal Chemicals LLC

It is becoming a widespread norm that refineries are processing an increasing quantity of heavy and sour crudes than ever before. The depleting oil reserves of light sweet crudes have resulted in the need for deeper exploration. Considerable finds in the Orinoco belt of Venezuela, California and Canada has increased the need of refineries to cater to the heavy crude market. Heavy crudes present a host of different issues in the processing world and refiners world over are aligning their refining strategies to cater to the newer type of crude oils. During the recent economical problem, several refineries have evaluated processing such "opportunity crudes" to improve declining refining margins.

The significance of this has been documented in terms of crude processing within the desalter, crude column overhead corrosion concerns and high temperature corrosion due to a high Total Acid Number (TAN) amongst other downstream concerns. Furthermore, the number of models used by refineries for determining crude compatibility has increased significantly. It has been found that most of the refineries that continue to foul their pre-heat exchangers in the crude unit very diligently conduct compatibility tests to determine potential incompatibility issues. If this routine is so rigid, then why does fouling still occur?

Generally as the crude density increases, the amount of sulfur and iron within the crude also increases; such crude oils have higher solids and asphaltene content. It is known that at the temperatures incurred within the crude preheat train asphaltenes foul at a rapid rate. The precipitation of these asphaltenes is accelerated by several factors although the crude blend in itself is compatible.

The ultimate consequence of the above situation is an increase in energy required to meet the same furnace outlet temperatures - often, there can be a limitation in maintaining these temperatures without achieving adequate fouling control in the heat exchangers. The need to fire more in the furnaces also increases emission levels.

This paper shall discuss:

- Preheat configurations and their general operating parameters from experience
- Fouling mechanisms of heat exchanger trains within the crude unit
- Foulant analysis results of deposits within heat exchangers
- Methods used to minimize fouling (including models, additives, inserts)

The understanding of the above points is a fundamental necessity for processing heavy crudes with complex characteristics and hence the purpose of this paper.

It has been accepted globally that fouling is occurring within crude preheat exchangers. Numerous studies and strategies have been implemented in order to control this fouling, yet the result is the same; fouling is still occurring.

There are a number of incentives for minimizing fouling of crude preheat train: decreased maintenance costs, reduced energy costs, increased capacity, confidence to purchase opportunity crudes and environmental benefits.

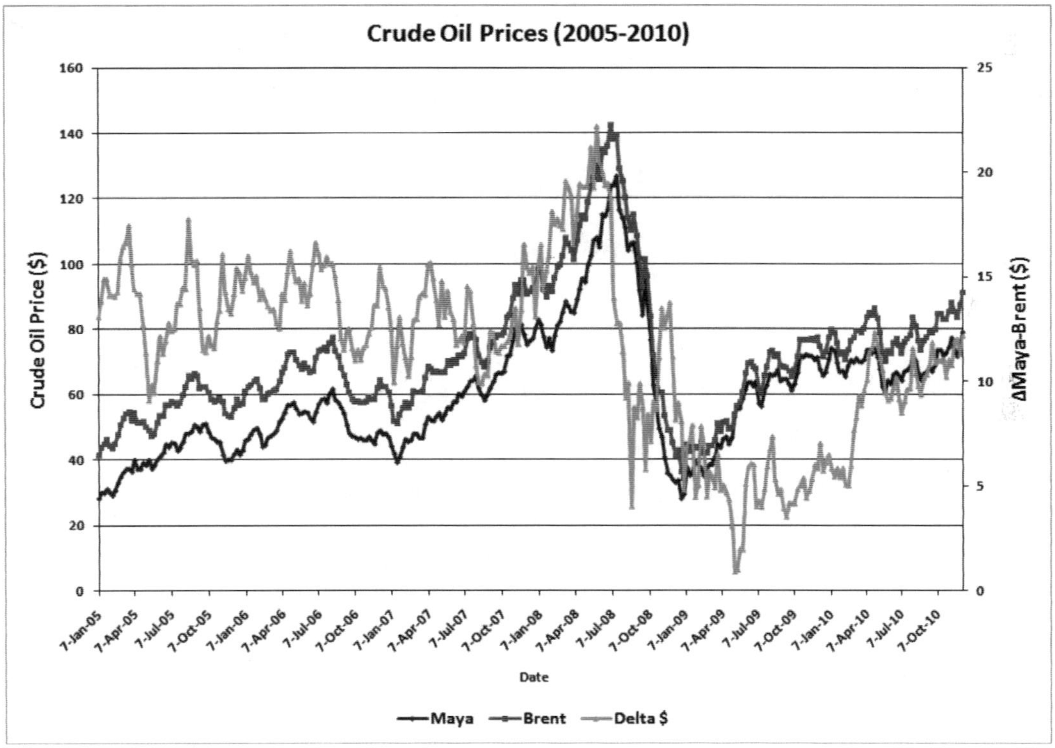

Figure 1: Crude Oil Prices (2005-2010) (U.S.EIA)

The above graph shows the trend for prices of Brent Crude (U.K., 38° API) and Maya Crude (Mexico, 22° API), (U.S.EIA). It is evident that the cost incentive from a purchase perspective is stronger for the opportunity crude, Maya, as it is \$7-\$12 cheaper per barrel than Brent Crude. However, along with the cost incentive, the following problems are ever-occurring due to fouling:

- Loss of energy (due to heat transfer problems) and its effects on cost and the environment. This refers to the additional fuel required for the furnace due to the reduced heat recovery in the preheat train as exchangers foul. The use of more fuel leads to additional production of CO_2
- Increased pressure drop
- Loss of throughput (reduced production)
- Replacement of equipment
- Cleaning and disposing of toxic wastes
- Increases safety hazards (fires)
- Maintenance costs.

Heat Exchanger fouling has been noted to account for 0.25% of the gross national product (GNP) in the highly industrialized countries (ESDU, 2000). Furthermore, the crude oil distillation unit is the largest consumer of thermal energy among distillation processes in chemical and petrochemical industries (Panchal & Huang Fu, 1998).

Steps to control fouling are implemented from the initial stages of refinery design. Licensors of preheat configuration technology incorporate fouling factors into the designs of their assemblies to limit the possibility of fouling occurring. TEMA (Tubular Exchangers Manufacturer's Association) tables are used to help design heat exchangers, however although the TEMA tables were originally considered to be rough guidelines, they are now often treated as accurate values causing considerable errors (IHS, 2007). Dorf Ketal has experience with a number of configurations given by different licensors. Three configurations are given below and Dorf Ketal's experience is discussed.

Configuration 1	
Cold Preheat Train	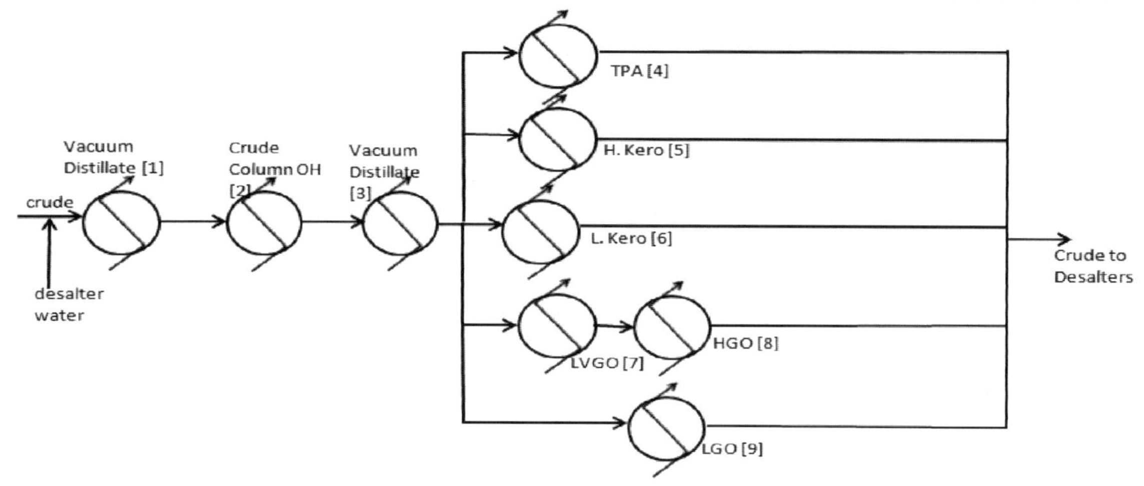

Hot Preheat Train	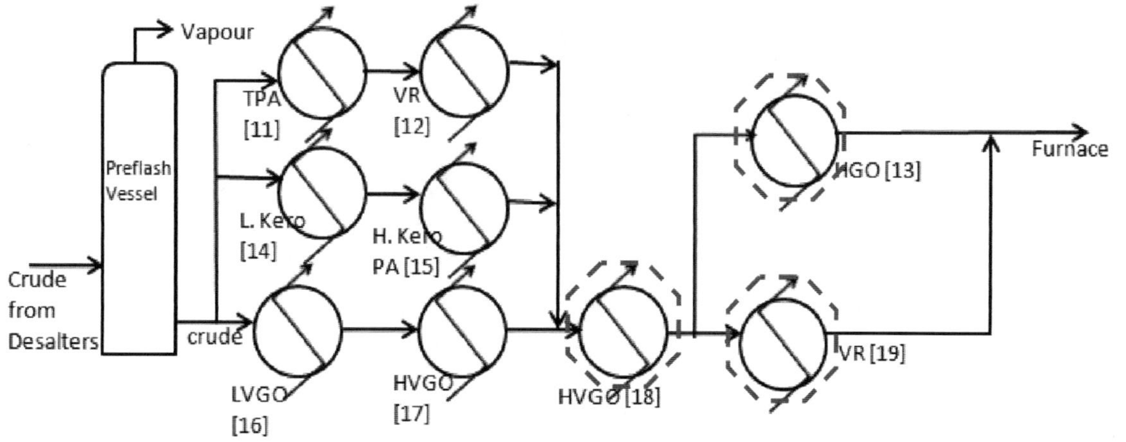

Configuration 2	
Cold Preheat Train	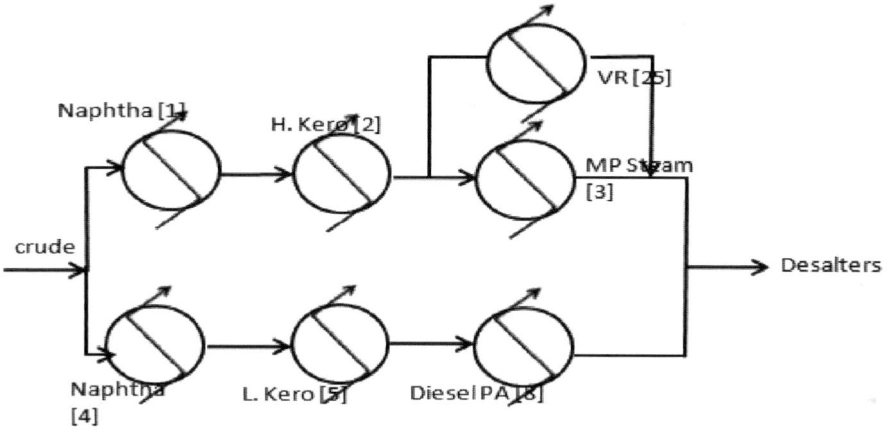
Warm Preheat Train	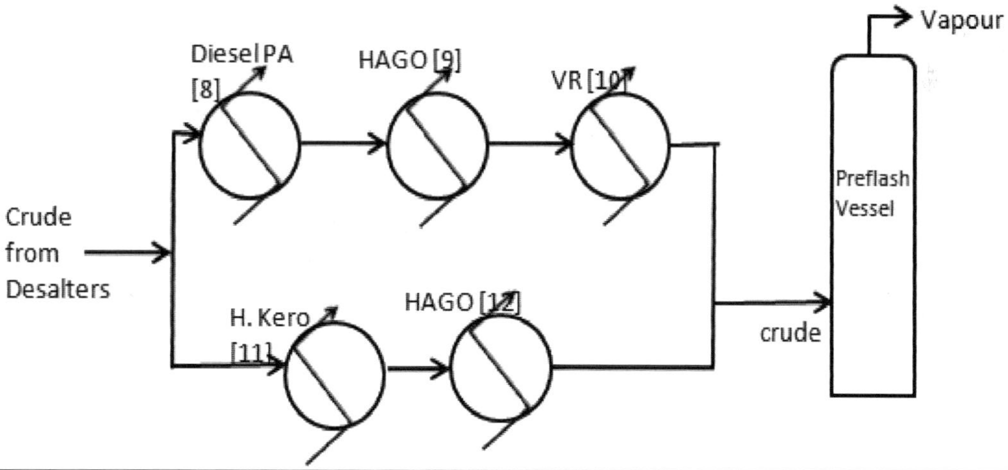

Hot Preheat Train

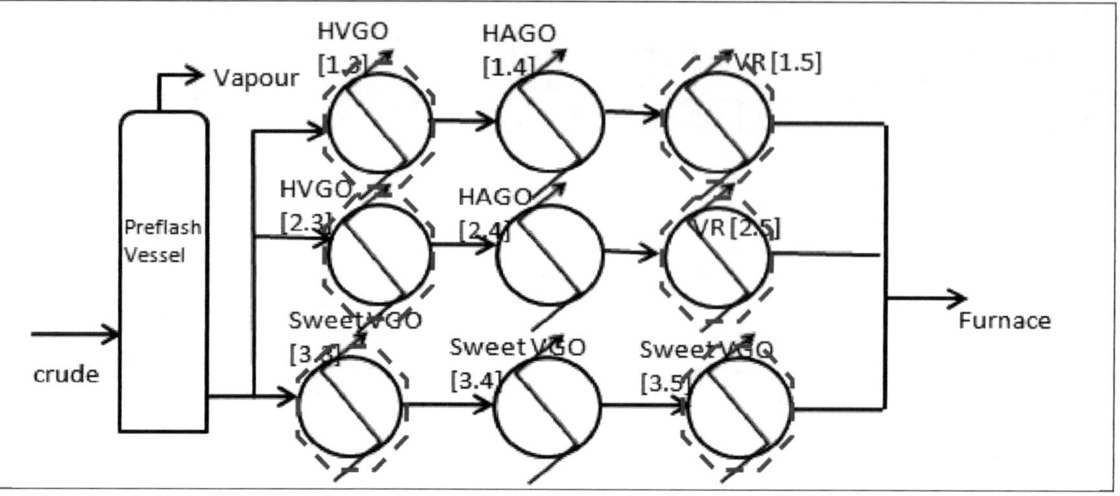

Configuration 3

Cold Preheat Train

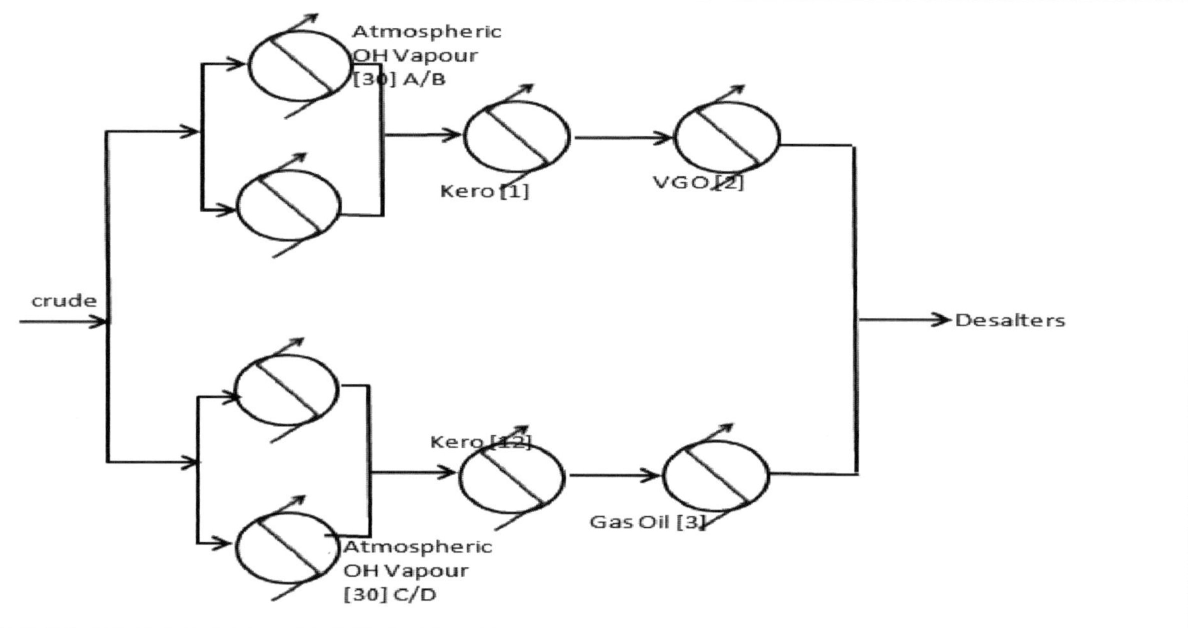

Hot Preheat Train

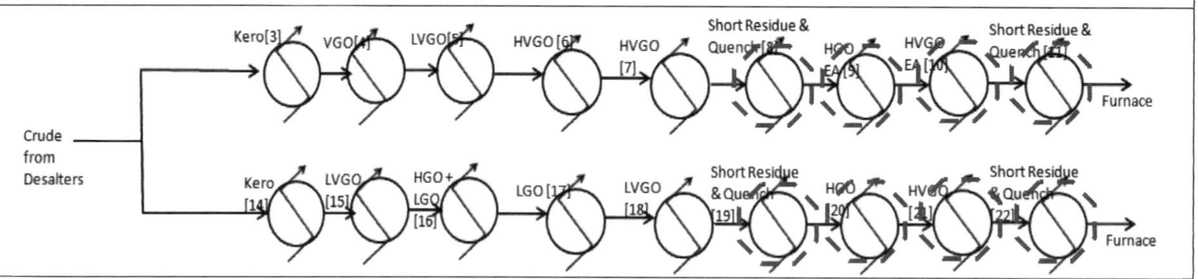

 Most Fouling-prone Heat Exchangers

Table 1: Configuration Characteristics

Parameter	Configuration 1	Configuration 2	Configuration 3
Hot Preheat Inlet Temperature		180°C	115°C
Heating Media	HVGO, HGO, VR, TPA, H. Kero PA, LVGO	HAGO, HVGO, VR	Kero, VGO, LVGO, HGO, HVGO, SR
Furnace Inlet Temperature (FIT)	270°C	265°C	270°C (Train 1); 275°C (Train 2)
Most Fouling Prone HX	Final 3 HX (HVGO, HGO & VR)	1 and 3 series (VR, HVGO and sweet VGO)	Final 3 HX (temp above 170°C)
Maximum Heat Recovery HX		HVGO (Penultimate HX Train 1); LGO (Train 2)	

Heat integration is optimized such that the streams receive the amount of energy required whilst limiting the fouling rates. Despite this measure, there is a long way to go before fouling is completely controlled. The second step is taken by the refiners themselves, by investing in crude compatibility software and models to ensure that the crude blend does not hinder operations. This is of interest especially when crudes of a heavy nature are being processed. These models give an indication of the probability of asphaltene fouling and generally do not predict the likelihood of fouling due to inorganics in the cold preheat train.

The fouling mechanisms in the cold preheat train, warm preheat train and hot preheat train all vary due to the temperatures experienced at each section.

Fouling that occurs within the cold preheat unit is initiated even at almost ambient temperatures due to the impurities that enter the crude unit with the crude. This type of fouling is known as inorganic fouling and is attributed by mostly corrosion products, silts and clay, other inorganic salts, sodium, magnesium and calcium oxides.
In detail, the inorganic constituents of the crude oil may be there as a result of:

1. Organic molecules containing hetero atoms. For example sulfur is often a constituent of crude oil in the form of mercaptans or other organo-sulfur compounds
2. Organo-metallic compounds since crude oil usually contains vanadium and nickel that may be present in combination with complex organic structures
3. Corrosion products that have become detached from upstream equipment and are trapped within the deposits produced by chemical reactions (Bott, 1995).

Depending upon the size of the particles, these solids are divided into 2 categories namely basic sediments and filterable solids. Particles above 20 micron size, which can be easily centrifuged out during standard BS&W test are called as Basic Sediments, whereas smaller particles are called Filterable Solids. Filterable solids are non-polar compounds and are coated with crude oil. Generally desalters are designed to give 50 to 60% filterable solids removal efficiency. The rest of the solids gets carried over to downstream equipments and settle down at low velocity areas of the equipment; this shall be discussed in detail later.

Corrosion that occurs within the heat exchanger and/or corrosion products within the crude oil represents a resistance to heat transfer hence contributing to fouling. The reactions that occur involve the production of ferrous and ferric ions.

$Fe \rightarrow Fe^{2+} + 2e$ Ferrous Ion (Eq. 1)

$Fe^{2+} \rightarrow Fe^{3+} + e^{-}$ Ferric ion (Eq. 2)

Iron sulfides in particular are heavy enough that they tend to settle out and foul various refinery equipments like Heat Exchangers/Trays. Table 2 shows the analysis results of a foulant within the cold preheat train with a heating media of naphtha. As is evident, the inorganic content is particularly high with 67% and 33% inorganic on the tube and shell side respectively. Iron specifically is of a high value at 3611ppm and 1479ppm.

Table 2: Cold Preheat Foulant Analysis

Cold Preheat Exchanger: Crude/Naphtha		
	Tube Side	Shell Side
Test	Result (ppm)	
pH	5.1	6.25
Chloride	9528	3062
Fe	3611	1479
Ammonia	2214	553
Sulphate	170	350
Ash	67.00%	33.00%
Ash content rechecked	66.20%	
Fe after acid soluble	49%	

Fouling in the hot preheat train is somewhat different as it is mainly organic. Fouling rates increase in hot preheat exchangers with injection of caustic downstream of the desalter. The extent to which each species contributes to fouling, depends upon crude properties, operational parameters etc. The properties of the deposit depend on the conditions present at that location within the heat exchanger and the amount of time the deposit has been in that location. This is one of the main contributing factors to lack of efficient fouling models available as it is difficult to simulate fouling changes over surface area with time. We know that it is extremely unlikely that the deposit will be uniform throughout its thickness due to chemical reactions since as the deposit thickens there will be a redistribution of temperature that could change the properties of the deposit.

Crittenden (Crittenden, 1984) has listed some relevant facts:

1. Liquids containing molecules of high molecular weight have a greater tendency to form deposits compared with lighter fractions
2. Branched chains have a greater tendency to deposition (i.e. increased deposition rates)
3. Thermally decomposed liquid streams are particularly susceptible to gum formation and deposition
4. The presence of oxygen (and its concentration) has a marked effect on fouling rates

5. Some sulfur and nitrogen containing compounds, particularly those that break down under the conditions prevailing near the heat exchanger surfaces to produce free radicals, promote deposition
6. Dissolved metal complexes and metals can increase deposition through catalytic action. Copper is said to be particularly deleterious
7. Desalting of crude oil streams does not always eliminate deposition of inorganic material.

Deposition of asphaltenes occurs through a series of degradation stages as crude oil is heated has been described by Eaton and Lux (Eaton & Lux, 1984). The process involves oxidation and polymerization.

$$\text{Saturated Hydrocarbons} \xrightarrow{\text{Inorganic Acids}} \text{Unsaturated Hydrocarbon} \xrightarrow{O_2} \text{Organic Acids} \quad \text{(Eq. 3)}$$

$$\text{Organic Acid} \xrightarrow[\text{Metals}]{} \text{Resins and Asphaltenes} \xrightarrow[\text{Hot Metals (Wall)}]{} \text{Coke-Like Deposit} \quad \text{(Eq. 4)}$$

Another reaction involving asphaltenes is given below and occurs in all hot preheat systems

$$\text{Asphaltenes} \xrightarrow{\text{Stress}} \text{Destabilized Asphaltenes} \xrightarrow{\text{Aggregate}} \text{Precipitation} \quad \text{(Eq. 5)}$$

Time is also necessary to allow these reactions to proceed to any degree (Crittenden, 1988).

Organic fouling is attributed by asphaltenes (poly nuclear aromatics) and waxes/heavier paraffins, coke (carbon), polymeric hydrocarbon (gum) and organometallic compounds which may exist in crude oil in various quantities and forms. The toughness of the precipitate deposited from the crude oil has a lot to do with whether there is asphaltene present in the crude oil even in minute quantities. Asphaltene, which is a highly polar compound, could act as a glue and mortar in hardening the deposits. Asphaltene is defined by chemists as the part precipitated by addition of low–boiling paraffin solvent such as n-pentane, and benzene soluble fraction whether it is derived from carbonaceous sources such as petroleum, coal, or oil shale. Asphaltenes readily destabilize when subject to changes in pressure, temperature, pH and solution environment. This can also occur when oils are blended and processed. On heating above 300–400°C, Asphaltenes are not melted but decompose, forming carbon and volatile products. Asphaltenes are mainly composed of sulfur, nitrogen, oxygen and carbon. Not much is known of the chemical properties of asphaltenes. Asphaltenes are lyophilic with respect to aromatics, in which they form highly scattered colloidal solutions. Specifically asphaltenes of low molecular weight are lyophobic with respect to paraffins like pentanes and petroleum crudes. It is demonstrated that while asphaltenes are the major compound responsible for various arterial blockage cases, in the petroleum industry its interaction with other heavy organics is a major factor whether depositions will occur or not (Mansoori).

Given below is the structure of a typical asphaltene molecule.

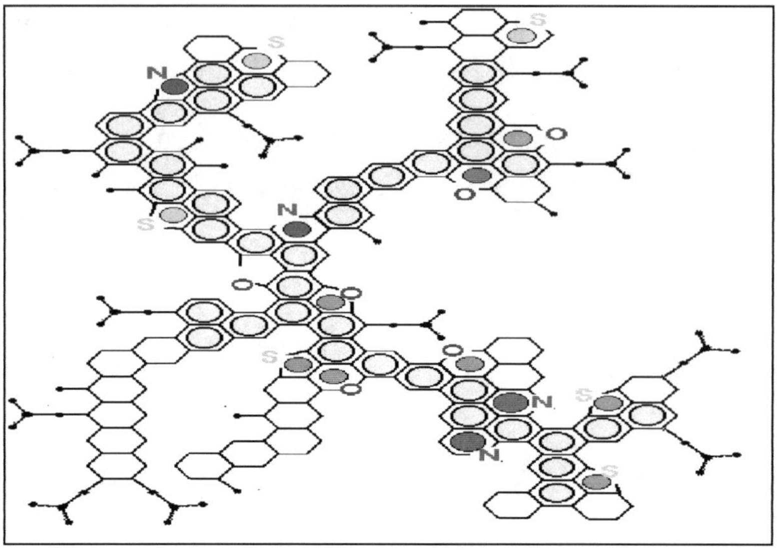

Figure 2: Typical Asphaltene Molecule Proposed for Maya crude (Mexico) by Altamirano, et al. [IMP Bulletin, 1986]

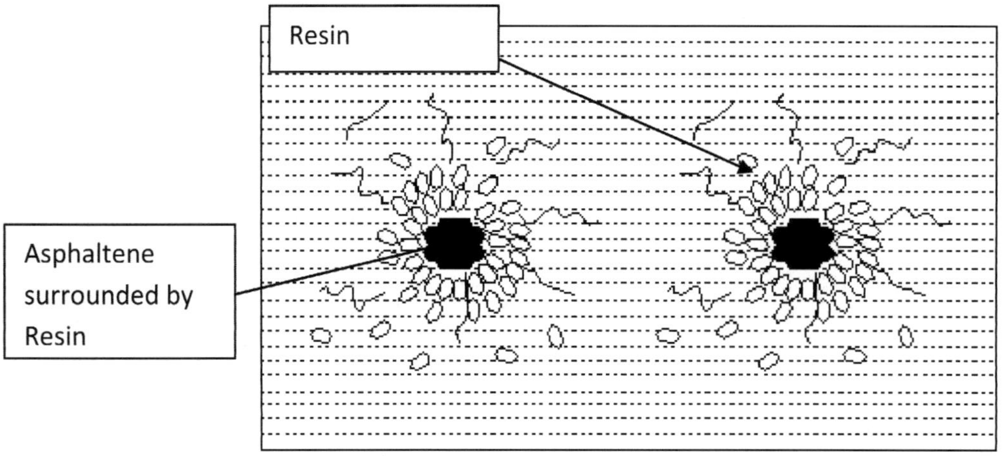

Figure 3: Physical Structure of Crude Oil at Storage Temperatures

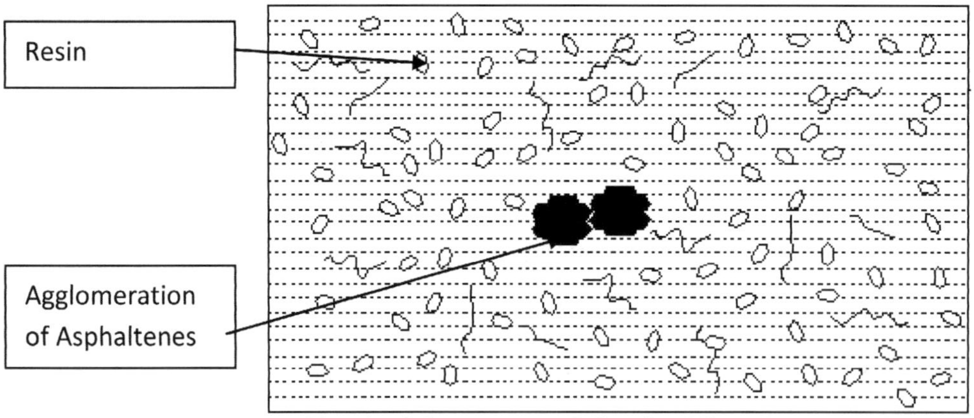

Figure 4: Physical Structure of Crude Oil at Process Temperatures

At these process temperatures, the concentration of the paraffin compounds increases and resins are lost from conversion to asphaltenes. The naphthenes become aromatic and the aromatics condense to form asphaltenes which precipitate out.

Wiehe (Wiehe, 2005) stated that when insoluble asphaltenes are deposited on a hot heat exchanger surface they may form coke, depending on the time and temperature. Above 370°C this reaction is almost instantaneous and below 200°C it does not happen in many years. The order of movement is such that the high temperature environment destabilizes asphaltenes which then aggregate to form large particles and precipitate. These precipitates deposit on the surface of the hot tubes and upon further heating, dehydrogenate to form coke. The significance is that once coke forms, it cannot be re-dissolved in the oil or any solvent. This is the reason that the fouling of the highest temperature heat exchanger in the crude unit is usually the greatest even though the asphaltene solubility is more at higher temperatures. Table 3 shows the analysis of a number of foulants within different hot preheat exchanger circuit. Dichloromethane (DCM) solubility gives an indication as to the amount of foulant that is asphaltenic. It is evident that asphaltenes dominate the properties of the foulant. The heptanes solubility is limited, as a property of asphaltenes is that they are insoluble in heptane.

Table 3: Hot Preheat Exchanger Foulant Analysis

Foulant Results (%)	Sr.1	Sr.2	Sr.3	Sr.4	Sr.5	Sr.6
DCM Solubility	11.3	11.36	14.3	18.1	21.3	18
Heptane Solubility	6.9	7.6	10.66	12.8	16.5	16
Loss on Drying	2.48	4.98	1.3	7.48	4.3	10.71
Ash Content	-	-	-	16.1	17.6	14.7
Asphaltene Content	4.4	3.76	3.64	4.9	4.59	-
Coke	-	-	-	59.7	57.8	-

There are a number of mechanisms that contribute to asphaltene precipitation as below (King, 2009):

- CO_2 - both from outgassing and acid shift of pH
- Acid - both unspent and spent acid can have significant effects
- pH - usually from CO_2, mineral or organic acid
- Turbulence - mixing of crude oil streams is a major culprit
- Chemical Shift that upsets micelle - loss of light ends, addition of alcohols and mutual solvents
- Stored, static conditions, where oil becomes 'dead' and allows the asphaltene platelets to aggregate. Stored crudes used as kill fluids are often a major cause of asphaltene problems.

Another significant contributing mechanism is that of coking catalyzed by metals. This is the reason why fouling is aggravated at the start-up of a unit and following cleaning when metal coils are exposed fully. Following this, thermal degradation of the asphaltenes begins as discussed earlier.

Generally the rate of fouling has been seen to increase with the increase in surface temperature as per the Arrhenius-style equation below (Crittenden, 1984):

$$\frac{dR_f}{dt} = A\exp(-E/RT_s) \qquad \text{(Eq. 6)}$$

where activation energy, E and proportionality factor, A are determined from experimental data and T_s is the surface temperature.

Recently, rules and regulations have become more stringent with regards to emissions and hence it is more vital to control fouling within the preheat train. Accelerated fouling will result in a need for increased fuel firing to achieve the necessary temperature required for the same yields. This, in turn, results in increased CO_2 emissions. According to studies by ESDU, crude oil fouling accounts for around 10% of the total CO_2 footprint of a refinery (IHS, 2007).

There are a number of methods undertaken to minimize fouling all of which contribute in some manner to achieve the desired results and run lengths.

- Oil Compatibility Model: this model helps determine whether crudes when blended in a certain ratio with each other or in a certain order are compatible or not. Some crudes have phase-separated asphaltenes already present e.g. Isthmus, which under any blending condition shall cause problems within the crude unit desalters. Other crudes however, have good properties as stand-alone crudes but when blended cause issues like asphaltene deposition. Dr. Irvin Wiehe has conducted extensive research in development of this model, which is now recognized globally. Dr. Wiehe's method involves determining two key parameters through laboratory studies; solubility number and insolubility number. Based on a calculation involving these numbers, it is possible to determine the likelihood of fouling of a crude blend.

- Another method to mitigate fouling is by constructing inserts within the heat exchanger tubes (e.g. HiTran, Spirelf, Turbotal™ (Bories & Patureaux, 2004)). These inserts offer enhanced heat transfer and fouling resilience but with increased pressure drop for a similar flow rate (Wilson, Polley, & Pugh, 2005). It has been noted that fouling resistance is a linear function of time, however after a certain time fouling occurs in an asymptotic or equilibrium manner which indicates a naturally occurring decrease in fouling rate.

- Ebert and Panchal have devised a model that can predict the fouling tendency of the exchangers of a crude unit preheat train. This model assesses the fouling threshold velocities at both ends of individual exchangers, heat exchanger temperatures and duties, identification of possible exchanger rearrangements within the network to mitigate its fouling. The deficiency in the model arises when trying to determine the change in fouling along the heat exchanger over time. For analysis at a single location at a specific point in time, the model is very efficient, but for real-time analysis it is limited. The Ebert and Panchal Model (1995) is stated below (Panchal, Kuru, Liao, Ebert, & Palen, 1999):

$$\frac{dR_f}{dt} = \alpha \,\mathrm{Re}^{-0.66}\,\mathrm{Pr}^{-0.33}\exp(-E/RT_f) - \gamma\tau_w \qquad \text{(Eq. 7)}$$

where α, E and γ are constants to be determined from the experimental data (Re=Reynolds Number; Pr=Prandtl Number; τ_w= wall shear stress) Following the introduction of this model, there have been a number of modifications to incorporate a greater number of factors that contribute to fouling within a preheat train.

- Another methodology is the Colloidal Instability Index, CII, (Asomaning & Watkinson, 1999).

$$\mathrm{CII} = \frac{\text{Asphaltenes + Saturates}}{\text{Resins + Aromatics}} \qquad \text{(Eq. 8)}$$

  CII is a measure of the instability of the crude blend. Using the ratio above, it is possible to determine the level of fouling expected from the incoming crude blend and so the CII value is kept within a set limit defined by the refiner. Theoretically CII should be kept less than 7 for stability, however in practice it is recommended to keep CII<2 (Hong & Watkinson, 2007). It is worth noting that this equation does not hold valid for situations where Resin and Aromatic content is negligible.

- EXPRESS® program helps identify rogue heat exchangers within a heat exchanger network (IHS, 2007) but does not help mitigate fouling as such.

Following this in-depth study into possible fouling mitigation methods, the final route to analyse is the use of chemical additives i.e. antifoulants. It is inevitable that at a reduced crude oil cost, the properties of the crude oil will not be so favourable towards refinery operations.

The mechanism of thermal destabilization of asphaltenes and subsequent precipitation from the crude phase triggers at a temperature of approximately 170°C as mentioned before. With antifoulants one can ensure limited asphaltene precipitation at this temperature and above such that the asphaltenes remain in a dispersed phase.

Chemistry of an antifoulant:

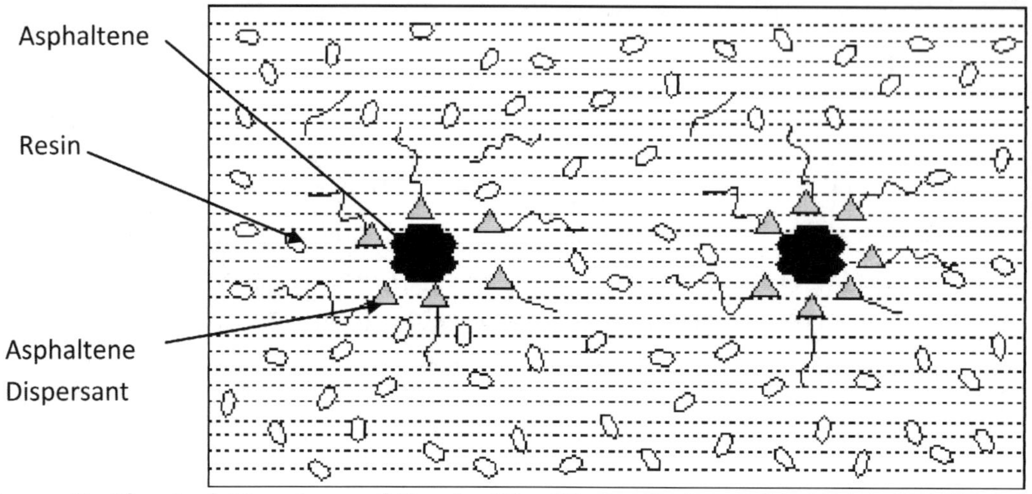

Figure 5: Physical Structure of Crude Oil with Asphaltene Dispersant at Process Temperatures

Tests have been conducted under microscopes to determine the behavior of asphaltenes with antifoulants. Figures 6 and 7 show 10x magnification of crude oil in paraffinic solvent.

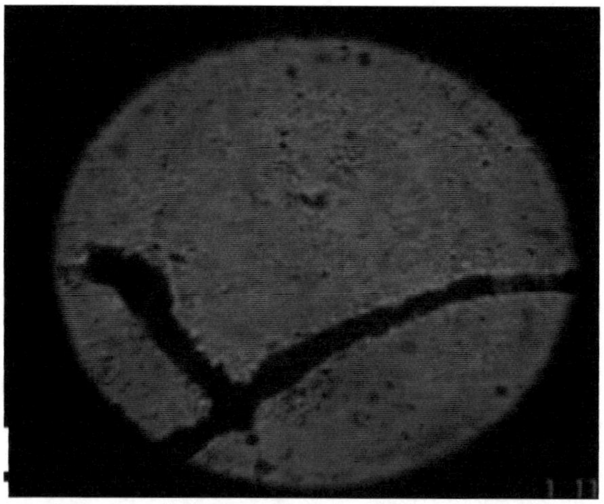
Figure 6: Asphaltene Agglomeration Initiation

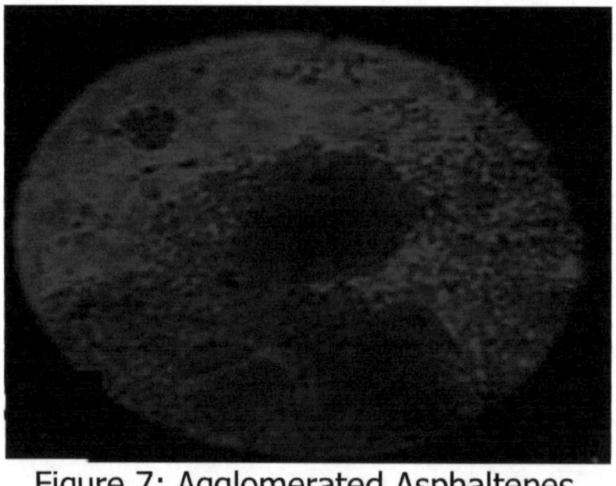
Figure 7: Agglomerated Asphaltenes

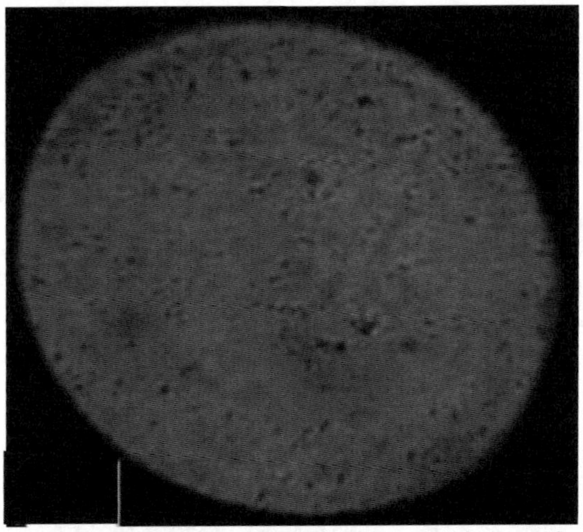

Figure 8: Dispersed Asphaltenes with antifoulant

Figure 9 shows asphaltene precipitation over time using Near Infared Light. Without dispersant the asphaltenes agglomerate and within 6 minutes, an evident phase separation is noted. With dispersant, the asphaltenes remain uniform throughout the crude.

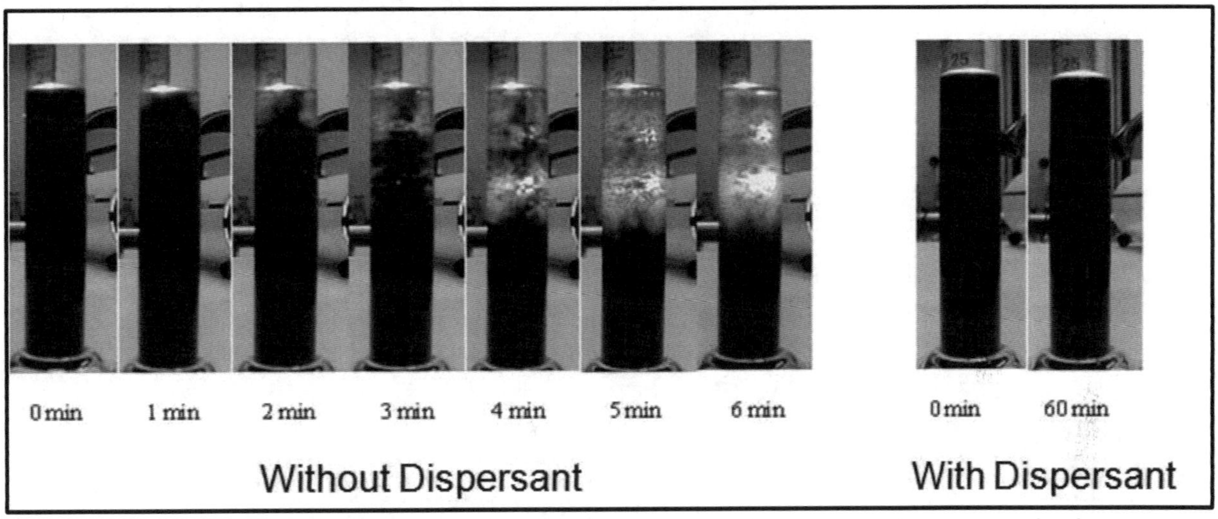

Figure 9: Antifoulant keeping Asphaltenes in dispersed phase vs. Blank

Monitoring of any antifoulant is necessary to ensure maximum benefits. To determine the fouling nature of crudes processed and then determine the dosage of antifoulant that is most beneficial to the system, Hot Liquid Processing Simulator (HLPS) is used. This instrument is basically a miniature shell and tube heat exchanger. If fouling occurs and material gets deposited on the tube surface, the heat resistance will increase and consequently the outlet temperature decreases. This equipment is used to determine the fouling tendency of a crude blend, along with allowing optimization of antifoulant dosage for desired results.

The performance of antifoulant and its beneficial effects on fouling is discussed in a number of case studies attached later in the paper.

- Trace quantities of impurities such as dissolved oxygen or suspended corrosion products add markedly to deposit formation. The influences of flow velocity, bulk and surface temperatures and particulate concentrations are demonstrated through experimental results, and compared to expectations from simple models. Through understanding of the key steps in the deposition processes a rational mitigation strategy can be formulated.

- Another contributing factor to crude preheat performance is the performance of the desalter. Desalter operation and appropriate chemical selection is key to ensuring minimal salt and solid carry-over to crude preheat trains. There are a number of key operating issues that contribute to desalter operation. Wash water addition of the correct quality and quantity is necessary as well as sufficient mixing and potential difference across the desalter for water agglomeration. Temperature is also important especially when crude blend viscosities are of a varying nature. Since crude blends are rarely of a consistent nature, timely operation changes are necessary along with chemical additives that can consistently give good performance regardless of the blend change.

CASE STUDY: A Refinery in South-East Asia

It is known that crudes containing high asphaltene content result in increased fouling of heat exchangers and further downstream equipment where temperatures are elevated. The issue of fouling can be very costly due to a number of hindrances caused to the refiner: fuel usage increases, operation is interrupted, maintenance costs increase and there are production losses.

In view of this, it is necessary to prevent heat exchanger fouling in the most cost effective and environmentally friendly way.

Dorf Ketal executed a trial of antifoulant in the hot preheat train. In August 2009, a heat exchanger, which has two sets of bundles, was due to be cleaned. The heat exchangers had not been cleaned prior to the trial so excessive foulants within the heat exchangers remained. Through use of antifoulant, the refinery was able to sustain heat exchanger operation without cleaning up until turnaround.

As stated, one heat exchanger within the hot preheat train was cleaned to maintain the profile across the vacuum column. This heat exchanger is a shell and tube crude-HVGO exchanger. Figure 1 shows the performance of the heat exchanger before and after cleaning. Following heat exchanger cleaning with the earlier treatment program, the duty achieved was at its peak immediately following heat exchanger cleaning and over the course of 4 months it did not achieved this duty again. Following heat exchanger cleaning in August 2009, the duty was approximately 17.65MW, the same duty achieved following heat exchanger cleaning in April 2009. However, the duty exceeded this initial duty on a number of occasions over the 3.5 months prior to turnaround. If a trend line is taken for the two periods, it is seen that following heat exchanger cleaning in August, the duty is almost constant as oppose to the evident decreasing trend from April.

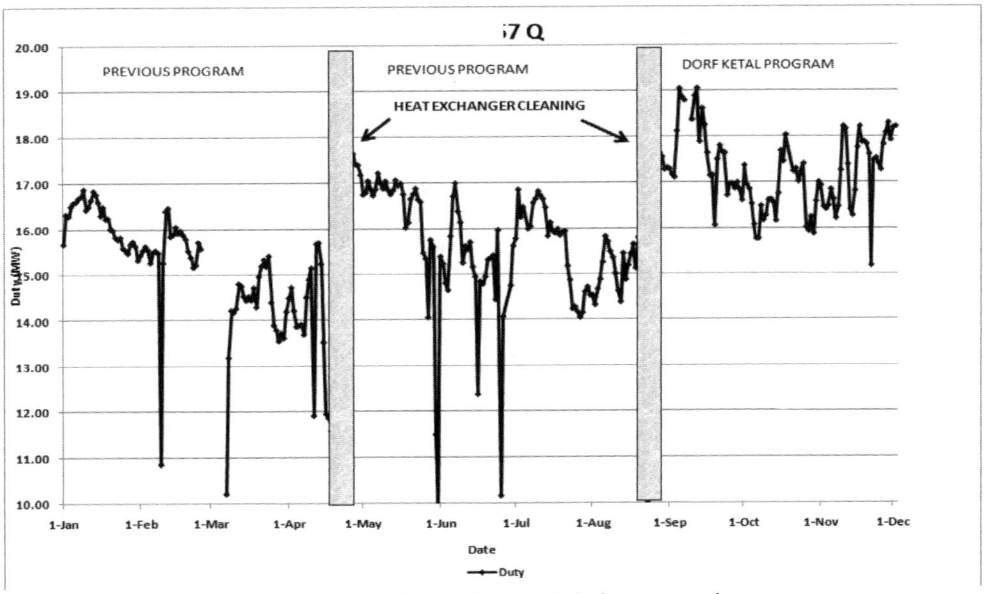

Figure 1: Duty of Crude-HVGO heat exchanger

There are a number of factors that one must consider when determining the performance of any antifoulant program. The condition of the heat exchangers prior to treatment contributes heavily, and so it is important to analyze the heat exchangers thoroughly before, as well as during, any new treatment program. Secondly, the quality of the feed processed contributes to the performance. If heavy, asphaltenic or incompatible crude blends are processed then these are properties that can lead to accelerated fouling within the hot preheat train. Finally the approach enthalpy to any heat exchanger affects the performance of that heat exchanger, and hence the hot preheat train as a whole depending on the duty contribution of that particular exchanger.

Considering the refinery was approaching a turn around, it could afford to process more challenging crudes than usual since all the units were soon to be opened and cleaned. This correlated with the time of the Dorf Ketal Treatment program, yet the results achieved were better than the earlier program. As mentioned, the four largest heat exchangers were monitored most closely.

The improvement in duty drop was clear and allowed the heat exchangers to perform until turn around without being cleaned.

Figure 2 shows how the duty of the heat exchanger stabilized following the introduction of Dorf Ketal antifoulant, whilst the approach enthalpy remained almost constant. This heat exchanger lies immediately downstream of the desalter. If the duty of the heat exchanger is extrapolated to the time until turnaround, cleaning would have definitely been necessary, hence incurring further costs to the refinery. It can be taken that each heat exchanger cleaning occurs at a cost of USD 16,000 (approximately). This figure does not account for reductions in throughput and slippage of products due to inefficient heat integration.

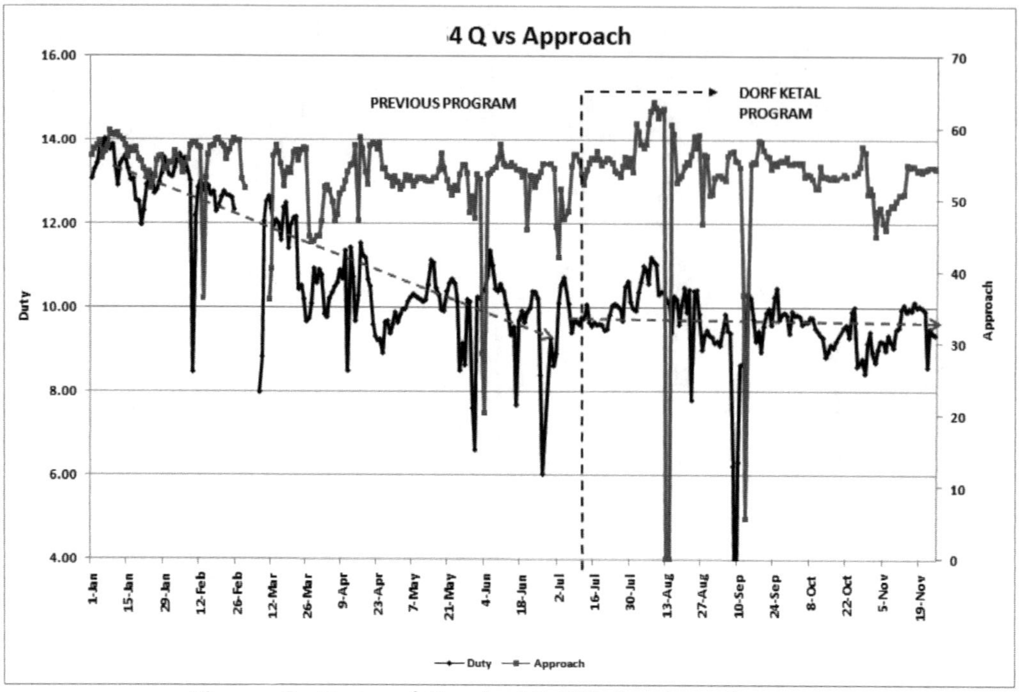

Figure 2: Duty of Crude-MVGO heat exchanger

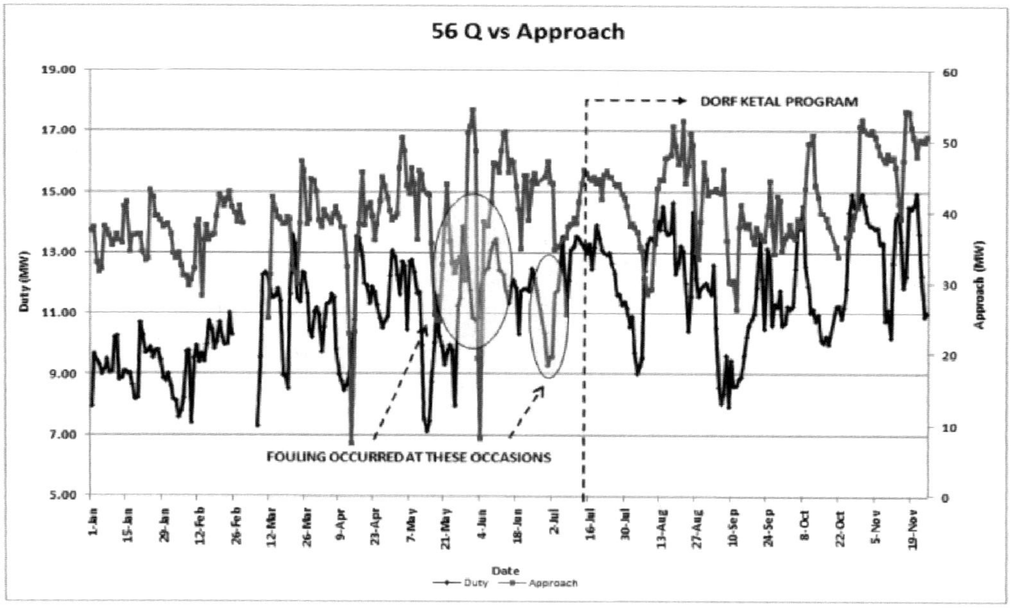

Figure 3: Duty of Crude-VR heat exchanger

The approach enthalpies have been included in Figures 3 and 4 to show how heavily the duty of these exchangers is influenced by the approach enthalpy and how often the approach changes. This also shows how often the crude diet and yields change within the refinery.

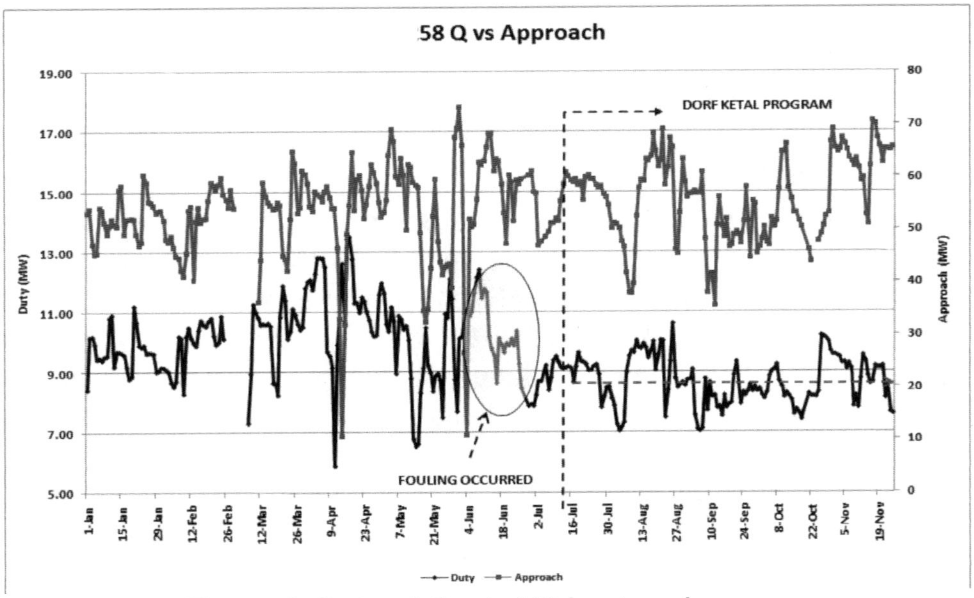

Figure 4: Duty of Crude-VR heat exchanger

Through use of the antifoulant, fouling rates due to organic fouling decreased significantly as seen in the four major heat exchangers. The refinery was able to achieve results that were completely unforeseeable due to the advanced technology and chemistry of the antifoulant treatment program.

CASE STUDY: An Asian Refinery

This particular refinery is renowned for processing a variety of crudes like Arab Light, Arab Extra Light, Arab Heavy, Maya, Sooroosh, Nowrooz, Khafji, Kuito etc. When blending crudes with such a wide spectrum of properties, it is all the more imperative to have fouling mitigation strategies in place.

Fouling in the preheat units encompass several aspects. In the feed exchangers fouling is related primarily to the type of crude (asphaltic, paraffinic) and to the history of handling prior to processing. The deposits typically found in crude feed exchangers consist of both organic and inorganic compounds. Inorganic deposits are found at temperatures between 300^{0}F and 500^{0}F (150^{0}C-360^{0}C). As the feed temperature rises above 300^{0}F (150^{0}C), water is evaporated and the inorganic salts reach saturation. These salts then begin depositing on exchanger surfaces. On the other hand, organic deposits, primarily asphaltic compounds, deposit as a result of blending of crudes, recycle of highly paraffinic streams or settling once they can no longer be held in solution. Asphaltenes that are soluble in benzene are lower molecular weight compounds naturally occurring in the crude while those not soluble in benzene represent the high molecular weight materials together with products of oxidation formed during transportation.

The particulates and corrosion products such as iron oxides or sulfides are collected on the surface of the organic deposits and also settle out in the low velocity areas. The fact that these inorganic materials are dispersed through most deposits makes it difficult to initially determine the sequence of fouling.

As a solution Dorf Ketal implemented a water injection facility upstream of the desalter to limit salt deposition. Where heavy salt deposits are found, fresh make-up water for the desalter can be added prior to the preheat exchanger to minimize deposition. Where sediment is a problem, improvement of crude oil handling is indicated. Older tanks should be cleaned and additional efforts made to segregate tanker ballast from crude oil. Where sediment or sludge is nominal, a dispersant can be applied so long as the additive does not interfere with demulsification in the desalter.

Good operation of the desalter will prove to be a valuable technique to limit downstream fouling by salts, as well as being the best means for corrosion control. Downstream of the desalter there are no truly effective means for removing salt deposits or preventing their deposition. Once the water of solution is evaporated, any entrained salts will drop out on exchanger surfaces. Water washing at this point, while on-stream, will only result in the movement of the salts to a higher temperature point where the water is again evaporated.

Organic fouling should be controlled by the use of Antifoulants. Ideally, antifoulants should be stable at high temperatures. The antifoulant helps in reducing the catalyzing effect of metals like Fe and prevent FeS deposition. The superior technology additives also reduce coking reactions on the metal surface.

Dorf Ketal makes use of number of tools to determine asphaltene flocculation within a crude unit preheat system which is a vital contribution to satisfactory results.

A mentioned, at this refinery, the crude basket consists of paraffinic crudes as well as asphaltenic crudes. These crudes when blended together above certain percentage causes incompatibility, which results in fouling due to Asphaltene precipitation in crude unit, preheat exchangers. The untreated system of the refinery used to lose 8–9°C of temperature in preheats exchangers/unit due to fouling. Heavy fouling was seen predominantly in the hot preheat with fuel consumption increasing at 10% per month. The refiner used to clean 1-2 exchangers every month to sustain unit operations. This led to throughput loss and raised several safety and environmental concerns. Following this performance, Dorf Ketal's preheat Antifoulant was tested in the lab and injection full-time began. Please find attached Furnace Inlet temperature trend of the system with and without usage of the Antifoulant.

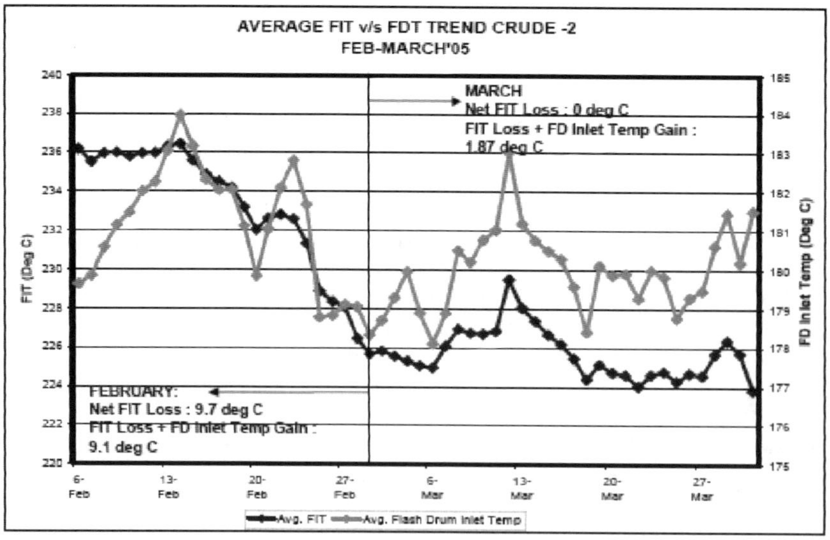

Figure 1: Crude Preheat Temperature-before and after Antifoulant usage

There was a significant return on investment with antifoulant usage. Lab screening results yielded a 75% reduction in fouling with the use of preheat antifoulant. With the application of Dorf Ketal's preheat antifoulant on field, superior fouling control was seen with reduction in fouling factor and stable furnace inlet temperature. Control in preheat fouling generated fuel savings of the order of 1 million USD in one year.

References

1. Asomaning, S., & Watkinson, A. P. (1999). Deposit Formation by Asphaltene-Rich Heavy Oil Mixtures on Heat Transfer Surfaces. (T. Bott, Ed.) *Understanding Heat Exchanger Fouling and Its Mitigation* , 283-290.

2. Bories, M., & Patureaux, T. (2004). Preheat Train Crude Distillation Fouling Propensity Evaluation by the Ebert and Panchal Model. In H. Muller-Steinhagen, P. Watkinson, & M. R. Malayeri (Ed.), *2003 ECI Conference on Heat Exchanger Fouling and Cleaning: Fundamentals and Applications, 27.* Santa Fe, New Mexico, USA.

3. Bott, T. R. (1995). *Fouling of Heat Exchangers.* Amsterdam: Elsevier Science and Technology Books.

4. Crittenden, B. D. (1984). Chemical Reaction Fouling in Fouling and Heat Exchanger Efficiency. *Inst. Chem. Engrs. Course* .

5. Eaton, P., & Lux, R. (1984). Laboratory Fouling Test Apparatus for Hydrocarbon Feedstocks. (J. W. Suitor, & A. M. Pritchard, Eds.) *Fouling in Heat Exchange Equipment , 35.*

6. ESDU. (2000). *Engineering Services Data Unit, Heat Exchanger Fouling-Case Study.* Retrieved December 10, 2010, from ESDU: http://www.uk.ihs.com

7. Hong, E., & Watkinson, A. P. (2007). Precipitation and Fouling in Heavy Oil-Diluent Blends. *Proc. 7th Intl Conf on Heat Exchanger Fouling and Cleaning.* Tomar, Portugal.

8. IHS. (2007). *IHS News Service.* Retrieved December 12, 2010, from IHS: http://uk.ihs.com/news

9. King, G. E. (2009, March 14). *GEK Engineering.* Retrieved December 22, 2010, from GEK Engineering: gekengineering.com/downloads/free_downloads/Asphaltenes_Depositon_And_Removal.pdf

10. Mansoori, G. A. (n.d.). *Asphaltene Deposition and its Control.* Retrieved December 20, 2010, from University of Illinois at Chicago: http://tigger.uic.edu/~mansoori/Asphaltene.Deposition.and.Its.Control_html

11. Panchal, C. B., & Huang Fu, E. P. (1998). Effects of Mitigating Fouling on the Energy Efficiency of Crude Oil Distillation. *AIChE Spring National Meeting, Session 81.* New Orleans.

12. Panchal, C. B., Kuru, W. C., Liao, C. F., Ebert, W. A., & Palen, J. W. (1999). Threshold Conditions for Crude Oil Fouling. In T. R. Bott, L. F. Melo, C. B. Panchal, & E. F. Somerscales, *Understanding Heat Exchanger Fouling and its Mitigation* (pp. 273-279). New York: Begell House.

13. U.S.EIA. (n.d.). *US Energy Information Administration.* Retrieved December 29, 2010, from US Energy Information Administration: http://www.eia.doe.gov

14. Wiehe, I. A. (2005). Petroleum Fouling: Causes, Tools and Mitigation Methods.

15. Wilson, D. I., Polley, G. T., & Pugh, S. J. (2005). Ten Years of Ebert, Panchal and the 'Threshold Fouling' Concept. In H. Muller-Steinhagen, M. R. Malayeri, & A. P. Watkinson (Ed.), *ECI Symposium Series: Proceedings of 6th International Conference on Heat Exchanger Fouling and Cleaning-Challenges and Opportunities. RP 2.* Kloster Irsee, Germany: The Berkeley Electronic Press.

Using Simple CFD Models to Identify Efficient Baffle Arrangements for Shell-and-Tube Heat Exchangers

Mario Alonso Vidal, A..R. Uribes Ramirez & G.T.Polley
University of Guanajuato, Mexico

Chemical reaction fouling is principally controlled by two factors: wall temperature and local fluid shear stress. Computer programs for the design of shell-and-tube exchangers generally calculate a mean heat transfer coefficient for the tube bundle. In reality both heat transfer coefficient and fluid velocity vary significantly from one location within the bundle to another location. This variation is a function of bundle geometry.

Commercial computer codes for the analysis and design of shell-and-tube heat exchangers generally allow the engineer to look at a very wide range of geometry. For instance, the prediction methods could cover baffle cuts ranging between 15 and 45% of shell diameter and virtually any baffle spacing to shell diameter ratio.

Designers working before the advent of these programs took a more conservative approach to baffle configuration.

It is possible to use CFD to determine the detailed flow field within a tube bundle. However, such models are difficult to develop and to modify. Furthermore, the solution of such models demands either a large amount of computer power or computation time.

In this paper the authors have analysed the performance of a rectangular channel containing a row of tubes. The CFD modelling of such a system is relatively straightforward and the computer power needed to obtain a solution to the model is low. Most significantly, experimental data on the distribution of heat transfer in such a system is available for one geometry operating at two different velocities.

Background

When using a shell-and-tube heat exchanger to heat crude oil many engineers place the oil on the tube-side of the heat exchanger. However, since the fouling on these duties increases with increasing wall temperature assigning the crude to the shell-side can result in reduced fouling rates for at a given heat flux the shell-side wall temperature is lower than the tube-side wall temperature because the surface area is significantly larger (typically 28 % greater for the thickness of tube used in many applications).

The "down-side" is that unless the baffle arrangement is selected properly stagnation areas can occur within the tube bundle. Then, because of low heat transfer coefficients and low shear stresses in these regions excessive fouling can occur.

There is a need for simple guidance on which baffle arrangements are efficient and which should be avoided.

The provision of such guidance can be achieved in a number of ways. A large experimental programme could be instigated; a large programme of work using sophisticated CFD modelling could be instigated; or, the use of idealised models (such as the photographic studies reported by Knudsen & Katz) could be used. In terms of effort and expense the latter approach is preferred.

Data from the literature

In the 1970's a simple rectangular test section was used at the UK's National Engineering Laboratory to obtain visual information on the two phase flow across a row of tubes (Figure 1).

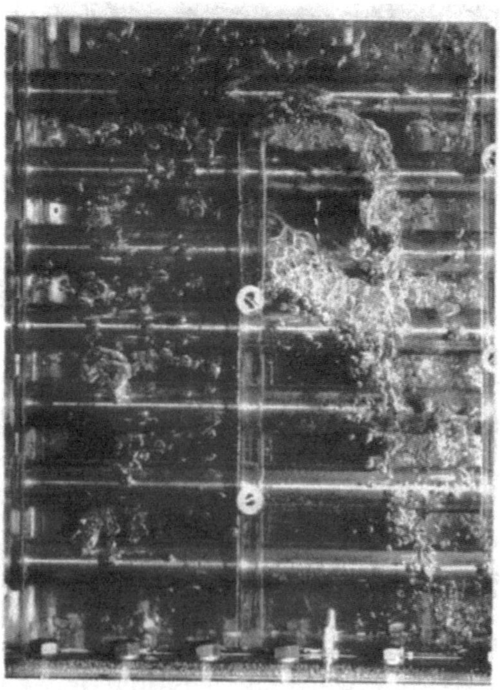

Figure 1. NEL Test Section

This facility also allowed the measurement of heat transfer coefficients at differing locations along the tubes and in differing tube rows. In addition to looking at two-phase flow measurements were taken when only water was pumped through the test section.

The results of these tests are re-produced in figures 1 and 2.

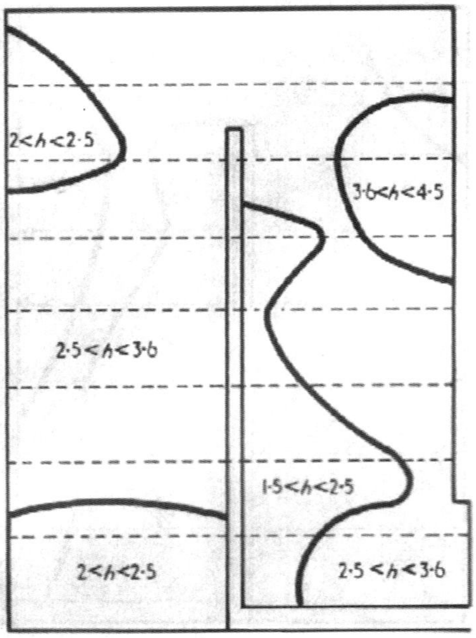

(a) WATER FLOWRATE 182 kg/m²s

Figure 2. Heat transfer distribution at moderate mass flux

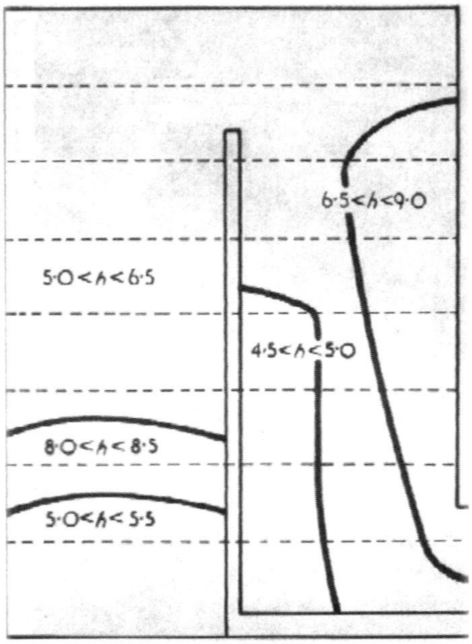

(b) WATER FLOWRATE 537 kg/m² s

Figure 3. Heat transfer distribution at high mass flux

A CFD Model has been set up of the test section used in these studies. The model has been used to determine the flow distribution through the apparatus.

Heat transfer coefficients have been calculated using the correlation derived during the extensive study of tube bundle behaviour undertaken at the University of Delaware in the nineteen sixties:

$$\alpha = jCp\,\overset{*}{m}\,\mathrm{Pr}^{-\frac{2}{3}}$$

where:

$$j = a\,\mathrm{Re}^{b}$$

Parameters for an in-line bundle were used.

The comparisons between predicted and measured heat transfer distributions are shown in Figures 4 and 5.

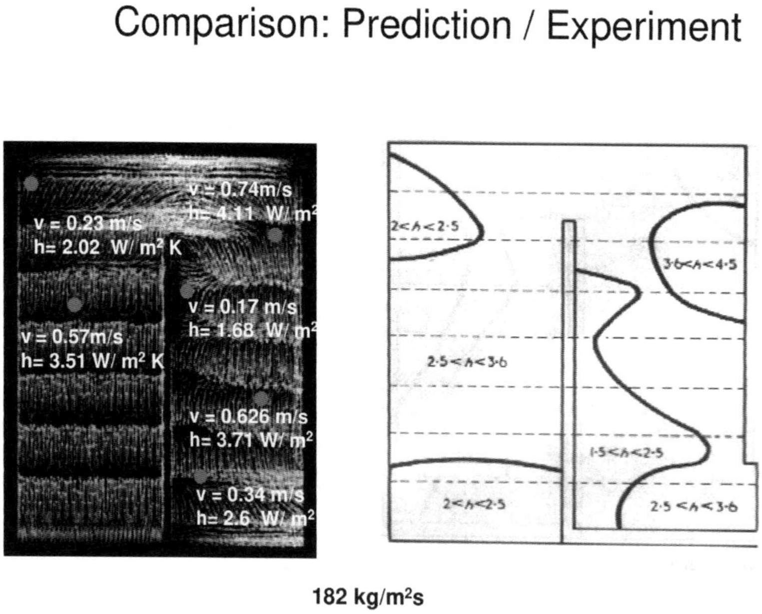

Figure 4. Heat transfer distributions at moderate mass flux

The comparisons indicate that the flow distributions are modelled quite well.

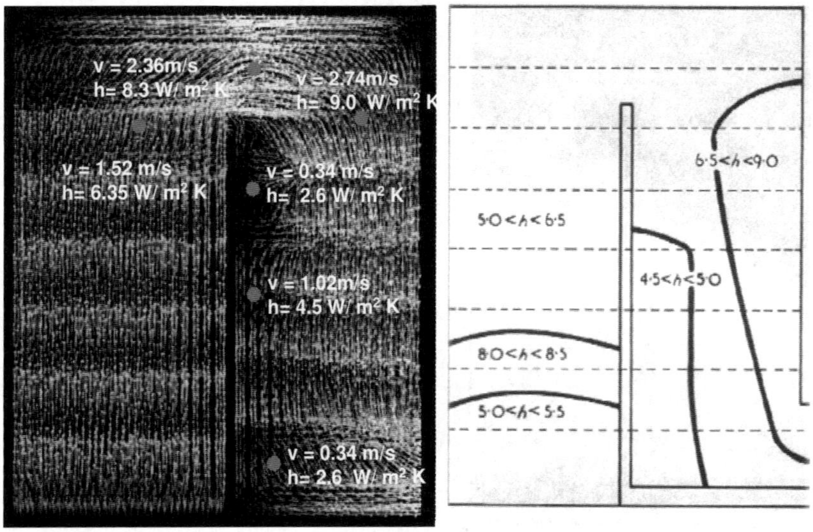

537 kg/m²s

Figure 5. Heat transfer distribution at high mass flux

Predictions of Baffle Behaviour

CFD models were then extended to look at the behaviour of single tube rows having various baffle separations and baffle cuts.

Results for a 15% cut with baffles spaced at 1.5 times the bundle height are shown in Figure 6.

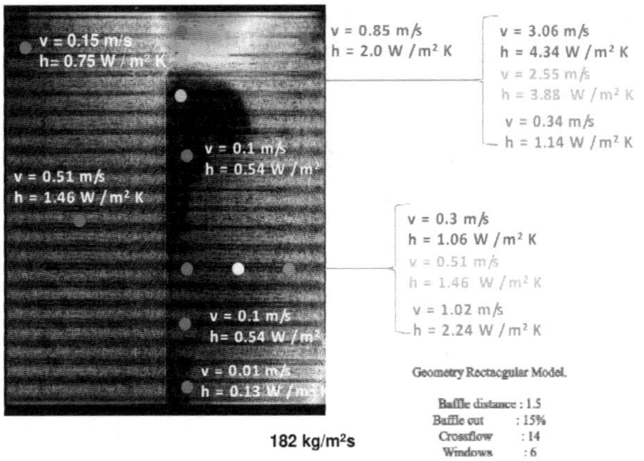

Figure 6. 15% Window – wide spacing

The velocity for a uniform distribution (shown in the first section of bundle) is 0.5 m/s corresponding to a heat transfer coefficient of 1 460

W/m^{2}K). In the window much higher velocities are observed. However, immediately behind the baffle there is a low velocity region whilst at the bottom of the baffle is a region of very low velocity.

Throughput has a marked effect upon these distributions. At high mass flux (537 kg/m^{2}s) good velocities are maintained throughout the bundle with the exception of a small area at the base of the baffle.

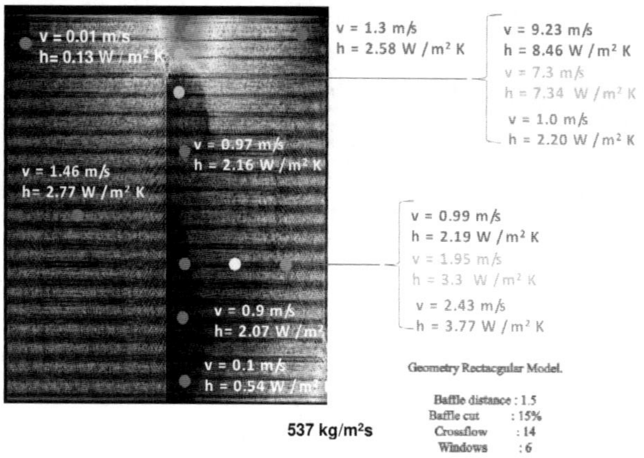

Figure 7. Same Geometry at High Mass Flux

In Figure 8 we observe the effects of tightening up the baffle spacing. Here the ratio of baffle spacing to bundle height is 0.5 (one third of the value used above). The mass flux is maintained at 182 kg/m^{2}s.

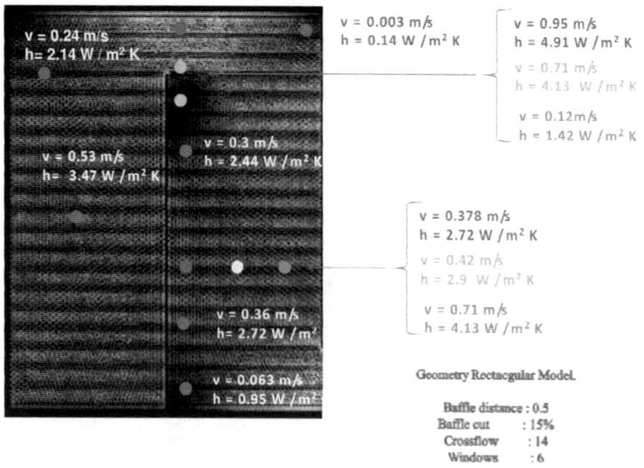

Figure 8. Tighter baffle spacing.

Here the flow is still skewed towards the second baffle. However, reasonable velocities are achieved throughout the bundle with the exception of small areas at the tip and at the regions where each baffle meets the edge of the duct.

Studies on the performance of baffle cuts of 25% showed similar results.

Studies on the performance of baffles having 40% cut showed large low velocity regions behind baffle and in the corners (Figure 8).

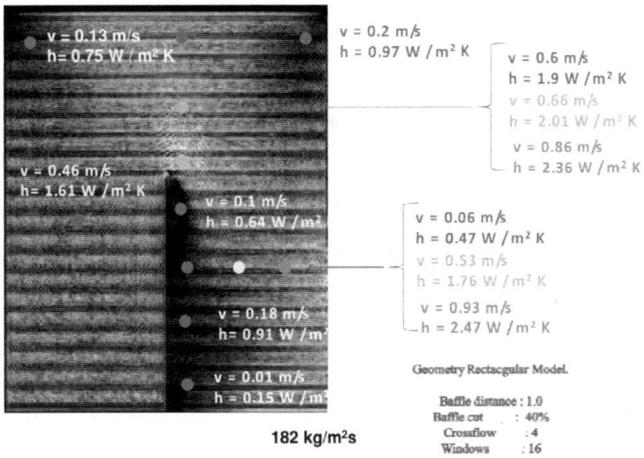

Figure 8. Flow Distribution with 40% cut

Further Work

These are the initial findings in this study. Our continuing work will:

1. Look at scaled up sections in order to ensure that there are no scale effects
2. Examine flow distributions with at lower mass fluxes and with liquids of higher viscosity
3. Develop a radial model covering a full shell-and-tube design, to confirm the validity or otherwise of the approach adopted above.
4. Develop a clear guide to the identification of effective baffle geometry.
5. Examine ways in which exchangers having poor baffle design can be improved.

(113d) Numerical and Experimental Investigation of Dry Particulate Fouling

Kiran Kumar Sathyanarayanarao Subbarao[1], Camilo Rindt[2] and Anton van Steenhoven[2], (1) Mechanical Engineering, Section energy technology, Eindhoven University of Technology, The Netherlands, Eindhoven, Netherlands, (2)Mechanical Engineering, Eindhoven University of Technology, The Netherlands, Eindhoven, Netherlands

Particulate fouling is one of the major forms of gas side fouling in heat exchanger systems. Numerical modeling of fouling has been an evasive task. One of the reasons for this is attributed to the lack of detailed experimental data. In-situ fouling experiments in real gasifier systems involve too many lumped parameters and are not very suitable in numerical modeling of the process. On the other hand, controlled lab-scale experiments that have been reported are meager and those that have been published are performed either at low temperatures or at very high temperatures (>1000 ^{o}C). In this direction, a high temperature controlled fouling experimental facility has been built to elucidate the effect of various parameters on fouling like: gas phase velocity, gas phase temperature, particle types and properties, tube shape and geometry. Experiments with different seeding particles and target geometries are performed under various temperatures and velocities. It is noted that the process of fouling is greatly influenced by the gas phase velocity i.e. shear due the gas phase, temperature, surface roughness of the tube and the presence of condensable species in the gas. As inertial impaction is reported to be the major mechanism involved in the particle deposition, a numerical model based on Elastic-Plastic deformation of impacting particles and removal by shear is presented and compared with the experimental observations.

See more of this Session: Fouling: Troubleshooting and Prevention

See more of this Group/Topical: Topical 7: 14th Topical on Refinery Processing

Fouling in Condensate Preheat Exchangers: A Case History

Safdar Rashid Toor
Operation Department Qatar Petroleum Refinery
P.O Box No. 50033 Masieed Industrial City State of Qatar, Ph: 009744188126, Fax: 009744772960

Abstract: Principal drive of this research paper is to indentify and determine the root causes of unexpected fouling aspect at various heat exchangers, which is hampering refining operational activities at large and leading toward massive revenue losses. Hereby, the fouling is categorized into various types as Epstein[3] suggested. Chemical reaction fouling is the formation of deposits at the heat transfer surface by chemical reaction, in which the surface itself is not a reactant. Chemical fouling involves chemical reactions and may increase exponentially with temperature. Crude oil organic fouling mechanism is a complex process and is a type of chemical reaction fouling - primarily due to destabilization of the asphaltene component of crude oil. A 1 mm film thickness of fouling can result in an increase of 10 °C in the radiant section and 30 °C in the heat flux section. Polymeric fouling was analyzed in a condensate refinery (feed preheat exchangers). Please refer to the following undertaken study.

Introduction:

A lot of research has been carried out to develop and understand fouling problems but in every case a new problems were faced and new solutions are recovered. Due to fouling in refineries about 6% of the energy contents of each barrel processed consumed by itself in refineries with globally production of 82-85 million barrels and equivalent of Exxon or Shell entire production[1].Fouling is transformation process and it starts with the transforming of deposited material after it has been formed on heat a surface[2]. In 1981 Epstein described the mechanism of fouling and pointed that subject of the aging in fouling was poorly developed[3]. Müller-Steinhagen[4] estimated the cost of heat exchanger fouling in the UK as 2.5 billion $ and 14 billion $ in **USA**. Transformation of fouling is very important because it determine the physical and chemical nature of deposits and consequently behavior of heat exchanger, dynamic of fouling process. Fouling is transformation process and starts with particulate deposition and followed by precipitation, chemical reaction, coking, corrosion and biological fouling. In heat exchangers it is undesirable since it reduces the effectiveness of heat exchanger and increases the operation difficulties such as high delta pressure and reduced flow. Such type of problems was faced newly established condensate refinery in **Arabian Gulf**. The effect of fouling can be categorized in four broad areas.

1. **Processing cost**: in case of fouling in heat exchangers processing cost will increase due to reduction in total heat transfer area due to high demand of utilities, 'ie' include pumping cost and more fuel. The adversely effect the environment due to excessive generation of CO_2.

2. **Production Loss**: for the cleaning of fouled heat exchangers production loss is there and approximately it will cost 2$ per US barrel margin.
3. **Capital expenditures**: if fouling is not mitigated then operators needed to provide large surface area heat exchangers and this will increase the capital cost.
4. **Maintenance cost**: it will include staff and cleaning and disposal of cleaning agents.

The significance of fouling in the preheat exchangers can be judged from the estimation of Panchal and Huang Fu[5] that a loss of 1^0 C in the temperature of the crude oil entering in the furnace in a 100 000 barrel/day crude oil refinery equates to 450 kW and an energy cost of 40 000 US dollars per annum and Baudelet and Krueger[6] estimate that an increase of furnace inlet temperature of 1 deg C is equivalent to 1 tone of fuel per day, which may translate into additional costs of $100 000 per year. The cost will depend on fuel costs and throughput.

Fouling in Crude Preheat Exchangers:

Fouling is categorized into various types and it has been discussed in many papers. Espteins[3] pointed out in his paper, these are crystallization, particulate, chemical reaction, corrosion and biological fouling. He classified fouling as initiation, transport, attachment, removal and ageing of the fouling layers. Esptein[3] define the chemical reaction fouling as the formation of deposits on the heat transfer surface due to chemical reactions. Chemical reaction fouling is that, in common with other chemical reactions, the rate of reaction increases exponentially with temperature. The formation of polymeric material during transport of the crude through the heat exchanger was originally thought to be an important factor. The formation of polymeric fouling was expected to be very sensitive to the presence of oxygen, as indeed it is. However, the change in the concentration of the polymeric material as the crude passes through the heat exchanger is small compared to the concentration of the material entering the heat exchanger and the formation process is not, therefore, dominant, as is confirmed by recent work.

Reaction Mechanism of fouling:
The Initiation and formation starts with following reactions;

Formation: R^*, RO, RO_2^* 1

Propagation: $\begin{cases} R^* + O_2 \rightarrow RO_2^* \\ RO_2^* + RH \rightarrow RO_2H + R^* \end{cases}$ 2

Termination: $\begin{cases} 2R \rightarrow R\text{-}R \\ 2RO_2 \rightarrow R\text{-}O\text{-}O\text{-}R\text{-}O_2 \\ R^* + RO_2^* \rightarrow R\text{-}O\text{-}O\text{-}R \end{cases}$ 3

Case History:

A condensate refinery was commissioned in 2001 in the **Arabian Gulf** and the propose of this refinery was to separate Naphtha, Distillates and residue products. The unit was designed to process 27000 BPSD of NF (North Field) condensate. Condensate feed is pumped from two storage tanks, 2101FA/FB to preheat exchanges 6101C to 6103 (figure 1) .Fouling has been observed in preheat exchangers over the years. High ΔP was observed in 6102C TPA exchanger (figure1).

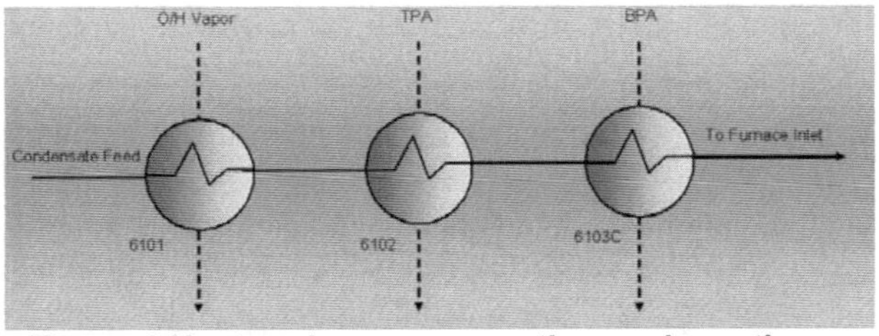

Figure.1 Heat exchangers operations schematic

Condensate heater is also having tube skin temperature limitation and due these limitations refinery can not be run more than 1 & ½ year instead of this was designed to run four year continuously. You can judged the condition by analyzing the following trend (figure 2) how quickly preheat exchanger are getting fouled and consequently heater tube skin temperature is abruptly increased in short time (1.3 years). The allowable temperature in radiant section of heater is 343^{0}C (clean condition) .

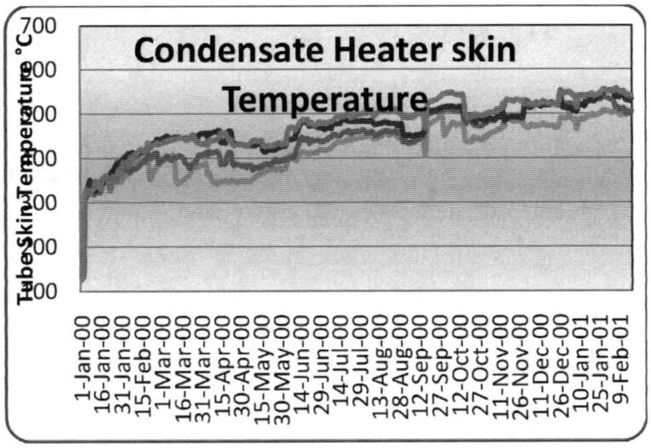

Figure 2 Temperature Variations in Heater Tube skin.

To find out the nature of fouling problem in condensate feed preheat exchangers, analysis of condensate feed was done from the NGL (NGL plant is feeding condensate to the refinery) to condensate feed tanks (2101FA/FB) with test method of ASTM D381 and ASTM D873.

Table 1

Sample	Existent Gum (mg/100ml)	Potential Gum (mg/100ml)	Heat Induced Gum with O_2 at 150^{0}C (mg/100ml)
NGL Tank T8 inlet	22	37	74
NGL Tank T 8 outlet	36	40	83
Condensate Feed in	41	47	94

Table 2

TEST	Condensate Feed	
Total filterable solids	5	ppm
Karl Fischer water content	0.1	wt%
Sp Gr @ 20^{0}C	0.754	g/ml
Bromine number	0.77	g Br2/100g
Total acid number	<0.05	mg KOH/g
Existent gum	13	mg/100ml
Heat induced gum	15	mg/100ml
Potential gum	18	mg/100ml
Basic nitrogen	<5	ppm
Mercaptan sulphur	1263	ppm
H_2S	<1	ppm
Pyrrole nitrogen	<1	ppm
Carbonyl content	<1	ppm
Total calcium ,ca	0.6	ppm
Total chromium , Cr	<0.2	ppm
Total copper ,Cu	<0.2	ppm
Total iron ,Fe	<0.2	ppm
Total Manganese, Mn	<0.2	ppm
Total sodium, Na	0.4	ppm
Total phosphor, p	0.3	ppm
Total Nickle, Ni	<0.2	ppm
Total vandium, V	<0.2	ppm
Total zinc , Zn	<0.2	ppm

Table 3

Test	Condensate feed before preheat train stabilization	Stabilizer Column RD line T-8	Unit 61 Feed after storage tank
Existent gum mg/100ml	21	6	6
Potential gum mg/100ml	38	36	51
Total iron, Fe ppm	0.6	0.3	0.3
Total phosphor p, ppm	1.9	0.8	0.8
Mercaptan sulphur ppm	1755	1319	1424

Based on lab analysis from three main points (as provided in table 1,2.3) NGL, intermediate storage tank 2101F/FA and unit feed, the possibility of polymerization is there due to mercaptan in feed and existent gum show polymerization already there and potential gum indicates problem will be more sever when contacted on hot surface of heat exchanger.

Conclusion: Polymeric fouling was analyzed in a condensate refinery due to the presences of mercaptans and the ingress of O_2 in the intermediate storage tank 2101F/FA. As indicated in table 1, there was potential gum existing there previously and a propagation reaction (refer to the reaction Mechanism given in the reaction mechanism section) started with a fouling rate exponentially increasing when the condensate contacted the hot heat exchanger surfaces. The problem can be mitigated with avoidance of O_2 by blanketing of intermediate storage tanks 2101F/FA.

Reference:

1. S.Macchieeto; Fouling in crude oil preheat trains: A systematic approach to an old problem. Proceeding of international conference on heat exchanger fouling and cleaning VIII-2009 Jun 14-19, 2009 Schladming, Austria.
2. D.I.Willson; Aging: Looking back and looking forward. Proceeding of international conference on heat exchanger fouling and cleaning VIII-2009 Jun 14-19, 2009 Schladming, Austria.
3. N. Esptein; 1983, thinking about heat transfer fouling a 5 x matrix. Heat Transfer Engineering, Vol. 4, pp. 43-56.
4. MÜLLER-STEINAGEN, H. (1995) Fouling of heat exchangers surfaces. Chemistry & Industry, No. 5, pp. 171-175.
5. PANCHAL, C.B. HUANG FU, E-P. (1998) the effects of mitigating fouling on theenergy efficiency of crude oil distillation. AIChE Spring National Meeting, New Orleans, March 8-12, Paper 81C.
6. BAILEY, A. GIBBARD, I. (1998) Improving heat exchanger reliability with heat transfer enhancement. Block Persee. Symp. on Innovative Approaches for Improving Heat Exchanger Reliability, November 2-4, Houston, Texas (Gulf Publishing Company)

Influence of pre-heat train performance and fired heater design on fuel efficiency of Crude Distillation Units

A. Morales Fuentes, G.T.Polley & M. Picon Nunez
University of Guanajuato, Mexico

Fouling of the heat transfer surface of fired heaters providing the boil-up for crude oil distillation columns can have a highly detrimental effect not only on the fuel economy of the pre-heat train but in some cases on the economic efficiency of the refinery as a whole. For instance, fouling on the crude-oil side of the heater generally leads to the formation of coke. In extreme cases, particularly with heaters used on the feed to vacuum distillation units, coke can be carried over from the heater into the column itself and form deposits that affect the efficiency of the column.

The modelling of the fouling on the crude side of the fired heater is complicated by the interaction between fouling on the cold-side and the temperature profile on the hot-side of the heater which then affects fouling rates in other locations within the heater.

A further major complication is the effect that the fouling on the crude side has upon the flow resistance presented by the furnace tubes. The thicker the deposit the greater is the pressure drop through the tubes. The higher the pressure drop through the tubes the higher is the pressure immediately after the pressure let-down valve at the entrance to the heater. This, in turn can result in less flash across the valve and poorer heat transfer in the tubes situated in the convective section of the unit. Poorer heat transfer leads to higher wall temperatures and higher fouling rates.

The work reported in this paper involves the combination of a simplified model of a furnace coupled with modelling of both the thermal and hydraulic behaviour of the crude oil side of the heater incorporating a consideration of the effects of fouling on this behaviour.

Furnace Model

Furnaces come in several configurations as presented by Berman (1978). The use of a convective heat recovery section following the radiant chamber is common to most configurations.

It is assumed that the heat transfer to the tubes positioned in the "fire box" is solely by radiation and a uniform radiant heat flux can be used.

Since the quantity of heat transferred in the radiant section is known, the temperature at which hot gas leave this section and enter the convective section are easily calculated.

In the "convective section" it is assumed that the heat transfer is solely by convection. The temperature of the hot gas is reduced as it passes over each individual row of tubes. The gases are assumed to be thoroughly mixed between tube rows.

Heat Demand Curves

Heat transfer on the crude-side is greatly enhanced once a two-phase flow has formed. It increases with increase in the vapour mass quality of the stream. Calculation of the vapour mass quality requires knowledge of the vapour-liquid equilibrium. This knowledge is best provided in the form of heat demand curves. One curve relates the fraction of feed vaporised with the quantity of heat transferred to the oil. The second curve relates fraction of feed vaporised with saturation temperature.

Unfortunately, these curves are dependant upon pressure which varies along the flow path.

The approach adopted here was to use the facilities provided in the ASPEN+ program to develop a series of curves at differing pressures (between the pressure at the inlet to the let-down valve and the distillation column pressure). Values at intermediate pressures were then obtained by interpolating between these curves.

Heat Transfer Equations

Single phase convective heat transfer was calculated using the Dittus-Boelter equation.

Two-phase convective heat transfer was calculated using the momentum-heat transfer analogy proposed by Chen (1966).

$$\frac{\alpha_{TP}}{\alpha_L} = \left(\frac{\Delta P_{TP}}{\Delta P_L}\right)^{0.44}$$

α_{TP} and α_L represent the two phase and the heat transfer coefficient that would have been encountered if the liquid phase flowed alone respectively.

Nucleate boiling heat transfer was ignored.

The vapour phase mass transfer resistance has a major effect upon surface temperature. This effect was calculated using the Bell & Ghally method (1972) described by Collier & Thome (1996).

Pressure Drop equations

Standard equations were used for the estimation of single phase pressure drop.

The Chisholm equation:

$$\frac{\Delta P_{TP}}{\Delta P_L} = 1 + \frac{C}{X_{tt}} + \frac{1}{X_{tt}^2}$$

where

$$C = \left(\frac{\rho_L}{\rho_g}\right)^{1/2} + \left(\frac{\rho_g}{\rho_L}\right)^{1/2}$$

was used for the calculation of two-phase pressure drop.

A uniform core model (Yap et al, 2004) was used for the calculation of the effect of fouling on pressure drop.

Fouling Model

A generalised form (Polley 2011) of the Ebert-Panchal Model (1997) was used for the calculation of local fouling rates within the fired heater:

$$\frac{dR_d}{d\theta} = \frac{A}{\alpha}\exp\left(\frac{-E}{RT_f}\right) - \gamma\tau_w$$

With wall shear stress calculated from the local pressure drop.

Calculation Procedure

Since, the objective of the study is to determine the effect of inlet conditions on the fouling occurring within the fired heater the analysis has been simplified by assuming that the fuel consumption is not adjusted as the outlet conditions deteriorate. In practice we could expect the firing to increase as fouling progresses in order to provide fixed boil-up for the column.

In this analysis it has been assumed that the crude oil enters the heater through the row positioned at the exit from the heater as shown in Figure 1.

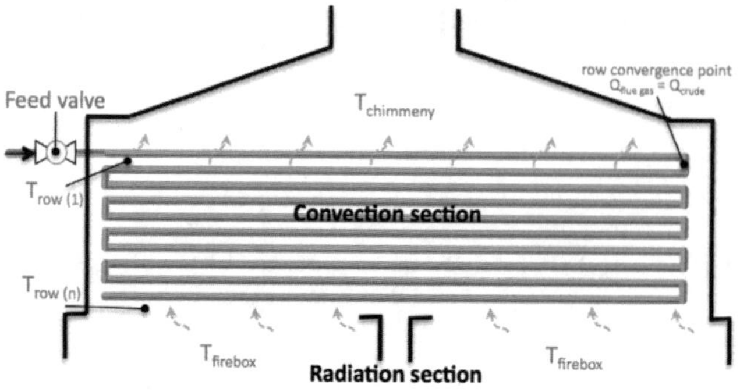

Figure 1. Contacting Arrangement in Convective Section

It is assumed that the following parameters are known: crude oil flow rate, temperature at exit from pre-heat train, pressure at exit from pre-heat train, pressure at exit from fired heater, radiant heat flux in heater, fuel consumption and heat flux in the radiant section.

Flue gas flow rate is determined from a heat balance and the addition of 15% excess air.

In this analysis it has been assumed that the crude oil enters the heater through the row positioned at the exit from the heater.

From the fuel flow rate, and furnace geometry, the hot gas temperature in the radiant chamber is determine in using a well stirred furnace model, as described by Shires et.al. (1994). This temperature is assumed to be the inlet condition to the convection section.

The first iteration loop relates to the temperature profile across the convective section. Starting from the chimney temperature (first guessed), heat balance between flue gas and crude oil in each row is carried out to determine the flue gas in the next row (closer to the firebox). Chimney temperature is obtained when convergence between last row flue gas temperature and firebox temperature match (calculated from the well stirred model mentioned above).

For operating time of zero:

A pressure at the entrance to the fired heater is assumed. A flash calculation is performed in order to determine the feed conditions at entry (temperature, vapour mass quality).

In the convective section of the heater the hot gas contacts the cold stream in a cross-flow manner. It is therefore assumed that each point along the tube encounters the same hot side conditions. The hot-side heat transfer coefficient is low. This results in a constant wall temperature condition.

The heat transfer along the first tube row in the convective section is calculated by numerical integration of the heat transfer and pressure drop equations. Local fouling rates are calculated.

The gas-side thermal changes are then bulked and applied in the analysis of the behaviour of the next tube row in the convective section.

The integration procedure is applied to each tube row in the convective section.

The integration is then conducted across each tube row in the radiant section. Here the heat flux at each location is assumed to be the same.

The pressure at the exit form the heater is then known and compared with the specified value. If there is a significant difference the inlet pressure is adjusted and the procedure is repeated.

A time interval is then applied and the change in local fouling resistances calculated from the predicted local fouling rates.

The analysis is then repeated until a specified operating period has been covered.

CASE STUDY

The fired heater was assumed to handle 75 000 barrels/day of light Arabian crude. The fouling parameters used in the study were those that best fitted the data reported by Knudsen et al. (1997).

The furnace geometry is given in Table 1.

The pressure at the end of the pre-heat train was 12 bar. The pressure at the exit from the fired heater was 3 bar.

Three inlet conditions (representing different levels of heat recovery in the pre-heat train were examined).

Table 1. Geometry of Fired Heater

	Convection	Radiation
d_i / d_o [m]	0.1347 / 0.1413	0.1453 / 0.1524
Lenth per pass [m]	16.64	18
Total Length [m]	166.4	324
Coils	8	8 (4 per firebox)
# pass per coil	10	18
# pass per row	1	1

At an inlet temperature of 250 °C the pressure at the inlet to the heater was so high that the feed was single phase. This resulted in high wall temperatures and high fouling rates in the first row of tubes (Figure 2). As time progresses, the build up of the deposit results in reduced surface temperature and reduced fouling rate.

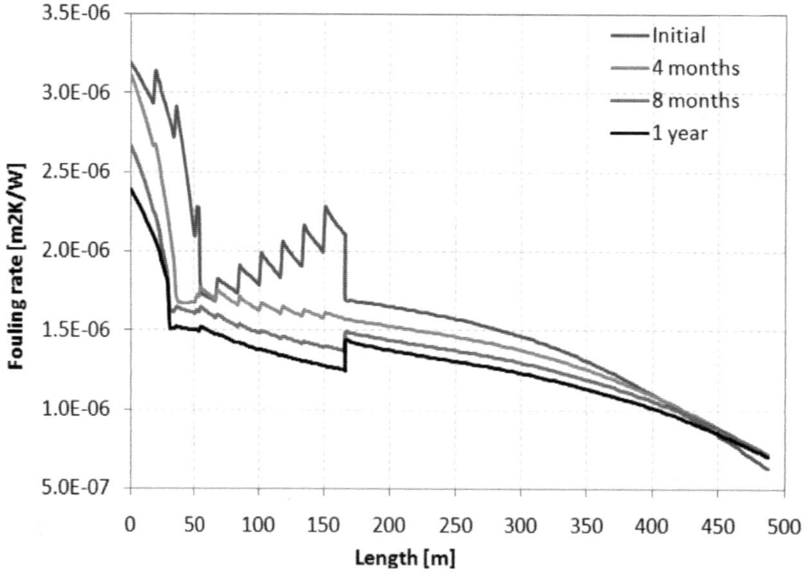

Figure 2. Fouling Rate Distribution: Pre-heat Train Exit of 250 °C

However, the deposit has a marked effect upon fuel efficiency. This is observed in the temperature profiles shown in Figure 2. The temperature at length zero being the exhaust temperature.

The pressure drop across the heater varied from 3.3 bar on start up to 3 bar after one year operation. This was due to a reduction in the quantity of oil being vaporised. This reduction would not occur if the firing was adjusted to control vaporisation.

The results for a pre-heat train exit temperature of 280 °C are shown in Figures 4 and 5.

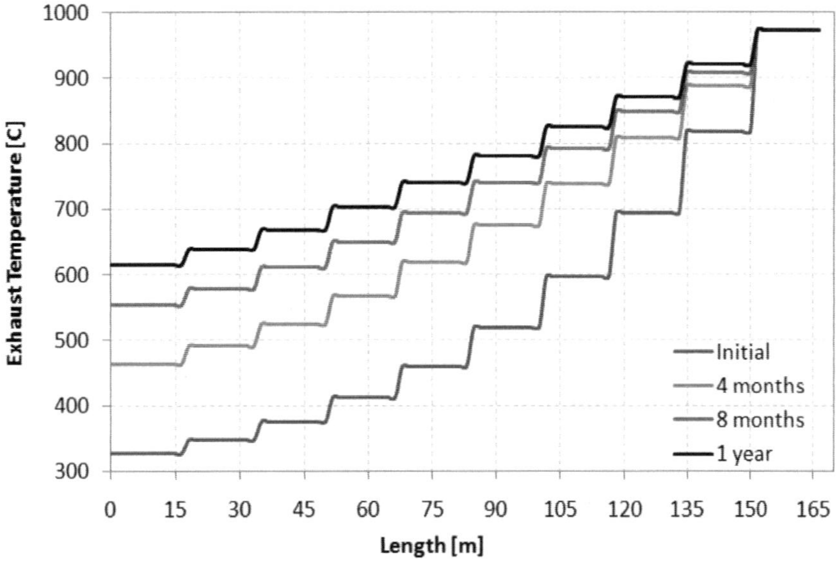

Figure 3. Exhaust Gas Temperature: Pre-heat Train Exit of 250 °C

At this temperature vapour is generated as the feed flashes across the let-down valve and the mass quality at heater inlet is around 4 %.

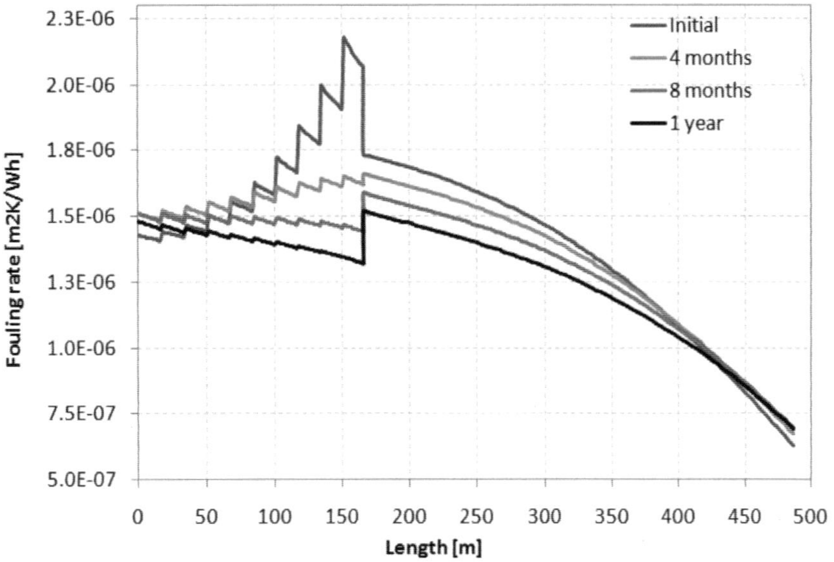

Figure 4. Fouling Rate Distribution: Pre-Heat Train Exit 280 °C

The fouling rates (Figure 4) in the convective section are about half the value observed for the lower pre-heat train exit temperature. The fuel efficiency of the heater is improved (Figure 5).

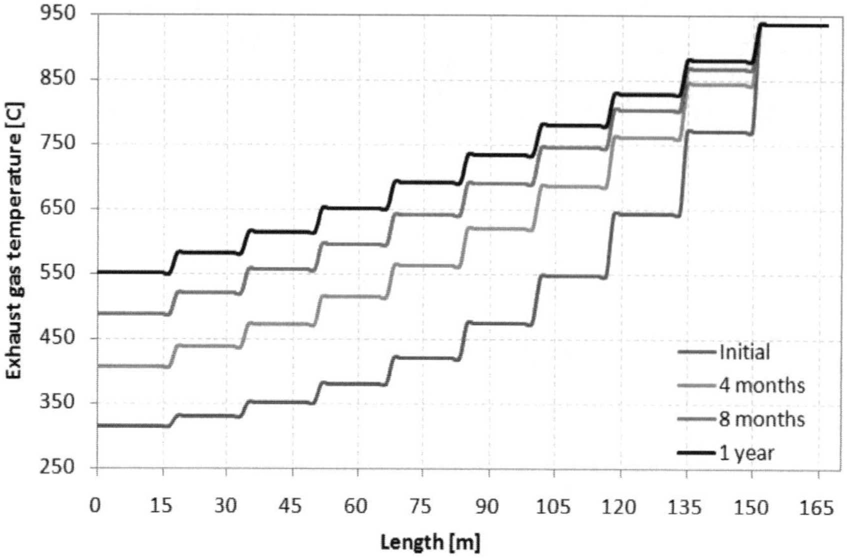

Figure 5. Exhaust Gas Temperature: Pre-heat Train Exit of 280 °C

The exhaust temperature after one year of operation is 550°C as opposed to the 620 °C observed at the lower inlet temperature. The pressure drop across the heater was around 3.5 bar throughout the operating period.

Finally, the results for a pre-heat train exit temperature of 300 °C are shown in Figure 6. Under these conditions the vapour mass quality at the inlet to the heater is just above 6%.

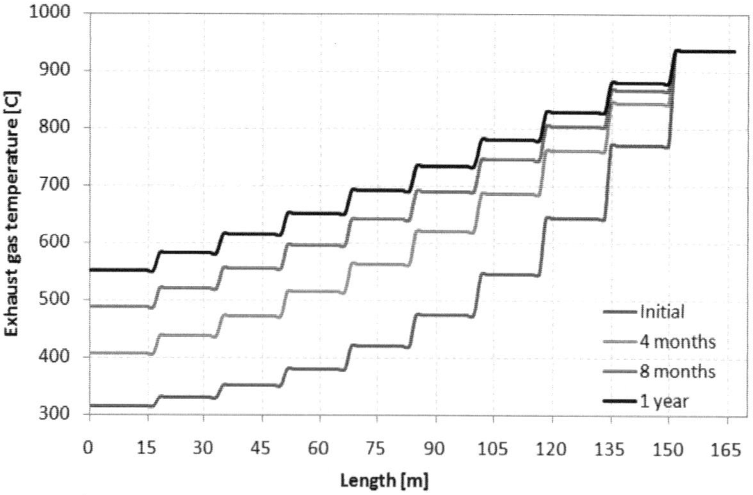

Figure 6. Fouling Rate Distribution: Pre-Heat Train Exit 300 °C

However, this increase in vapour mass quality has little effect upon fouling rates. Consequently, there is little effect upon fuel efficiency (the exhaust temperature after one year operation being 540 °C).

The pressure drop across the heater was around 3.8 bar throughout the operating period. Given the large tube diameters used in the heater fouling has little effect on pressure drop.

Conclusion

For the heater design considered in this study the key factor in setting fuel efficiency is shown to be the temperature at which the crude oil exits the pre-heat train. If this temperature is not high enough to ensure that a two phase flow is developed at the inlet to the heater then fouling rates in the convective section will be higher than necessary and fuel efficiency will be affected.

We do not need to set up this type of model in order to gauge what the minimum heater inlet temperature should be. We can assume a typical pressure drop across the heater (say 4 bar) and then conduct a flash calculation in order to determine the heater inlet condition. If a two phase flow has been generated excessive fouling is unlikely. If a single phase feed is present then excessive fouling can occur within the heater and steps should be taken to increase the heat recovery in the pre-heat train.

EXTENSIONS TO WORK

The work conducted here can be readily extended to include a consideration of the interaction of furnace design and fuel efficiency.

In the case of heater for use on vacuum distillation units it can be extended to provide estimates of oil residence times and if the fouling model adequately covers the chemistry of asphaltene precipitation to estimation of coke formation and possible carry over into the distillation column.

The behaviour of heaters feeding VDU's will also depend upon the vapour generation on entry. So, additional heating of feed prior to entry may provide a means of reducing coke fouling in the distillation column itself.

REFERENCES

Berman, Herbert L., 1978, Fired heaters I. Finding the basic design for your application, Chemical Engineering (New York, NY, United States),), 85(14).

Bell K. J., Ghaly M.A. "An approximate generalized design method for multi-component partial condensers" Chem. Eng. Prog. Symp. Series, No. 131, 69, 72-79, 1972.

Chen J.C. 1966, "Correlation for Boiling Heat Transfer to Saturated Fluids in Convective Flow". Ind. Eng. Chem. Process Des. Dev., 5 (3), pp. 322–329.

Collier, J.G., Thome, J.R., Convective Boiling and Condensation, Oxford Engineering Science Series, Oxford University Press

Ebert, W.A., Panchal, C.B., 1997, "Analysis of Exxon crude-oil slip stream coking data", in: Bott, T.R., et al. (Eds.), Fouling Mitigation of Industrial Heat-Exchange Equipment, Begell House, New York, pp. 451–460.

Knudsen J.G., Dahcheng L., Ebert W.A., 1997, "The determination of the threshold fouling curve for a crude oil", Engng. Found. Conf. Understanding Heat Exchanger Fouling and its Mitigation, Lucca, Italy

Polley G.T. 2011, "Models for chemical reaction fouling", Tutorial on Fouling & Design of Crude Pre-Heat Trains, AIChE Spring Meeting, Chicago, 2011

Shires, G. L., Bott, T. R., Hewitt, G. F., 1994, Process Heat Transfer, CRC-Press, Chapter 24 pp. 777

Yeap, B.P., Wilson. D.I., Polley, G.T., Pugh, S.J., 2004, "Mitigation of Crude Oil Refinery Heat Exchanger Fouling Through Retrofits Based on Thermo-Hydraulic Fouling Models", Chemical Engineering Research and Design, 82, pp. 53-71.

Extended Abstract – 2011 AIChE Spring National Meeting

More Refinery Hydrogen Through Efficiency, Recovery, and Process Improvements

Alan Zagoria, Ron Long, and Kathy Picioccio, UOP, a Honeywell Company

The Importance of Hydrogen

Hydrogen has played an important role in refinery operations, but for many refiners it now has a critical impact on the refinery profitability. Factors driving the growing importance of hydrogen include the introduction of low sulfur fuel requirements, benzene reduction in gasoline, stricter diesel cetane and aromatics limits, new marine fuel regulations, the processing of heavier crudes, increased bottoms upgrading, and the increased cost of energy.

Hydrogen Network Analysis and Improvements

The most successful hydrogen network improvement programs will draw the largest possible optimization envelope - and take advantage of all the "knobs" that are available to turn, including network connectivity, increased hydrogen production capacity, target hydrogen partial pressures, process changes in producers and consumers, catalysts, operating procedures, revamped and new purification capacity, PSA product bypass, hydrogen plant feeds, compressor modifications, ability of LP models to accurately represent hydrogen availability constraints, and header pressure control system improvements.

Hydrogen has a significant impact on process performance and profitability. Hydrogen partial pressure, a parameter under the operator's control, can increase catalyst life, increase throughput, increase conversion, improve product quality, or process more-profitable feeds. These benefits may be much greater than those from simply reducing hydrogen production / purchase costs. Establishing and faithfully maintaining the target hydrogen recycle purity of key hydroprocessing units is an important component of effective hydrogen management.

Methodology

It is important to first clarify the objective before analyzing the hydrogen network. In general the objective should be refinery profitability rather than minimizing H2 losses or reducing H2 purchase costs. In essence hydrogen network optimization requires a) identifying the constraint that most limits profitability b) identifying ways to alleviate the constraint and selecting the best option and c) repeating these steps until a constraint is reached that can not be cost-effectively relaxed or alleviated.

Tools

Hydrogen network pinch analysis enables identification of the theoretical minimum hydrogen requirements for a given network achievable with unlimited modification of the existing network connectivity including addition of major compression. This target and associated pinch purity provide useful guidance to the analyst, but must be tempered with practical considerations (viability of changing piping, flowrates, hydraulics and unlimited modification of scrubbing and compression equipment). Pinch analysis is supplemented with a detailed refinery-wide model of the hydrogen network. One of the critical features of this model (as opposed to a spreadsheet model) is that it simulates the non-linear relationships between charge rate, makeup rate, makeup purity, and recycle purity (hydrogen partial pressure) and purge rate. Potential modifications can be tested such as changing a makeup source, trading off purity versus recovery in an existing purifier, adding new purification or changing the target hydrogen partial pressure in a hydrogen consuming process unit. This model can be used by the refiner for operations and project planning.

Detailed process and purification models enable consideration of changing operating conditions, equipment, or catalyst to increase hydrogen yields in catalytic reformers; the impact of changing hydrogen partial pressure on catalyst life, product quality, conversion and hydrogen consumption; and the impact of changing purifier feed composition on product purity and recovery.

Full-Powered Optimization

We have optimized grass roots refineries before individual process unit design bases are set. In that situation we have many more degrees of freedom available and much greater potential to optimize. All parameters are available for optimization and we can be that much more effective in minimizing capital and optimizing refinery margin. For example, in the design work we did for a South American refinery, we had full freedom to route and recover streams since design pressures and makeup purities were not yet fixed. We were able to set flash drum pressures, makeup purity and PSA feed composition design bases. This enabled us to cascade flash gas from one unit to makeup of another, and recover flash gas from a second unit in a PSA included for reformer netgas. Initially, since no external fuel could be purchased in this refinery we expected that we would have to feed the hydrogen plant with expensive naphtha. After preliminary energy optimization of the whole refinery we determined there was an excess of fuel gas available. Not only could we utilize lower value fuel gas as hydrogen plant feed, but we could isolate hydrogen-rich fuel gas streams and utilize them for this purpose. The hydrogen content enabled a significant reduction in energy consumption in the hydrogen plant furnace over any other hydrocarbon feed.

Optimizing Beyond the Hydrogen Network

Broadening the optimization envelope even further, it is possible to consider optimizing more than just the hydrogen system at one time. There are often additional benefits when several systems are optimized together. For example, when designing a new catalytic reformer and hydrocracking unit, it is possible to consider the hydrogen and LPG recovery systems at the same time. As stand-alone designs each unit would have its own LPG recovery system, and produce moderate purity hydrogen to be purified/recovered (reformer net gas and hydrocracker flash gas). The reformer net gas contains the recoverable LPG and hydrogen while the recoverable hydrogen and LPG are found in the purge and flash gas streams of the hydrocracker.

LPG recovery is much more efficient in the hydrocracker than in the reformer because the gas stream has a lower hydrogen concentration. And, it is much more cost-effective to process both hydrogen streams in one PSA. It is possible to improve the two systems together by designing a single PSA that takes both hydrogen streams as its feed and sending the PSA tail gas to the existing LPG recovery system in the hydrocracker. Since the PSA concentrates LPG in the tail gas, its LPG recovery requires less energy and capital than if it was recovered separately in the reformer. An additional benefit of the integration is that a PSA product bypass can now be integrated with the hydrocracker to enable optimizing the hydrocracker makeup purity – a degree of freedom that would not otherwise exist.

Hydrogen Network Improvements – Results

Our hydrogen management studies generally identify \$2MM to \$20MM per year in benefits.

Figure 1 Results of Recent Hydrogen Management Studies

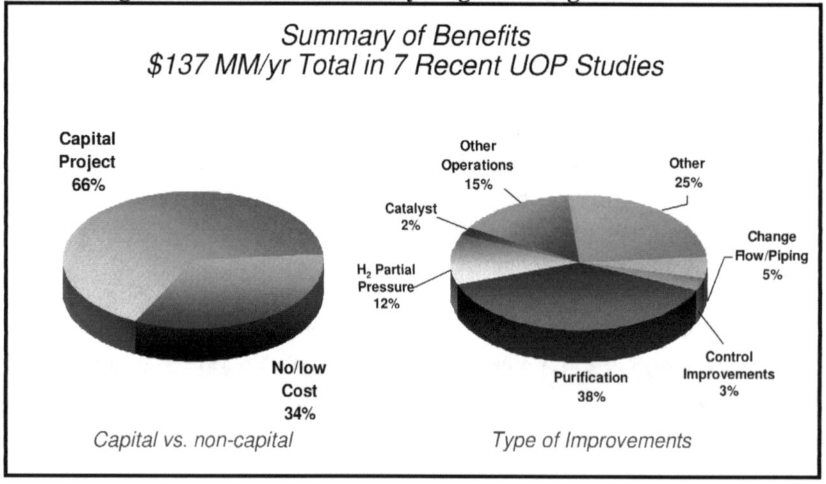

In our most recent seven studies (Figure 1) we identified a total of $137MM/year in benefits. A third of the opportunities identified were no / low cost changes and the remainder required capital projects (all with simple payback of less than 2 years). While 38% of the improvement opportunities came from adding or improving hydrogen purification capacity, a much broader scope of evaluation is required to achieve these benefits. Operating changes and better management of target hydrogen partial pressure were important as well as many other issues.

Hydrogen Recovery and Purification Options

PSA hydrogen purification is a cyclical process in which the impurities in a hydrogen-containing steam are adsorbed at high pressure and subsequently rejected at low pressure. The hydrogen product is at just slightly below the feed pressure and is typically upgraded to 99.9+% purity with hydrogen recoveries of 60 to >90%.

Membrane purifiers separate a gas mixture by the differences in permeation rates of various gases through the polymeric membrane. The more permeable gas (hydrogen) is enriched on the permeate side of the membrane, while the less permeable gas enriches on the feed side of the membrane. The hydrogen product stream (permeate) is produced at a significantly lower pressure than feed.

Because of its high product purity and low hydrogen pressure drop, PSA purification is favored in most cases. Feed streams below 40% hydrogen in the feed are very unlikely to be economically recoverable. Recovering streams between 40% and 55% may be attractive depending on site-specific economics. Streams above this value are likely to be economically recovered. The membrane process is the most economical process for high pressure purge gas upgrading.

In addition to adding a new PSA or Membrane unit, there are often opportunities to improve the performance of existing units. PSA and membrane units are often revamped to increase hydrogen production, recovery and/or purity. These revamps can be as simple as replacing adsorbents or as complex as adding an additional unit. Frequently refiners will elect to perform the revamps in phases where each phase adds additional capacity.

Figure 2 Selection of Purification Technology

Variable	*Polybed™ PSA*	*Polysep™ Membrane*
Product Purity	**✓ 99-99.999 mol-%**	**up to 98 mol-%**
Remove CO_x, H_2S, H_2O	✓	—
High Product Pressure	✓	—
Economy of Scale	✓	—
Feed Pressure	**<1,000 psi (70 kgcm²)**	**Favored@ > 1,000 psi**
Feed H_2	**Preferred > 50 mol-%**	**Pref. > 70 mol-% Min. 15%**
H_2 Recovery	**70- 90%**	**70- 97%**
Ease of Expansion	**Easy**	**✓ Very Easy**

Process Improvements to Enhance the Hydrogen Network

Catalytic reforming is a significant source of hydrogen in almost every refinery. Recent designs of new UOP CCR Platforming™ units optimize reactor pressures to maximize yields of valuable liquid products and hydrogen. Existing catalytic reforming units can be revamped to produce more hydrogen. Revamps can focus on increasing capacity, increasing severity or reduced pressure. Frequently the hydrogen yield can be improved by reloading with a more selective catalyst. Also, UOP has developed an integration of new Platforming and PSA hydrogen purification units to substantially increase hydrogen and LPG recovery.

Diesel hydrotreating, VGO hydrotreating, and hydrocracking catalyst breakthroughs enable achieving desired target product properties and yields while minimizing hydrogen chemical consumption (suppressing unwanted hydrogen-consuming reactions).

Hydrogen Optimization – Sustaining the Benefits

Optimizing the refinery hydrogen network on a daily basis is a challenge. Daily operating targets must be optimized to reflect the day-to-day changes of the refinery. Operations must know which are the critical hydrogen network operating parameters to monitor and manage. Ideally, one person is responsible for the network as a whole and can manage the network to maximize overall refinery margin rather than individual units.

Operations must understand and monitor the key network constraints and know what adjustments they can make to increase hydrogen network efficiency by pushing closer to a real network constraint. For example, in a cascaded system, they might work to minimize the make-up purity of a given unit (maintaining target recycle purity with purge) until the make-up compressor is at full capacity. Managing the compressor spillback in this way minimizes hydrogen losses (upstream PSA unit) under all refinery operating conditions.

Operators should be aware of the value of hydrogen, the cost of sending hydrogen to fuel, and the penalty for operating too conservatively. Running a PSA unit at too high a product purity, and operating a diesel hydrotreater at greater than target recycle purity are good examples of safe but inefficient operation.

Refinery Hydrogen Management – more than a balancing act

Sanjiv Ratan and Mariano Pagano - TECHNIP

ABSTRACT

Modern day refining landscape with its increasing clean fuels global mandate, compounded by growing proportion of "opportunity crudes" processing and widened bottom-of-the-barrel strategies has led to extensive need for hydrogen (H_2) and thus calls for its usage optimization. The substantial energy and investment costs associated with on-purpose hydrogen generation can significantly affect the profitability of a refinery operation. Thus advanced and creative solutions are being sought for ensuring and enhancing refining profitability beyond just satisfying the H_2 balance and for minimizing the cost burden of compliance to the clean fuels requirements world-wide.

Further, the increased H_2 usage against emphatic CO_2 management as well as capital efficiency targets against volatile oil pricing have pushed the need for advanced options and concerted techniques in establishing cost-effective H_2 balance and its added value in optimizing the refinery operations. Such needs, in turn have provided growing impetus to develop and apply innovative and smart hydrogen management methodology for optimizing the refinery H_2 networks.

HYDROGEN MANAGEMENT

Hydrogen management entails optimal utilization and distribution of hydrogen in relation to the various H_2-consumers and H_2-producers in the refinery, and to fulfill hydrogen balance effectively. Key objectives are cost-effectiveness, operational flexibility and reliability, and utilization of the various hydrogen containing refinery off-gases plays a significant role in having increased hydrogen availability.

Though hydrogen management in refineries has been practiced for several years, however today's high-complexity refining and processing, coupled with the need for extensive (energy) integration and need for enhanced cost-effectiveness require wider framework and more advanced methods and systems to achieve the objectives of H_2 optimization and refinery objectives in terms of overall margins, efficiency and environmental compliance.

Addressing such needs, an advanced integrated H_2 management program called HyN-DT™ has been developed by Technip based on integrated H_2 recovery and ROG utilization, H_2-pinch analysis and advanced LP programming using PIMS platform. It also includes the option of 'poly-generation' based on residue/petcoke gasification in order to arrive at the most appropriate and cost-effective refinery hydrogen network as well as overall energy management, including future CO_2 capture capability. It has been successfully applied for a grass-root refinery master plan and also for several other refinery hydrogen optimization studies.

As each refinery is different in configuration and its H_2 network, the advanced integrated hydrogen management solutions can be tailored to specific case in order to effectively analyse, assess and optimize the H_2 usage against objective targets and maximized returns for existing as well as expansions and grassroot refineries. Further, implementation of such solutions requires know-how and engineering expertise to effectively execute the findings, with absolutely no compromise on safety, reliability and operational flexibility.

ADVANCED H2-MANAGEMENT TOOL

The advanced hydrogen management and refinery optimization tool HyN-DT™ (Hydrogen Network Design Tool), has been developed by Technip for meeting refiner's needs towards establishing efficient, cost-effective and optimized hydrogen network architecture, including underlying minimization of C-footprint. It is structured as a "virtual console" to obtain a simulated overview of the refinery H_2 balance for identifying and ranking applicable solutions and options based on the various hydrogen consumers and producers in a refinery, in order to achieve increased efficiency and profitability.

Such a console is based and integration of commercial available software, refinery Linear Programming (LP), rigorous kinetics modelling simulator, data import and reconciliation tool to optimize the make-up and/or 'on-purpose' hydrogen demand. It woks out a smart H2 balance between the appropriate hydrogen sources and sinks which are connected to the data network, while also addressing the energy integration and economics for the refinery-wide operations.

The program platform simulates new or existing refinery hydrogen systems and identifying best potential hydrogen containing refinery off gas (ROG) streams for hydrogen recovery, assessing the capacity of dedicated H_2 producers and optimizing the hydrogen balance including application of advanced concepts of pressure and purity cascading. It entails a comprehensive design methodology (H_2 Network Analysis) involving all the associated H_2 distribution costs relevant to each H_2 recovery solution, in order to arrive at the Capex / Opex evaluation for each prospective H_2 network alternate.

From conducting optimization studies and modification schemes for the hydrogen network of existing refineries to initial configuration and assessment of new / grass root refineries, HyN-DT™ can help refineries in satisfying their hydrogen balance with significantly reduced investment in comparison to a non-H2 optimized overall operations. Its application has been very effective for a consolidated master plan of a grass-root refinery in Tatarstan as part of a full comprehensive design methodology and hydrogen network analysis involving all the associated costs relevant to each H_2 -recovery option and for identifying the most appropriate or optimum overall solution for the refinery.

HYDROGEN NETWORK OPTIMIZATION

The optimization analysis within HyN-DT™ basically involves:

- Defining of the actual refinery H_2 balance
- Investigate and optimize potential ROG hydrogen recovery and/or utilization
- Analyse options for upgrading/revamping the existing hydrogen production unit
- Evaluate additional new H_2 capacity
- Asses availability of the H_2 distribution system
- Specific refinery-wide energy assessment and integration
- Overall cost optimization, with optional C-readiness

The distributive recursion is extensively used in the refinery LP Model to structure the sub-models of the main process units, using the "base-delta" modelling techniques. The base delta structure of each process sub-model makes the model sensitive to variations of operating severity and to the feed qualities such as specific gravity, aromaticy factor, sulphur content etc, as well as H_2 make-up purity. A single base yields structure provides the level of conversion and product properties associated with a typical quality feed, on normal operating conditions. This base yields structure is customized and tuned to the particular needs of each refinery process unit. Multiple base yields are provided at alternative severities conditions.

The base yields structure of a process unit sub-model is an LP vector generated from a single run of a tuned simulator. Alternatively it can be derived from the preliminary yields received by the process unit Licensor. After this preliminary step, the feed streams are structured with recourse properties in the LP model to dynamically change the feed properties. The variation in the yields and products quality is achieved through "property drivers" which adjust the performance of the process unit in terms of changes in feed properties in relation to the base yields case. These property drivers represent the linear change in the yields and products properties as function of changes in the feed properties. These structures of the LP model make the LP model streams more reflective of the real refinery streams and also the performance behaviour of the unit operation..

The property drivers are generally calculated by use of rigorous kinetics modelling simulators or with a close collaboration with the Catalyst supplier or with the Unit Licensor. The common technique used to generate these property drivers adjustors, when rigorous kinetics modelling simulators are used, is the perturbation techniques: each feed property is changed independently (up and down) in the simulator and then it is calculated the linear shift from the perturbation cases in order to produce a new LP vector. The H_2 make-up purity is also configured as property driver thus each variation will affect the yields and products properties. One of the case overview displays of HyN-DT console is shown in Figure-1.

In the LP model H_2 consumers such as hydrotreaters / hydrocrackers are represented with reactor yields for the liquid products and with pools of lights hydrocarbon for the ROG and unit purge. The data required for each pool of lights are flow, composition and pressure. Automatic reports to import stream data from gas-chromatographic analysis or from rigorous kinetic simulation are available. All the complementary equipment such as Compressors, Pressure Swing Adsorption (PSA) and Membranes modules are simulated inside the refinery LP model to make the most reconciled model of the H_2 network. Combining the data from all hydrogen sources and sinks and relevant equipment, it is possible to structure a very detailed refinery H_2 balance.

Once the minimum 'on purpose' hydrogen generation requirement has been detected, the LP model will help to discover the bottlenecks to increasing the refinery profit due to the limitations of the H_2 network. The debottlenecking options of the base H_2 network through purifiers or coupling synergies of different purifiers can be investigated within the LP Model to define a shortlist of practical modifications which can positively impact the refinery profit.

The capital costs estimate has been introduced in the hydrogen management console as "flexible penalty cost" with the purpose to combine H2-pinch analysis with Economic Analysis. Refinery LP Modelling normally accounts for the operating costs (Opex) of the refinery processes to forecast the Gross Refinery Margin. Based on Technip's cost-estimating knowledge and global procurement database, it also includes, beside the Opex costs, the investment cost analysis and capital expenditure (Capex) related to the different H_2 recovery and production solutions.

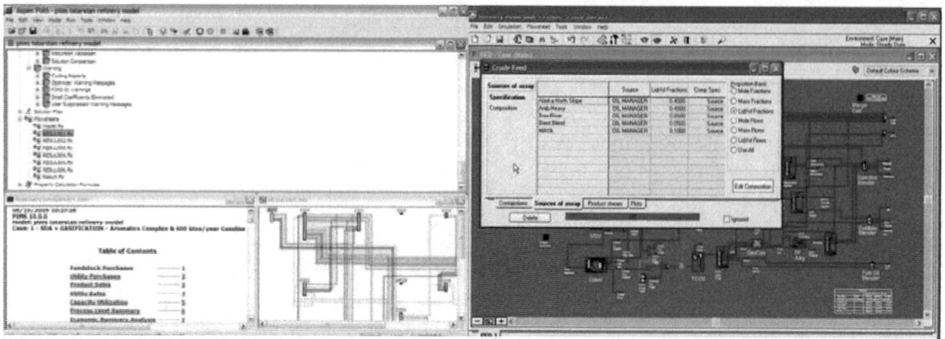

HYDROGEN BALANCE

USERS	Mass Rate	Nm3/h
H2 for Naphtha HDT	6.39 t/d	2985 Nm3/h
H2 for Kero HDS	2.25 t/d	1052 Nm3/h
H2 for Diesel HDS	17.12 t/d	7994 Nm3/h
H2 for Hydrocracker	508.22 t/d	237297 Nm3/h
H2 for ARO Complex	17.00 t/d	7936 Nm3/h
Tot	551.0 t/d	257264 Nm3/h
PRODUCERS	**Mass Rate**	**Nm3/h**
CCR-H2 (High purity)	-151.93 t/d	-70937 Nm3/h
H2 from Gasification	-241.96 t/d	-112977 Nm3/h
HC-H2 PSA	-38.86 t/d	-18146 Nm3/h
Hydrogen Generation Unit	-118.23 t/d	-55204 Nm3/h
Tot	-551.0 t/d	-257264 Nm3/h

Hydrogen Unbalance	0.0 t/d	0.0 Nm3/h

(+/-; shortage/overproduction)

STOP BLINK

Naphtha Warning

NO SHORTAGE

SYNGAS DISTRIBUTION

Default

Sweet Syngas distribution	T/Day	DEFAULT SPLIT		
Feeds		%	%	% wt
Sweet Syngas to Power	2943.0 t/d	52.65	52.65	52.65%
Sweet Syngas to Hydrogen	2646.3 t/d	47.35	47.35	47.35%
Sweet Syngas to Fuels	0.0 t/d	0.00	0.00	0.00%
Tot	5589.3 t/d		100.00	100.00%

REFINERY FUELS BALANCE

REFINERY ENERGY BALANCE			
REFINERY FUELS AVAILABILITY	MMKcal/h	503.24	
REFINERY FUELS DEMAND	MMKcal/h	755.84	
DELTA (+import/-excess)	MMKcal/h	252.60	
FUELS IMPORT	LHV		
Feeds	Kcal/Kg	T/Day	MMKcal/h
NATURAL GAS	11700.00	518.16 t/d	252.60
Fuel Oil M 100	10680.00	-	-
FUELS EXPORT	LHV	OTHERS	
Feeds	Kcal/Kg	T/Day	MMKcal/h
Refinery Fuel gas (flare)	12464.55	0.00 t/d	0.00
Fuel Oil M 100	10680.00	0.00 t/d	0.00

Figure-1: One of HyN-DT console displays

The "H_2 Production Factor" represents the Capex and Opex to produce 1 additional Nm^3 of H_2 from the Hydrogen Generation Unit while the "H_2 Recovery Factor" represents the Capex and Opex to recover one Nm^3 from the ROG. The level of CO_2 generation is simulated in the LP Model to account the CO_2 costs when the project is executed in a CO_2 trading scheme. The comparison of H_2 Production and Recovery factor can helps refiners to rank different H_2 network solutions, evaluate the investment escalation for solution at increasing H_2 recovery and finally identify the trade-off among H_2 production and H_2 recovery for a specific economic environment. The challenge is to minimize the capacity of a new H_2 unit whilst maintaining the same H_2 availability. The following two specific penalty factors have been introduced in the LP Modelling:

$$H2\_PRODUCTION\_FACTOR = \frac{CAPEX(HGU) * \Delta CAPACITY}{(PLANT\_LIFE) * H2\,Recovered} + OPEX(HGU) + CO2credits$$

$$H2\_RECOVERY\_FACTOR = \frac{CAPEX(HDTs) * \Delta PRESSURE}{(PLANT\_LIFE) * H2\,Recovered} + \frac{CAPEX(H2\_RECOVERY\_equipment)}{(PLANT\_LIFE) x H2\,Recovered}$$

$$+ OPEX(H2\_Recovery) + CO2credits$$

CASE STUDY

Background

A master plan study was conducted for a grass-root refinery in Tatarstan as Nizhnekamsk Integrated Complex (NIC) project. The Crude oil supply was characterized by high sulphur content (3.8 % wt) and high density (23.4 API°) to be converted into a variety of high add value products, with special attention given to application of bottom of the barrel (B-o-B) processes and their integration for achieving *"Zero fuel oil refinery"*. The major objectives of NIC are summarized as below:

Production Targets:

- Deep oil processing refinery able to process 7 MMTY of carbonic crude;
- Maximizing the depth of oil processing and reduce the HC wastes;
- Maximized production of light and medium distillates (i.e. Diesel and Kerosene/Jet Avio A1) complying to Euro V specifications;
- Minimized production of fuel oil, with the challenging target of "Zero Fuel Oil Refinery";
- Minimization of Natural Gas import;
- Production of aromatics (i.e. Benzene and Paraxylene);
- Production of iso-butane for Petrochemical downstream (i.e. synthetic rubber production);
- Production of raw material for Steam Cracker (LPG, petrochemical Naphtha and Raffinate);

Efficiency Targets:

- Rationalization of internal available fuels and network
- Maximization of heat recovery & heat integration
- Hydrogen optimization through advanced H2 management

Specific Requirement for Hydrogen Optimization:

- Reduce the refinery hydrogen system costs
- Implement cost-effective ROG integration and recovery measures
- Minimize H_2 burned as fuel or sent to flare
- Reduce utility consumption (fuel and power)

HyN-DT program was successfully applied for meeting several of above targets,

Hydrogen Balance

The products slate from the refinery is driven for meeting current and future environmental restrictions on sulphur emissions,. The level of desulphurization, beside the high level of bottom-of-the-barrel processing using both full conversion hydrocracker and the SDA combination requires a large amount of hydrogen of ~ 2.7% by weight on total crude input. The major producers and consumers are listed in Table 1.

After a preliminary study of 6 alternatives a complex scheme involving one world scale full conversion hydrocracker, plus SDA (i.e. Solvent De-Asphalting) combined with TAR gasification was selected. The gasification of heavy residues has an important part to play in the development of deep conversion refinery schemes and minimization of B-o-B residues. Various compressors' configurations were analysed to determine the optimal configuration considering availability, Capex and Opex. The scheme with a common 3-stage centrifugal compressor resulted with the highest system effectiveness among all cases analyzed. This solution was further justified by the fact that about 90% the overall refinery H_2 consumption was consumed for hydrocracking.

Table 1

Refinery H2 Consumption/Production (as pure H2)

Unit	FRESH H2 MAKE-UP			OFF GAS/PURGES		
	Flow Rate Nm3/h	Purity % Vol.	Pressure Barg	Flow Rate Nm3/h	Purity % Vol.	Pressure Barg
NHDT	2,985	80÷99.9	24.1	1,426	67.8	5/23
KHDT	1,052	80÷99.9	24.3	92	16.1	5/24
DHDT	7,994	80÷99.9	58	460	81.5	5/51
HYD. CRK.	237,297	99.9	155	20,795	90.6	31
ISOMAR	1,821	80÷99.9	13.5	neg.	-	-
TATORAY	6,117	80÷99.9	32	neg.	-	-
SRU	neg.	99.9	-	neg.	-	-
CCR (PSA)	-70,937	99.9	25	5,199	58.7	0.3
POX (PSA)	-112,977	99.9	25	15,228	53.3	4
HPU (PSA)	-73,352	99.9	25	used for SR internal firing		

The syngas from POX/TAR Gasification unit was sent partially (50%) to a combined cycle power plant and the remainder portion to a CO shift unit for pure hydrogen production. For the gasification section the number of gasifier trains and the specific capacity of each gasifier train had been optimised and agreed with the Client/Licensor upon fixing the processing priorities and once the Capex/Reliability trade-off have been fully investigated; in particular the design capacity, operating turndown and finally the POX/TAR gasification reliability are critical for defining number and capacity of each gasification string.

An additional feature of the flow scheme was the inclusion of a CO_2 removal step before the PSA unit in the CO shift section of the POX/TAR Gasification train. The shifted gas however contains about 33% vol. CO_2 which both inflates the cost of the PSA as well as reducing its efficiency. The key issue is however the tail gas LHV, which without a CO2 removal unit would have a heating value of 2300 Kcal/kg. Since this is the determining factor it is clear that it was not necessary to remove all the CO_2 by any means and a slip-stream unit for partial CO_2 removal was quite adequate. Figure-2 shows case analysis overview for one of the selected cases.

Base Case

The base case was developed considering the H_2 balance when all the refinery units run under steady state conditions and normal operating modes. Only one H_2 network at high H2 purity (PSA quality of > 99.9 % vol.) was considered.

Alt Case 1

Provision of an additional H_2 network at lower purity (i.e. 90 % vol).. In this case, more natural gas was needed to replace the heat duty of the H_2 recovered from ROG.

Alt Case 2

Additional PSA in the Hydrocracker unit to effectively recover H_2 from the offgas ex-HP Cold Flash Drum, thereby reducing the capacity of HGU.

Alt Case 3

Extension of case 2 with recovery of H_2 from the PSA tail gas in the CO shift section of the POX/TAR Gasification. The tail gas has still a decent quantity of hydrogen at moderate purity of ~ 53%, and the proposed system to recover H_2 from this stream is the combination of a three stages compression train followed by membrane modules based on the optimization of feed to permeate pressure ratio as well as other variables being H_2 recovery, product H_2 purity levels for effective cascading with the Hydrotreaters and catalyst volumes.

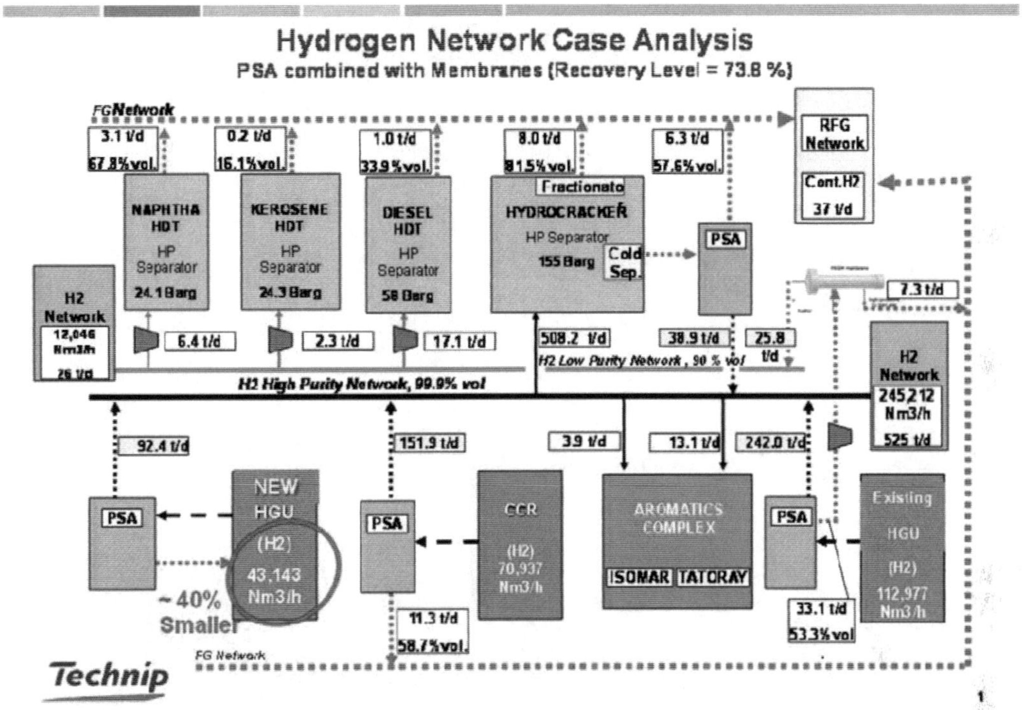

Figure 2: HyN-DT Case Simulation Overview

Table-2 provides an overview of the three cases based on HyN-DT simulations and assessment. These cases analysed in the study were selected at incremental level of H_2 recovery thus involving a decremented capacity for the "on purpose" HGU. From complexity point of view the Case 2 which foresees one H_2 network is preferable; moreover for case 2 the hydrogen availability and reliability is high and comparable with the base case thus ensuring the best compromise among H_2 recovery and H_2 availability.

Table 2

H2 Recovery Solutions - Operating Features Overview

		BASE CASE	CASE 1	CASE 2	CASE 3
H2 Network Complexity		-	Moderate	Low	High
Piping modification		-	Moderate	Low	High
Impact on HDTs HP equipment		-	Moderate	-	Moderate
Additional PSA		-	-	1	1
Additional Compressors		-	2	-	8
Additional Membranes modules		-	-	-	1
H2 availability		-	High	High	Average

Economic analysis

The analysis is based on Opex for the HGU and Capex based on a plant life of 20 years, to arrive at the overall unit cost of hydrogen. The methodological approach to the economic analysis of the H2 management is strongly depended by the project typology & environment constraints. As general indication for grassroot refinery projects the "flexibility & availability" of the H_2 for the complex is predominant against minimizing Capex for the selection of the most appropriate H_2 network, while for revamping or capacity augmentation projects, Capex of the H_2 recovery system plays the major role.

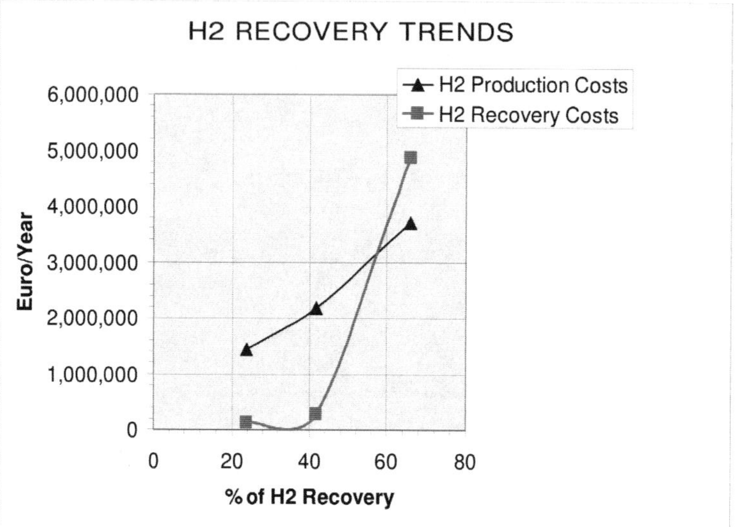

Figure 3: Overall H2 Recovery Optimization

Additional H_2 recovery solutions were investigated as sensitivities to analyse intermediate recovery levels of H_2 from the ROG, with a break event point at ~56 % of the overall H_2 recovery (see Figure-3) for NIC Complex (the solution is an intermediate condition among cases 2 and 3): beyond this recovery level the Capex and Opex of H_2 recovery & purification exceeds the hydrogen generation costs. Therefore is preferable to increase the capacity of the HGU.

CONTRIBUTION-BASED INTEGRATED HYDROGEN MANAGEMENT

HyN-DT interface with HGU can be extended by integrated ROG utilization based on the H2-contribution from such integration. Three major routes as described below (represented in Figure-4) can be applied.

1 **Low Contribution**: ROG mixed upstream the PSA of HGU (as common PSA)
2. **Medium Contribution**: Direct use of ROG as (part) feedstock for HGU
3. **High Contribution**: Dedicated recovery PSA with its purge gas to either fuel net or optional (extended) integration as feed for HGU.

The underlying driver of these options and routes being that any hydrogen burned as fuel is a loss of its 'asset' value above its heating value. Accordingly, by integrating ROG in a HGU, not only does most of its hydrogen content get recovered, but also improves the economics since the price attached to the ROG is often lower than the base hydrocarbon feedstocks used for HGU. Figure-4 shows the three routes for ROG integration with HGU.

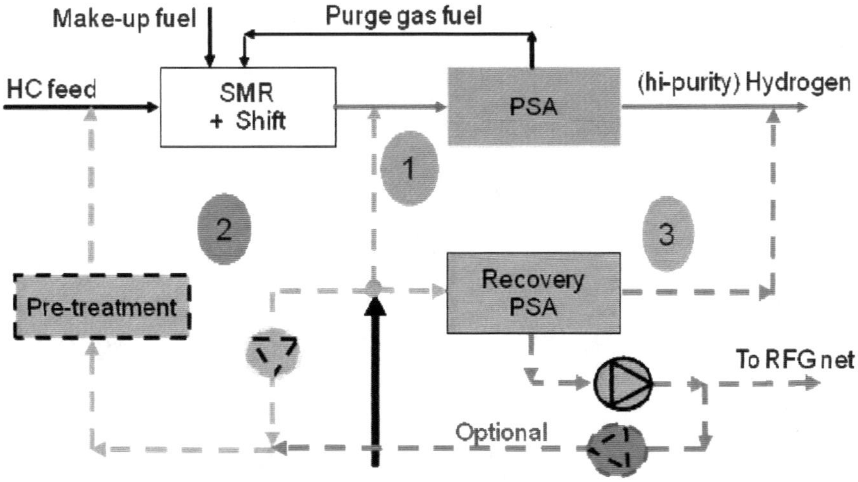

Figure-4: ROG Integration Routes with HGU

Within these proven routes, several variations are usually encountered or possible based on the case-specific attributes. These are summarized in Table-3 which can help in general assessment overview and provide some broad pre-screening guidelines.

Table-3: ROG Integration Summary and Guidelines

Application	**Attributes #**	**H_2 contribution**
Direct hydrogen end-user	2, 3, 4	High
Common PSA recovery integration	3, 5	Low
Direct feed for generation / reforming	3, 4, 6	Low-High
Dedicated recovery PSA	1, 3	Medium-High
Recovery PSA and extended integration of its purge gas as reformer feed	1, 3, 4, 5	Medium – high
Membrane or PSA-membrane Hybrid	1, 3	low-High
Cryogenic or Hybrid with PSA /Membrane	1, 7	Low
Retain for direct use into RFG fuel net	1, 3, 4	-- none

 # Attribute list

1. Hydrogen fraction in ROG
2. Cost-effective end-use hydrogen purity
3. Available pressure (and required hydrogen pressure)
4. Level of (possible) contaminants
5. Ratio of Reformed gas to ROG
6. Relative price of Hydrocarbon feedstock versus ROG
7. C_2+ credit and streams' composition

Case History

A case history is described below for utilizing ROG (with average 40 vol % H_2) as primary feed for hydrogen generation based on medium level contribution. The available quantity (averaged over time) was too large for common PSA and not enough to justify dedicated recovery PSA. For full hydrogen generation, it needed to be supplemented by Natural gas (NG) under advanced front-end control for steam/carbon ratio and capacity. The comparative economics is presented in Table-4.

Table-4 ROG Integration as SMR feed (45 mmscfd H2)

		Case-1 ROG as partial feed	Case-2 NG feed only
Feed ROG	kg/h	6708	--
Feed NG	Kg/h	3763	14427
Fuel NG	kg/h	1011	1667
Export steam	t/h	29.5	36.8
Power	kWh/h	1470	690
Operating cost savings[1)]	US$/h	188	Base
Additional TIC investment [2)]	MM $	4.1	base
Pay-out	**Years**	**2.5** [3)]	**Base**

1) ROG: 260 $/t (avg. 40 vol % H_2); NG: 180 $/t; Steam: 8 $/t; Power: $60 / MW
2) For compression and pre-treatment
3) Shorter if NG feed needs compression

Though the pricing of ROG was similar to NG in terms of LHV basis, the operating cost benefit due to potential (contained) hydrogen contribution was substantial (in the order of 7%). The payout of additional investment was less than 3 years, without downsizing the reformer and downstream section, which were still sized for NG feed as the controlling design case. Other inherent benefits for operating with ROG feed mix included relaxation on recycle hydrogen; NG feed flexibility, easier start-up and longer reformer tube life. Various ROG integration schemes have been implemented and proven.

Conclusions

Present-day refining needs call for advanced hydrogen management beyond just balancing the H_2 network in order to arrive at wider optimization. HyNDT™ platform console from Technip can prove to be an effective and powerful tool capable to identify trade-offs between hydrogen production, recovery and purity, as well as smart solutions based on purity and pressure cascading, energy optimization, steam-power synergy and ROG integration, along with necessary C-management in order to establish a roadmap to maximise refinery profitability.

Bibliography

1. Hallale, N, and F. Liu, "*Refinery hydrogen management for clean fuels production*", Advances in Environmental Research, 6 (2001), Elsevier Science Ltd" pp. 81-98.
2. Linnhof; B., *"User Guide on Process Integration for the Efficient Use of Energy,"* IChemE, Rugby, UK.
3. Ratan S., Wentink P. *Cost effective hydrogen from refinery offgases*, PTQ, 2Q 2001
4. Ratan S., van Uffelen R., *Curtailing refinery CO2 through H_2 plant*, PTQ Gas 2008

CHALLENGES IN USING REFINERY FUEL GAS FEEDSTOCK FOR STEAM REFORMING BASED HYDROGEN PLANTS

Peter V Broadhurst & Graham C Hinton

Johnson Matthey Catalysts, PO Box 1, Belasis Avenue, Billingham, Cleveland TS23 1LB, United Kingdom

Prepared for Presentation at AIChE 2011 Spring Meeting, March 14 - 17, 2011
14th International Topical Conference on Refinery Processing
Session 114: Hydrogen Management

ABSTRACT

Operators of steam reforming based hydrogen plants are seeking feedstock options to minimise operating costs and maximise operational flexibility. One option being considered increasingly by hydrogen plant operators associated with oil refineries is the refinery fuel gas (RFG) pool. RFG usually has a low value compared to other hydrogen plant feed options and so represents a particularly attractive feedstock option where there is excess RFG available.

RFG is a combination of refinery unit waste or by-product gases, often referred to as off gases. Off gases are sent to the RFG pool because these are difficult to process elsewhere in the refinery. Accordingly, incorporation of RFG into the hydrogen plant feed slate is not straightforward in terms of plant design and operation. The challenges may comprise

- High olefin levels;
- Significant sulphur levels;
- Substantial levels of higher hydrocarbons which may include naphthenes and/or aromatics;
- Less usual trace and minor components;
- Variability in the RFG composition as the blend of off gases changes with fluctuating operation of the contributing unit operations.

These factors impact substantially upon the design, catalyst and absorbent selection, and operation of the front end purification and the steam reforming section of the hydrogen plant. This is complicated further where the RFG is part of a complex multiple feed slate for the plant.

This paper will review the challenges presented by the use of RFG feed and the design and catalytic solutions that are available to allow the successful processing of an RFG feed.

Introduction

Operators of steam reforming based hydrogen plants want feedstock options to minimise operating costs and maximise operational flexibility. Consequently, new build hydrogen plants are often designed for a number of hydrocarbon feeds. It is common to have three or four feedstocks ranging from off gases through to naphtha requiring full operational flexibility across the range.[1,2] Operators of existing plants are also evaluating alternatives to the original feedstock slate and in some cases implementing changes which may necessitate modification of the plant operating conditions, hardware, equipment, catalyst selection and so forth.[3-6]

The quest for cheaper feedstock options is undoubtedly heightened by the significant increases in both natural gas and crude oil prices in the last few years. A feedstock option being considered increasingly by hydrogen plant operators associated with oil refineries is the refinery fuel gas (RFG) pool. Relative to imported natural gas and many other hydrocarbon streams and off gases in the refinery, RFG has a comparatively low value. Thus, RFG represents an attractive feedstock option where there is excess RFG available.

RFG is not widely used as a hydrogen plant feed. This is because its composition leads to a number of difficulties in processing in the feedstock purification and steam reforming sections of the hydrogen plant flowsheet. In this paper, we will explore these difficulties and the strategies for hydrogen plant design and operation, which may be used to allow processing of RFG as a feedstock. This will include some recently developed catalytic and control solution developed jointly by Johnson Matthey and Air Products & Chemicals.

2 RFG Composition and processing difficulties

RFG is a combination of refinery unit waste or by-product gases, often referred to as off gases. The off gases that are sent to the RFG pool are those which are cannot be processed elsewhere in the refinery either because of the composition or that there is an excess available. The off gases in the RFG come from various refinery unit operations (e.g. catalytic reforming, FCC, hydrotreating, coking), the availability and amount of which will depend on the refinery operation. Hydrogen containing off gases may be partly or fully used in hydrogen consuming units or may be treated to recover the hydrogen in a membrane or PSA unit such that a hydrogen lean gas is available to the RFG pool. Also, off gases or off gas blends with high olefin levels may be treated to recover olefins with the olefin lean gas going to the RFG pool. Thus, RFG can differ significantly between refineries. Examples of RFGs, which have been proposed for hydrogen plant feeds, are given in Table 1.

This shows the substantial variations in hydrogen from 11 – 35 mol%, methane from 26 – 65 mol% and olefins from 2.6 – 15.9 mol%.

RFG feeds often contain quite high levels of sulphur compounds. Up to 100 ppmv can be found and this can contain quite a significant proportion of mixed organic sulphur compounds. The level and speciation are necessarily dependent on the blend of gases going to the RFG pool.

Additionally, the RFG composition can fluctuate significantly in a given refinery as rates on different units change and particularly if a unit comes off line. The amount of off gas available to the RFG pool changes and so impacts on the composition of the blend, which comprises the RFG. This presents some control issues in some cases.

Table 1: RFG compositions considered for hydrogen plant feed

Mol%	Example 1	Example 2	Example 3	Example 4	Example 5	Example 6
Hydrogen	34.9	10.8	25.5	34.1	13.9	14.4
Methane	26.3	64.9	34.8	45.3	43.2	42.3
Ethane	11.0	13.4	23.5	8.2	12.2	13.7
Propane	9.5	2.5	5.2	3.2	8.3	6.8
Butanes	7.1	1.4	3.2	1.4	0.3	4.4
Pentanes	0.0	0.2	0.5	0.6	0.3	1.5
Hexanes	0.0	0.3	0.9	0.2	0.3	0.6
Ethene	3.8	3.1	3.4	1.5	7.2	6.6
Propene	2.1	0.5	0.6	0.9	8.3	6.7
Butenes	3.0	~	~	0.2	0.3	0.3
Pentenes	0.0	~	~	~	0.1	trace
Nitrogen	1.9	1.8	1.9	3.7	4.5	2.2
Argon	~	~	~	trace	trace	~
Oxygen	trace	~	~	~	~	~
Carbon monoxide	~	0.3	0.3	0.3	1.1	trace
Carbon dioxide	0.4	0.8	0.2	0.4	~	0.5
Total	100.0	100.0	100.0	100.0	100.0	100.0

In terms of incorporating RFG into the hydrogen plant feed slate, the aspects that may cause difficulties can be summarised.

- high olefin levels;
- Variability in the RFG composition as the blend of off gases changes;
- High hydrogen levels;
- Significant sulphur levels;
- Substantial levels of higher hydrocarbons which may include naphthenes and/or aromatics;
- Less usual trace and minor components.

Not every RFG will present each of these difficulties and each case must be considered separately. If a new build hydrogen plant is being considered, then it must be designed to include any RFG feed case(s). When considering using RFG on an existing plant, the extent to which there is a problem will be influenced by original design basis.

3 Feedstock purification section

RFG feeds can cause various issues in the feedstock purification section.

3.1 High olefin levels

Olefins need to be removed from the hydrocarbon feed in a hydrogen plant to a level below 1 mol% to minimise possible olefin derived carbon formation in the steam reforming section. The hydrodesulphurisation (HDS) catalyst is also an effective olefin hydrogenation catalyst and removes olefins almost completely as long as there is sufficient hydrogen present. Thus, for RFGs with significant olefin levels, the hydrogen available in the feed and as recycle must be sufficient for the olefin hydrogenation reaction, other hydrogen consuming reactions and to provide the target hydrogen level specified at the HDS converter exit. Johnson Matthey recommends a different level of hydrogen be present in the HDS effluent, depending on the feed composition, and how 'heavy' it is. For the gases given as examples 5 and 6, insufficient hydrogen is present in the RFG on its own to hydrogenate the significant olefin content. This means that additional hydrogen must to be added, usually recycled to the purification section of the plant, of approximately 5 mol% of the RFG feed rate for example 5, and approximately 3 mol% of the RFG feed rate in example 6.

Olefins hydrogenate exothermically over the HDS catalyst and the temperature rise can be 20+°C (36°F) per mol% olefin depending on the heat capacity of the feed gas. The inlet temperature must be adjusted to ensure that the maximum HDS bed temperature remains below 400°C (752°F), however, using standard HDS catalysts, such as **KATALCO$_{JM}$**^{TM} 41-6T or **KATALCO$_{JM}$** 61-1T, the inlet temperature needs to be above 300°C (572°F) to provide sufficient activity for reactions to initiate. Given the need for some operating margin inside these limits, this limits the olefin that can be processed to a few mol% in a once through reactor system.

To process higher olefin levels, a re-circulation system is usually employed around the HDS reactor (see Figure 1). The reactor effluent is olefin free so the recycle dilutes the inlet olefin level to an acceptable level. This approach has some disadvantages. Depending on the inlet olefin level and therefore the extent of dilution required, the recycle can be substantial. This increases capital expenditure in a number of ways; the recycle flow usually increases the vessel size to stay within design space velocity design limits and additional equipment is needed such as a recycle cooler and circulator.

Figure 1: HDS converter with recirculation

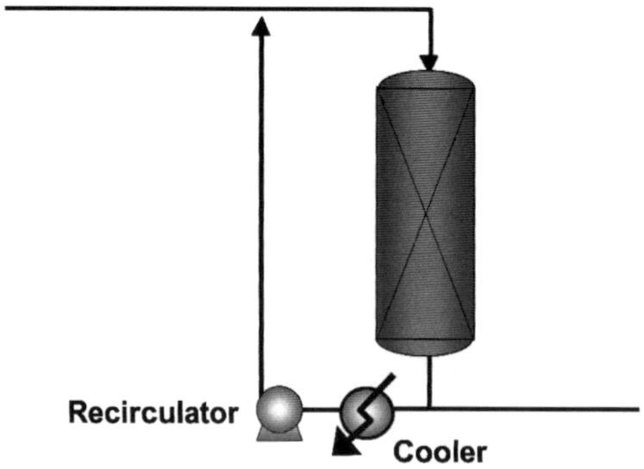

Without a recycle system, there is limited flexibility in terms of the olefin level using standard HDS catalyst types, thus, approaches that widen the operating envelope without installation of a recycle will be beneficial. One such approach which Johnson Matthey has recommended is to use a higher activity catalyst, such as **KATALCO$_{JM}$** 61-2T and its successor, **KATALCO$_{JM}$** 61-2F. This has a higher active metals loading and allows operation at inlet temperatures down to ~250°C (462°F) with the same exit temperature limit of 400°C (752°F). Thus, the amount of olefins that can be processed without installation of a recycle is increased to 7 – 8 mol% depending on the precise composition of the feed in terms of heat release and specific heat.

This concept has been extended which forms part of the technology claimed in recent patent applications filed jointly by Johnson Matthey and Air Products & Chemicals,[7] and which is available for licence. Detailed evaluation of a typical range of RFG feed compositions established that the extremely active pre-sulphided **KATALCO$_{JM}$** 61-2TS (and **KATALCO$_{JM}$** 61-2FS) allows the olefin hydrogenation reaction to strike below 150°C (302°F). This widens the operating temperature envelope to over 250°C (450°F) and allows processing of feeds containing well in excess of 10 mol% olefins.

It is imperative, however, that the consequences of the changes in operating temperatures are fully considered if **KATALCO$_{JM}$** 61-2T/F or **KATALCO$_{JM}$** 61-2TS/FS is to be retrofitted into an existing reactor to accommodate a higher olefin feed.

An alternative approach is to use a tube cooled converter.[8] The heat of reaction is removed on the shell side of the converter using water and the temperature in the catalyst packed tubes is maintained in the required range. The reactor is not truly isothermal as the heat release is sufficiently fast for the temperature to rise in the reaction zone within the tubes before the coolant brings the temperature back down. This approach has been commercialised in at least one plant.

3.2 Variability in the RFG composition

As refinery unit operations change rate, come off line or back on line, the RFG composition can fluctuate. This can make step changes in the gas going forward to the hydrogen plant and can cause control issues particularly where there is a substantial olefin content in the feed. If the olefin level increases sharply, it is possible that the 400°C (752°F) upper limit for HDS operating temperature might be exceeded. Also, if the inlet temperature has been lowered to well below 300°C (572°F) to allow for the hydrogenation exotherm and the olefin level decreases sharply, the exotherm will collapse which may lead to organo-sulphur slip. In both cases, the HDS inlet temperature needs to be adjusted quickly in response to the change in RFG composition.

A control system concept has been developed to allow this to happen.[7] There is a rapid feedback system from changes in the HDS bed exotherm into the upstream feed pre-heat such that the temperature in the catalyst bed can be controlled within acceptable limits.

Another possible problem is if the hydrogen level in the RFG decreases suddenly to a point where there is insufficient to hydrogenate olefins (and any traces of oxygen and the expected organo-S compounds) in the feed. Additionally, the recommended HDS exit hydrogen level should ideally be maintained. If a rapid decrease in feed hydrogen level occurs, another source of hydrogen, probably recycle hydrogen, needs to be rapidly established. If not, the consequences could be slippage of olefins and/or sulphur compounds to the steam reforming section with resulting carbon formation and/or poisoning.

3.3 High hydrogen levels

Some RFG blends may contain a significant amount of hydrogen. In some circumstances, this might provide a driving force for side reactions over the HDS catalyst if the RFG unusually happened to contain extremely low sulphur levels. The undesirable side reactions are methanation, which would also need a substantial amount of carbon oxides present, and hydrocracking, which would need heavier alkanes present. Both are very rarely observed and are associated with operation at transient or abnormal operating conditions (e.g. low flow; very low S; excess temperature). Methanation and hydrocracking activity can be significantly suppressed by pre-sulphiding the HDS, an effect which is also achieved if there is a reasonable level of sulphur in the feed.

3.4 Significant sulphur levels

RFG blends can contain significant sulphur levels of which much may be organo-sulphur compounds. As long as the maximum and typical levels are known and there is speciation of the sulphur, new plants can be design appropriately.

Existing plants may have been designed for a lower level of and/or less difficult organic sulphur compounds such that conversion in the existing HDS converter may not be possible. There are three retrofit options depending on the additional HDS capacity needed.

1) Replace the existing HDS catalyst with a higher activity version (e.g. replace **KATALCO$_{JM}$** 41-6T or **KATALCO$_{JM}$** 61-1T with **KATALCO$_{JM}$** 61-2F).
2) Install a combined product, which delivers HDS activity and H_2S absorption (e.g. **KATALCO$_{JM}$** 33-1) in the downstream H_2S removal beds.
3) Replace the existing HDS converter with a larger vessel.

If the total sulphur level is higher than the original design, the life of the H_2S removal beds will be shortened. If the design features lead-lag H_2S removal beds allowing change out on line, the consequence is more frequent change out. Selection of the highest capacity H_2S removal absorbents (e.g. **KATALCO$_{JM}$** 32-5) can lessen the change out frequency. The problem can be greater if the design has only a single H_2S removal bed designed to last between turnarounds where it is possible the increased sulphur level will not allow operation for the normal interval. Higher capacity absorbents may help but a retrofit of a second vessel may be required.

3.5 Substantial level of higher hydrocarbons

In the event that the RFG contains higher hydrocarbons, there may be an increased tendency for thermal cracking in the feed pre-heat coil. The carbon formed can carry forward and deposit on top of the HDS catalyst leading to increased pressure drop over time and some deactivation if the top catalyst becomes coated with carbon. If this problem is observed, it can be mitigated by replacing the hold down balls on top of the catalyst with a high voidage shape (e.g. **DYPOR** 604 – see Figure 2), which has more capacity for the carbon foulant before the pressure drop becomes critical.

Figure 2: DYPOR 604

3.6 Less usual trace and minor components

There are a number of other species which can be present in refinery off gas stream and which, therefore, could find their way into RFG. These include chlorides, arsines, oxygen, acetylenes and dienes.

Chloride can be dealt with by hydrogenation of any organo-chlorides over the HDS catalyst followed by removal of HCl using an HCl removal absorbent (e.g. **KATALCO$_{JM}$** 59-3). This must be placed upstream of the ZnO-based H_2S removal absorbent. In existing plants, it may be necessary to retrofit this into the purification section usually by putting it on top of the H_2S removal absorbent.

Arsines will be removed on the surface of the HDS catalyst and act as a poison. If present, levels are likely to be sub-ppm and the effect on the HDS catalyst is minimal. For a new plant design, the volume of HDS catalyst would be increased slightly to allow for the absorption of As and associated poisoned of the HDS catalyst at the bed inlet.

Oxygen will be removed by hydrogenation over the HDS catalyst. This is a very exothermic reaction and so the temperature rise and hydrogen consumption must be taken into account in addition to any from olefin hydrogenation.

Acetylenes and dienes may be present at low levels. There is some evidence that these may cause polymerise in the feed pre-heat or HDS converter leading to carbon deposition. This is unlikely to be a major issue as long as the level remains at low ppms such that any pressure drop rise and catalyst deactivation caused by the carbon is very slow.

4 Steam reforming section

Below are presented some of the considerations and impacts for the steam reformer and associated equipment.

4.1 High olefin levels

With proper design and operation, olefins should be converted in the feed purification section. However, levels of up to 1 vol% olefins can be accommodated in the feed to a steam reformer.

4.2 High hydrogen levels

The process duty in the reformer may be lower for feeds with high hydrogen levels, compared to other feeds, such as natural gases. This is because there is less endothermic steam reforming reaction in order to generate the same quantity of hydrogen. Additionally, since the feeds are more hydrogen rich than a natural gas, less feed flow may needed in order to produce the same quantity of hydrogen. However, even though the feed is more hydrogen rich than a typical natural gas, the average carbon number can be higher, meaning that the tendency for carbon formation within the reformer is also higher. This may necessitate either an increase in steam to carbon, a reduction in outlet temperature, or a change in the steam reforming catalyst type. This is discussed further in section 4.4 below.

All of these effects are illustrated below in Table 2. In this, the reformer heat load for constant hydrogen production has been calculated, assuming that no other process parameters are changed (inlet and outlet temperatures, steam to carbon ratio and reformer outlet pressure). The second row of the table shows how the molar flow of the RFG in question will compare to the molar flow of a typical natural gas for the same hydrogen production rate.

Table 2: Impact of varying feed composition on feed flow and reformer heat load

	Natural Gas	Example 1	Example 2	Example 3	Example 4	Example 5	Example 6
Reformer heat load	100%	93.4%	97.8%	94.2%	89.6%	99.9%	99.5%
Feed flow	100%	76.2%	88.4%	81.6%	109.7%	69.6%	64.5%
Average carbon no.	1.06	2.09	1.34	1.76	1.42	1.71	1.85

In a new plant design all these factors can be taken into account. If introducing such a feed to an existing plant, however, these factors and implications arising from them need to be taken into account and reviewed. The reduction in reformer heat load will have a number of consequences. The first, and most obvious, is that the firing on the reformer box must be reduced. This means that the bridge wall temperature will reduce, the flow rate of combustion gases will reduce and the amount of heat available to be recovered in the flue gas duct will be less. Hence, there is the possibility that some of the coils may become pinched leading to a reduction in steam production or in combustion air pre-heat temperature. Secondly, if the plant is a modern PSA hydrogen plant in which the PSA tail gas is used as the major fuel source on the reformer, it is possible that the trim fuel requirement is dropped to a level that there are control issues in maintaining the reformer outlet temperature as the PSA unit goes through its cycles. Additionally, if RFG is being used as the feed to the plant, it is likely that the trim fuel type will also have changed. This will necessitate that the suitability of the burners is checked to fire this new fuel type with good control and flame stability.

4.3 Significant sulphur levels

With proper design and operation, sulphur should be removed in the feed purification section.

4.4 Substantial level of higher hydrocarbons

In new plant designs, the selection of steam reforming technology and catalysts should take account of the RFG composition. In existing plant with non pre-reformer flowsheets, the impact of a heavier hydrocarbon component in the feed must be assessed in case a change in steam reforming catalyst is required to cope with the heavy component in the RFG. As mentioned above, this may necessitate an increase in the steam to carbon ratio for the plant, a reduction in reformer outlet temperature, or even a change in steam reforming catalyst type.

4.5 Less usual trace and minor components

With proper design and operation, these should be removed in the feed purification section.

4.6 Variability in the RFG composition

This presents possible problems for the steam reforming section if there is not sufficient control. If there is a sudden decrease of olefins causing the HDS exit temperature to decrease, slip of organo-sulphur is possible which would poison the pre-reformer or steam reforming whichever is first in the flowsheet.

If there is a sudden decrease in hydrogen, there may be insufficient to hydrogenate olefins and/or organo-sulphur. The resulting slip could lead to sulphur poisoning and carbon formation in the pre-reformer or steam reforming whichever is first in the flowsheet.

5 Conclusions

While RFG is an attractive economic option for hydrogen plants associated with refinery operations, its processing presents challenges in the feedstock purification and steam reforming sections of the flowsheet. These challenges result from the composition of the RFG and also from its variability.

In the feedstock purification section, the range of options for handling the high olefin levels often found in the RFG pool has been expanded in recent years by considering more active catalysts and heat exchange reactors. In conjunction, control systems have been developed to allow the HDS converter operating temperature to be adjusted as the feed composition changes to maintain the temperature within the acceptable range for successful operation.

In the steam reforming section, the impact of the change in feed and fuel type need to be modelled and understood. There may be minimal impact associated with the change, but conversely there may be changes required in operating parameters, catalyst type, heat exchanger performance, or even heat exchanger design, as well as consideration given to any necessary changes in control schemes.

6 References

1. "Hydrogen Market Review", B J Cromarty; Synetix Technical Paper, 1999.
2. "Hydrogen Technology Overview", S Ratan, Proceedings of 6th International Conference on Refinery Processing, AIChE Spring National Meeting, New Orleans, April 2003.
3. "Alternative Lower Cost Feedstock for Hydrogen Production", K Chlapik and B Slemp, NPRA Annual Meeting, San Antonio, 21st–23rd March 2004.
4. "The Study on the use of Butane Feed for the Repsol-YPF La Coruna Hydrogen Plant", A J Winsper and I C Irizar, ERTC 11th Annual Meeting, Paris, 13-15th November 2006.
5. "Feedstock Conversion for Indian Ammonia Plants. A Review of the Challenges", W J Cotton and V Singh, Proceedings of Fertilizer Association of India (FAI) Seminar 2003, New Delhi, December 2003.
6. "Taking Feedstock", P V Broadhurst and W J Cotton, Hydrocarbon Engineering, March 2005, p41.
7. "Process for Hydrogenating Olefins", R A Davis, N Macleod and G E Wilson, World Patent WO 2009/123909 A2, 8th October 2009.
8. "Process and System for Reducing the Olefin Content of a Hydrocarbon Feed Gas and Production of a Hydrogen-enriched Gas Therefrom", N Musich, R S Natarajan and H Klein, US Patent Application 20080237090, 2nd October 2008.

NEW PROCESS FOR NATURAL GAS LIQUID (NGL) AND OLEFINS RECOVERY FROM LOW PRESSURE OFF-GAS GENERATED IN REFINERIES & PETROCHEMICAL PLANTS

Rajeev Nanda, Technip, Houston, TX,USA

Rahul Singh, Technip, Houston, TX,USA

Abstract

There are various NGL (Natural Gas Liquid) recovery units in North America which are designed to recover hydrogen, olefins and NGL components like ethane, propane and butanes by processing the off-gas from various sources such as a refinery or Petrochemicals Complex. There is increased demand to install these units to recover valuable components, and, in return, provide clean gas for fuel.

The installation of the Off-Gas Processing units has both economic and environmental benefits as the main polluting components from the off gas are removed and valuable products are separated. The methane rich fuel gas produced from this unit without polluting or soot forming components is beneficial from an environmental point of view.

The industry is presently using the modified version of cryogenic NGL recovery process typically used in the natural gas processing industry. There are some disadvantages in using the modified conventional NGL process for off-gas. The existing plants have a history of downtime due to maintenance issues associated with the compressor, turbo expander and brazed aluminum heat exchangers. This is due to the nature of off-gas which can have significant impurities and variations in feed gas flow and composition.

This paper presents a new recovery process, specifically developed for Refinery, Upgrader or Petrochemical plant off-gas. This process is resilient, as it tolerates feed impurities and variations and is reliable.

Introduction

Off-gas is generated from several refining processes and is generally directly utilized as a fuel gas. Even till date many refineries are continuing to do the same. Apart from being a valuable source of hydrogen, the off-gas contains components such as ethylene, ethane, propene and LPG which are valuable chemical and petrochemical feed stocks than they are as fuels. The hydrogen requirement in petroleum refineries has witnessed an increasing trend worldwide in recent years. This is attributed to the growing demand for lighter and cleaner fuel. Also, the need to process heavier oil feedstock has resulted in scarcity of hydrogen. The oil refiners are forced to make effective use of existing facilities and implement economical methods to recover hydrogen from the hydrogen rich gases produced in the refinery itself. The up-gradation of the refinery off-gas streams to a high purity Hydrogen stream may be more economical than to install a new hydrogen production plant.

Recovery of valuable products from refinery off gas not only creates additional revenue for the industry but it also helps in reducing the carbon emissions. These advantages of refinery off gas processing and its economic viability has resulted in an increased demand to install these units. These units recover valuable Olefins & NGL and in return send the concentrated hydrogen streams for further recovery of Hydrogen by traditional technologies such as adsorption, membranes etc.

One of the traditional methods applied in industry for the refinery off gas processing is a modified version of expander based cryogenic NGL recovery process typically used in the gas processing industry for treating natural gas. There are some disadvantages in using the traditional turbo-expander process like plant downtime due to maintenance issues with compressor, turbo expander, brazed aluminum heat exchangers etc. This is mainly due to frequent changes in the feed gas composition and hydrate formation in the plant.

The higher temperature operation of the conventional turbo expander type plant to prevent the blue oil formation and freeze up of other solids in exchangers can significantly lower down the recovery of the valuable NGL and olefins from the refinery off gas.

The changes in the feed gas composition along with frequent fluctuations in flow rate results in inefficient operation of the plant. This often results in lower product recovery from feed gas in turbo-expander units.

This paper presents the process schemes to reduce or eliminate the currently experienced problems associated with the traditional turbo-expander process. The two new process schemes and their advantages are described in the ensuing sections. In these process schemes, indirect refrigeration loop consisting of turbo expander in mixed refrigerant loop is considered which gives high degree of operational flexibility with variation in feed gas composition. The refrigerant is a mix of the components available in the feed gas like methane, ethane, ethylene, propane, butanes etc. Also, the gas is compressed after first stage chilling and removal of some heavy components in refinery-off gas thereby reducing the fluctuations of molecular weight to the compressor.

Brief process description for the New Refinery Off-Gas Scheme – Base Scheme

Refinery off gas is the feed to the plant which arrives at around 100°F (37.7 °C) and 70 psia (4.8 bara). The bulk moisture in the feed gas is removed by pre-chilling the gas up to 70 °F (21.1 °C) in an exchanger which is heat integrated with the main NGL recovery plant.

The gas is dried in molecular sieve dehydrators. The dehydrated feed gas is cooled to around -70 °F (-56.7 °C) in a plate fin heat exchanger (E-100). The condensed hydrocarbon is separated in a low pressure separator (V-101). The condensed liquid is pumped to Demethaniser (T-100) operating at around 390 psia (26.9 bara) as a feed after heating it in E-100 up to around 30 °F (-1.1 °C).

The vapor separated from V-101 is compressed to around 355 psia (24.5 bara) in a compressor (K-103). As most of the heavy hydrocarbons are condensed in V-101 therefore there is very less variation in the molecular weight of the lean gas being compressed by K-103. The performance of the compressor remains optimum for wider range of variations in the feed gas composition. The compressed gas is chilled in E-100 to about -96 °F (-71.1 °C) and is fed to a separator, V-100. The liquid leaving the separator is rich in C2+ and is pumped to the Demethaniser (T-100).

T-100 operates at 395 psia (27.2 bara) at the bottom and 390 psia (26.9 bara) at the top. The T-100 overhead gas is chilled in E-104 to about -99 ° F (-72.8 °C). The hydrocarbon liquid formed as a result of cooling to a very low temperature is separated in a high pressure separator (V-105) and is pumped back to T-100 on the top-most tray.

The separated vapor from V-105 is combined with the vapor exiting V-100. The combined vapor exchanges heat in E-104, before it is sent out as the lean gas at about 92 °F (33.3 °C) and 135 psia (9.3 bara).

The C2+ is recovered from the T-100 bottom. The column bottom temperature is maintained at around 81 °F, which makes it possible to provide the reboiler duty by cooling the refrigerant after the final stage of compression.

The refrigeration duty required is provided by means of a closed loop turbo expander cycle. The refrigerant is made by mixing vapor and liquid generated within the plant at several locations. This allows refrigerant composition to be changed accordingly in case the feed composition changes significantly in future.

<u>(Figure 1) Simplified Sketch of New Refinery Off-Gass Process (Base Scheme)</u>

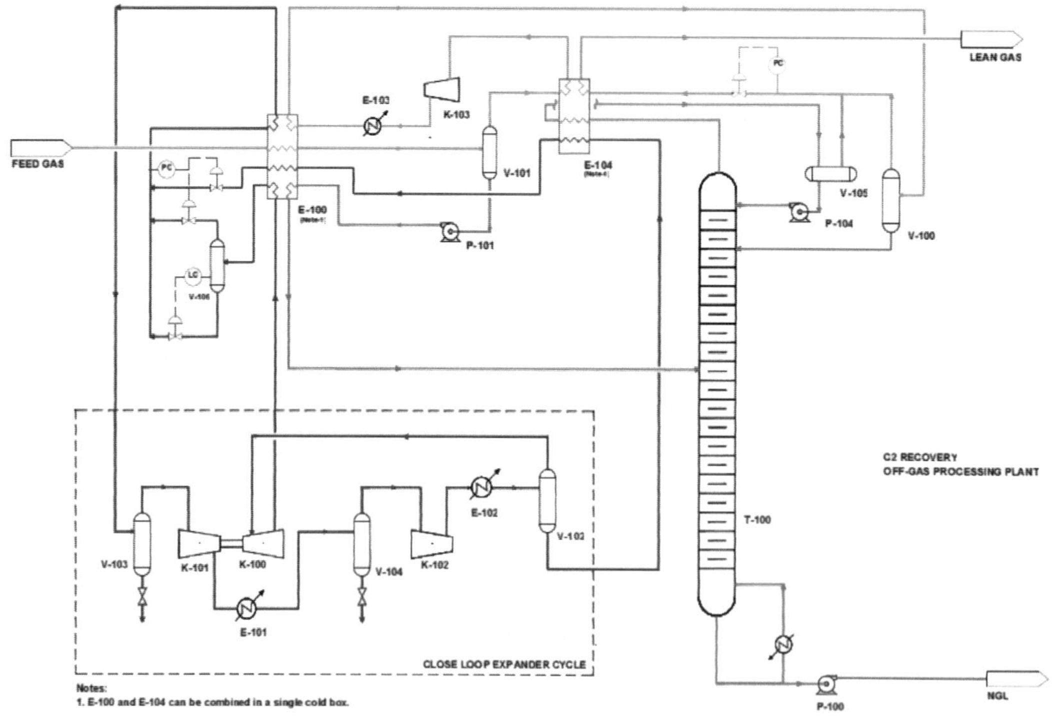

The base scheme is suitable for the recovering ethane and ethylene from refinery off gas. The product recovery is moderate, in range of 85% ethane recovery and is limited by the minimum operating temperature considered for designing the plant. This base scheme provides limited propane recovery while operating under ethane rejection mode. This scheme is good for projects requiring moderate ethane recovery and no specific requirements for propane recovery in ethane rejection mode of operation.

The deep ethane recovery is achieved by modifying the base scheme. The modification includes inclusion of a heavy reflux absorber. The heavy reflux in the absorber provides ethane recovery greater than 90%. Moreover, propane recovery of greater than 90% can be achieved under ethane rejection mode of operation. A detailed process description for deep ethane recovery scheme has been provided below:

Process description for the new refinery off-gas scheme – With Deep Ethane Recovery

Refinery off gas is the feed to the plant which arrives at around 100 °F (37.8 °C) and 85 psia (5.8 bara). The bulk moisture in the feed gas is removed by pre-chilling the gas up to 70 °F (21.1 °C) or lower in the exchanger (E-102). The pre-chiller (E-102) is heat integrated with the main NGL recovery plant.

The condensed water from the feed gas is separated in a filter separator before dehydrating the gas in molecular sieve dehydrators. The dehydrated feed gas is cooled to around -69 °F (-56.1 °C) in a plate fin heat exchanger (E-100A). The condensed hydrocarbon is separated in a low pressure separator (V-103). A part of condensed liquid is pumped to Demethaniser (C-102), a distillation column operating at around 390 psia (26.9 bara)., while remaining portion of the condensed liquid is sent as a heavy reflux to an absorber (C-101) to absorb C2+ from the compressed lean gas. The Heavy Reflux Absorber (HRA), C-101, operates around 340 psia (23.4 bara).

The vapor separated from the V-103 exchanges heat with the incoming feed gas in a brazed aluminum plate fin exchanger (E-100B) before being compressed to around 355 psia (24.5 bara) in a compressor (K-102). The inlet feed to the compressor is at around -65 °F (-53.9 °C). The compressed gas is cooled in steps, first in an air cooler (AC-104) and next in the plate fin heat exchanger (E-100A) to about -96 °F (-71.1 °C).

The compressed and chilled V-103 vapor is fed to the absorber (C-101) at the bottom. The Liquid leaving the absorber column is rich in C2+ and a part of the liquid is pumped to the Demethaniser (C-102) as a reflux at the top. The remaining portion of the HRA (C-101) bottom liquid is fed to the 15th tray of column C-102 after heating it in E-100A up to around 30 °F (-1.1 °C). The vapor leaving the top of C-101 at around -77 °F (-60.6 °C) is cooled to -99 °F (-72.8 °C) in E-100A to condense remaining C2+ content in the gas. The hydrocarbon liquid formed as a result of cooling to a very low temperature is separated in a high pressure separator (V-104) and is pumped to the Demethaniser (C-102) on the top-most tray.

The Demethaniser (C-102), operates at 395 psia (27.2 bara) at the bottom and 390 psia (26.9 bara) at the top. The C-102 overhead gas is mixed with the C-101 overhead gas.

The vapor separated from the high pressure separator (V-104) exchanges heat in the plate fin heat exchangers, E-100A and E-100B, before it is sent out as the lean gas at about 92 °F (33.3 °C) and 135 psia (9.3 bara). A part of this gas is taken for regeneration and is returned back and combined with rest of the lean gas before it is sent out as residue gas.

The C2+ is recovered from the C-102 bottom. The column bottom temperature is maintained at around 81 °F (27.2 °C) which makes it possible to provide the reboiler duty by cooling the refrigerant after the final stage of compression. The C2+ product is flashed to a lower pressure of around 140 psia (9.6 bara) which provides pre-chilling duty in the pre-chiller E-102. The C2+ product is compressed by K-104 and pumped with the help of product pumps P-104 A/B to the required B/L export pressure.

The refrigerant is compressed in a compressor K-103, to a pressure of about 510 psia (35.2 bara) and cooled in steps, first in an aircooler AC-102 and subsequently in Demethaniser C-102's reboiler (E-106), a shell and tube type heat exchanger. The cooled refrigerant downstream of reboiler (E-106) is a two phase mixture. The liquid formed is separated in refrigerant separator (V-105). The separated liquid is further cooled in E-100B to about -80 °F (-62.2 °C). The refrigerant vapor exiting V-105 is expanded in a turbo expander (K-101) to a pressure of about 220 psia (15.2 bara).

The expanded vapor then exchanges heat in the pre-chiller (E-102) and is further cooled to about -85 °F (-65 °C) in E-100A. The liquid formed as a result of cooling is separated in a vertical drum V-122. The liquid and vapor are flashed to a pressure of around 55 psia (3.8 bara) with the help of a control valve. The flashed vapor and liquid is mixed with the cooled refrigerant liquid from V-105. The cooling of refrigerant liquid from V-105 is done in E-100B and it is let down by means of a control valve before mixing. The mixing of refrigerant streams takes place at around 55 psia (3.8 bara) and the combined stream exchanges heat in the plate fin heat exchanger (E-100A) to provide the required refrigeration duty. The refrigerant is heated to about 90 °F (32.2 °C) and is fed to the turbo compressor associated with the turbo expander, K-101. The partially compressed gas from the turbo compressor is further compressed in refrigerant compressor, K-103.

The refrigeration duty required is provided by means of a closed loop turbo expander cycle. The refrigerant is made by mixing vapor and liquid generated within the plant at several locations.

(Figure 2) Simplified Sketch of New Refinery Off-Gas Process (Deep Ethane Recovery Scheme)

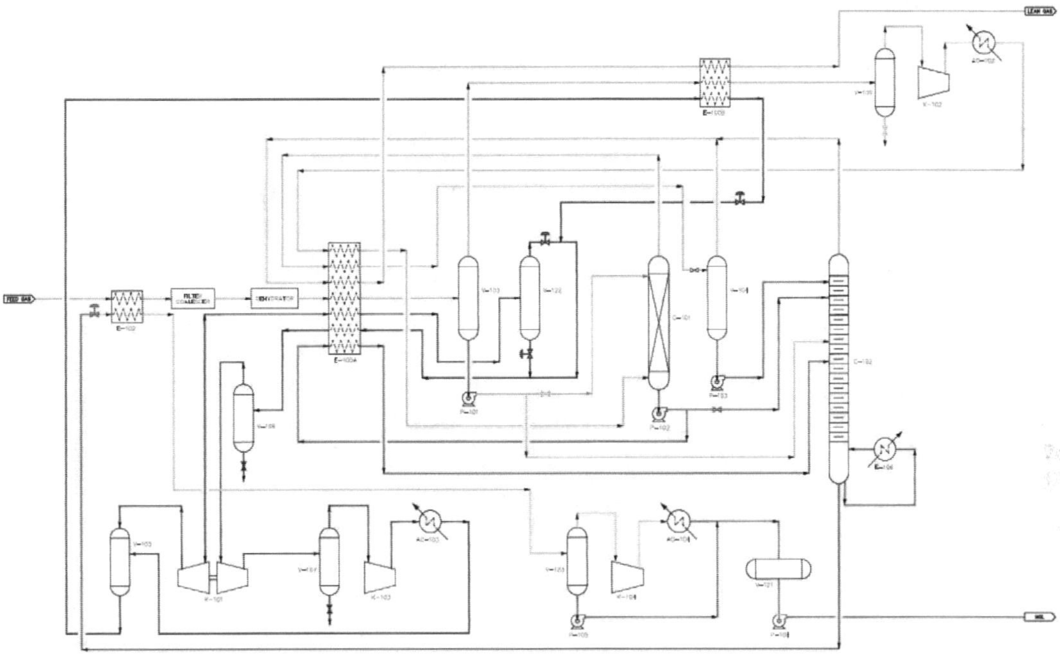

Advantages of New Refinery Off-Gas Process

I. Low Pressure Operation Of Inlet Facilities:

Generally, the conventional NGL recovery units for gas plants require feed gas compression up to a certain level (around 400 to 750 psig or even higher depending on the composition) in order to produce low temperature by expansion of the high pressure gas with the help of a turbo-expander. Also, the dehydration system becomes more economical with high pressure operation for a typical capacity natural gas plants

However, in case of low pressure refinery off gas plant (usually less than 150 MMSCFD capacity), the system can be optimized to achieve more economical solution.

The indirect refrigeration system can be used to process the low pressure refinery off-gas thereby avoiding direct compression of the inlet feed gas. In this case the feed gas can be processed by pre-chilling, followed by dehydration and then first stage chilling before going through compression step in order to remove most of the heavy components in the gas. The resulting gas is leaner and has fewer variations in molecular weight as compared to the feed gas. The low pressure feed gas holds more water at same temperature versus higher pressure gas. The pre-chilling of the gas helps to reduce the excess water before dehydration of gas thereby achieving similar water loading as that of the higher pressure gas operating at higher temperature.

The following can be observed from the typical adsorption isotherms available from any molecular sieve vendor for 3A sieves.

The water loading on mol sieves remains same for dehydrator operation at 315 psig (21.7 barg) & 115 °F (46.1 °C) versus 68 psig (4.7 barg) & 70 °F (21.1 °C).

The water content in the feed gas is 232 lb/mmscf at 315 psig (21.7 barg) & 115 °F (46.1 °C) as compared to 224 lb/mmscf at 68 psig (4.7 barg) & 70 °F (21.1 °C).

The lower design pressure also reduces the thickness of the dehydrators resulting in lower metal weight for the vessels (Refer to Table 1).

Table 1: Comparison of dehydrator system operation at various pressure levels

Parameters	Case 1	Case 2
	High inlet pressure	Lower inlet pressure
Feed Flow, MMSCFD	51	51
Temperature, °F (°C)	115 (46.1)	70 (21.1)
Pressure Psig (barg)	300 (20.7)	68 (4.7)
Inlet water, lb/MMSCF	232	224
Dehydrator bed size	12 X 21.6	12 X 20.9
Mol sieve weight, lbs	110,500	106,500
Steel Weight, lbs	82,975	28,223
Regen Gas Flow, MMSCFD	15.8	15.3
Regeneration Duty, MMBTU/H	6.66	6.43

In summary, the low pressure dehydration of the refinery off gas has following advantages:

a) Dehydrator regeneration at low pressure and temperature minimizes the risk of decomposition /cracking of olefins and polymerization of di-olefins thus preventing partial degradation of sieves and further moisture breakthrough during the drying cycle.

b) Operation far away from dew point curve thus reducing the risk of condensation and increasing reliability of the process.

c) Pre-chilling of gas upstream of dehydrator results in potential savings in metal weight due to low pressure design.

II. <u>Hydrate Tolerant Process:</u>

Warmer process conditions with lower pressure will result in lowering the chances of hydrate formation.

The Refinery off gas NGL extraction plants experience frequent shutdowns due to hydrate formation in the cold sections of the plant due to deterioration of molecular sieves and improper regeneration. This problem in many cases is attributed to the components in the feed gas from refinery. In refinery off gas units, the regeneration temperature is kept below a certain temperature so as to avoid the risk of decomposition /cracking of olefins and polymerization of di-olefins.

The hydrate formation issue for the refinery off gas NGL extraction units can be reduced by designing the plant warmer so as to tolerate at least 0.1 ppmv moisture breakthroughs in the dehydrated feed gas. It should be noted that most of the mol sieve suppliers will only guarantee less than 0.1 ppmv water content in the dehydrated feed gas from the dryers.

III. Better Adaptability to Significant Changes in the Feed Gas Composition

The refinery off gas composition varies significantly depending upon the various units operating at a particular time in the refinery. For example the H2 content in the refinery off gas may change from 15 to 30 mol %. The conventional NGL extraction plants will require adjustment of the external refrigerant to provide the required refrigeration duty. Often it is difficult to achieve the designed product recoveries in the conventional NGL extraction plant for operation with significant changes in the feed gas composition.

The new process scheme provides better adaptability to changes in the feed gas composition. The plant can be run with the same refrigerant for two different feed gas compositions with minor loss of product recoveries. Table 2 presents the comparative data for the plant operation under deep ethane recovery mode with two different feed gas compositions.

Table 2: Simulation Results for Plant Operation with Different Feed Composition		
Parameter	Feed Composition 1 15% H_2	Feed Composition 2 30% H_2
Feed Flow, MMSCFD	100	100
Feed Pressure, Psia (bara)	90 (6.2)	90 (6.2)
Feed Temperature, °F (°C)	100 (37.7)	100 (37.7)
Ethane % Recovery	90.7	90.1
Ethylene %Recovery	76.4	74.1
Propane % recovery	97.5	96.8
Propene % recovery (TPA)	96.6	95.7
C4+ % recovery	99.8	99.8
Minimum metal temperature in Heat exchanger °F (°C)	-119 (-83.9)	-119 (-83.9)

The figures shown in the table above are based on following typical Refinery Off-Gas (Feed Gas) Composition (mol %)

Feed Composition 1:

H_2: 15.0, O_2: 0.05, N_2: 0.6, CO_2: 0.05, CO: 1.1, C1: 43.2, C2=: 2.3, C2: 20.01, C3=: 4.1, C3: 10.9, C4+: 1.69, H_2O: 1. 0

Feed Composition 2:

H_2: 30, O_2: 0.00, N_2: 0.5, CO_2: 0.03, CO: 1.3, C1: 37.5, C2=: 1.8, C2: 15, C3=: 3.2, C3:10, iC4: 0.1, nC4: 0.4, 1-BUTENE: 0.1, iC5: 0.02, nC5: 0.02, nC6: 0.02, Mercaptans: 0.01, Water: saturated.

IV. Superior Performance Under Ethane Rejection Mode

NGL extraction plants are often required to operate under ethane rejection mode. The primary objective under ethane rejection mode of operation is to maximize propane recovery.

The new scheme allows for highly flexible operation under ethane rejection mode. Propane recovery greater than 90% can be achieved under this mode of operation.

V. Good Controllability of Residue Gas Heating Value

One of the requirements for refinery Off-gas processors is to maximize the NGL recovery while still achieving the required minimum heating value of the residue gas. This process has good flexibility with respect to selectively increasing the ethane production over propane recovered. A small amount of C3+ slippage in the residue gas can increase the heating value to the required value without loss of ethane recovery or significant NGL production. This allows good controllability of the heating value of the residue gas product. Hence it is easy to manage the future changes if warranted, in heating value specification of the residue gas.

VI. No feed gas compressor in inlet gas service and no turbo-expander in Off-Gas Service

One of the problems faced by many Off-gas processing plants is the frequent maintenance of the rotating machines such as Feed gas compressors, Turbo Expanders. The failure of expanders and the compressors in this severe service is attributed to the significant changes in feed flow rate and molecular weight. By nature the refinery gas is a waste gas whose composition depends upon the number and type of units working at a particular time in the refinery. The changes in the composition are significant to effect the operation of the rotating machinery. The unpredictability and significant variations in the feed composition & conditions makes it difficult to specify the rotating machinery for various feed composition to which the machinery will be exposed. During operation, the rotating machinery is subjected to severe off design cases, not anticipated during the design phase, leads to inefficient performance and downtime.

The new process discussed in this paper is efficient and mitigates this widely known industry problem. The feed gas is not compressed until it is pretreated and chilled to about -80 °F (-62.2 °C). At this low temperature mostly all the heavier hydrcarbons along with significant portion of C2+ condenses. The lean gas is then compressed to the required pressure. The compression of the lean gas ensures less variation in the molecular weight of the gas compressed. The compressors therefore operate close to the design point even if the feed gas composition varies significantly.

The new scheme provides following advantages as compared to conventional processes:

- Less downtime and maintenance as compression equipment is provided for leaner dry gas with significantly lower variations in molecular weight.
- Change in composition and flow surges will have minimum effect on plant operation.
- Power saving, due to efficient operation of the compressors near the rated point.
- No turbo expander in process gas stream results in higher process availability. Turbo- expander is provided on refrigerant side.

VII. External refrigerant contains components available in feed gas

The refrigerant used for closed loop expander cycle is generated within the plant. Various draw off points at different locations are provisioned in the plant. Streams in different ratios are mixed to generate the refrigerant required for operation. The start up of the plant at lower capacity is possible while using dehydrated feed gas as refrigerant. Subsequently, the dehydrated feed gas is replaced by the refrigerant once the plant is stabilized at lower capacity.

Refrigerant composition can be easily adjusted should the feed composition changes dramatically during the life of project thus resulting in efficient operation in off-design cases also.

No significant refrigerant storage is required as streams generated within the plant constitute the refrigerant.

VIII. Use of Standard, Reliable and Proven Equipment used in Industry in Similar Service.

This process scheme is new but the equipment used are standard and proven. The process is simple using equipment already being used in the similar services.

<u>Conclusions:</u>

It is evident that the new process developed certainly has following advantages over the traditional modified turbo expander process.

a. It is flexible as it can efficiently handle wide variations in feed gas composition and flow rate without the loss of product recovery or frequent adjustment in refrigerant composition.

b. This process is reliable as there is no turbo- expander directly in contact with the difficult refinery off gas. Also, no compression of inlet feed gas in most cases.

c. Better recovery and less down time.

d. Life cycle savings due to higher availability of the plant and better product recovery.

e. The process is safe as the warmer operation of the plant avoids risk of blue ice formation, freezing of undesirable components and hydrate formation.

f. Moreover there are no significant storage requirements for refrigerant inventory.

(114e) The Importance of Carbon and Hydrogen Management In the Development of Thermochemical Production of Biofuels

Lexie Kinion[1], Daniel Resasco[1], Lance Lobban[1] and **Richard G. Mallinson**[2], (1)University of Oklahoma, Norman, OK, (2)School of Chemical, Biological and Materials Engineering, University of Oklahoma, Norman, OK

Biomass has the potential to supply significant quantities of transportation fuels as increasing worldwide demand will place significant pressure on the supply. In the US, increasing imports of energy could make any imbalance especially acute, as has been demonstrated in the past. Additionally, pressure continues to build for reductions in carbon emissions.

The transportation fuels infrastructure has developed around hydrocarbons over the last 100 plus years, and there is intrinsic incompatibility with oxygenates that limit their incorporation into the existing infrastructure that will require significant capital investment to change.

A "big picture" problem is that biomass carbon is in a relatively highly oxidized state that has a low energy density. To increase the energy density, and reduce the oxygen content to increase compatibility, the biomass must be significantly deoxygenated. Using thermochemical methods, there are various paths to achieve this that result in some combination of a loss of carbon or the consumption of hydrogen. Theoretically, the highest possible yield retains about 2/3 of the original carbon. This would be achieved through reactions that lose only CO2 and require no hydrogen. Practically, almost any process yield will be lower and the optimal cost may be a combination.

Within a process, several paths may be chosen and evaluated. For example, decarboxylation removes two oxygens for each carbon. Reactions wherein CO is lost allows the CO to be used to generate hydrogen via the water gas shift reaction that may be used to further remove oxygen as water, with no additional carbon loss. Hydrogen may be generated by gasification of the biomass, or its char, by steam reforming of natural gas, or by renewable wind and solar.

These methods all have tradeoffs that must be considered from several perspectives. This study does not consider the cost, which of course ultimately controls whether a process will be built, but considers the main objective of the conversion of biomass, that is production of transportation fuels.

See more of this Session: Hydrogen Management
See more of this Group/Topical: Topical 7: 14th Topical on Refinery Processing

AUTHOR INDEX

AUTHOR INDEX